'This book provides invaluable insights into the nature of modern conflict, particularly within the context of Grey Zone warfare. The Editors and Contributors masterfully combine theory and practice, offering innovative solutions for small states faci
and Grey Zones: Design for Small States is an essential cc
and security studies. It's an indispensable source of knowle
within NATO and the European Union.'

Gene
Polish Military Represen

'This book provides a comprehensive and much-needed examination of modern conflict, with a particular focus on small states navigating the grey zone. *Modern War and Grey Zones: Design for Small States* is essential reading for anyone involved in crisis response and strategic planning. Its practical approaches to managing hybrid threats and fostering institutional resilience make it an invaluable resource for security professionals across Europe and beyond'.

Evgenia Karatari, *Deputy Head of Division,*
EU Situation Room, Crisis Response Centre (CRC),
European External Action Service

'*Modern War and Grey Zone: Design for Small States* addresses one of the most pressing issues in contemporary warfare: how smaller nations can confront and counteract the subtle, multifaceted threats of hybrid warfare. Through rigorous analysis and case studies, this book lays out strategic solutions that are both innovative and realistic. It's a must-read for anyone concerned with national security and defense in today's evolving global environment, which is full of hybrid warfare.'

Larry P. Goodson, *PhD, Professor of Middle East Studies,*
US Army War College, General Dwight D. Eisenhower
Chair of National Security (2004–2007, 2014–2017)

'Marzena Żakowska and David Last have edited an impressive collection that brings together leading experts to explore the complex dimensions of modern warfare, with a particular focus on Grey Zones and hybrid threats. Through its comprehensive structure, spanning topics from environmental insecurity and cybersecurity to violent nonstate actors and crisis management, the volume offers both theoretical insights and practical approaches. This work is a critical resource for scholars, practitioners, and policymakers seeking to understand and address the multifaceted challenges of contemporary conflict.'

Mikael Weissmann, *Associate Professor of War Studies,*
the Deputy Head of the Land Operations Division at the Swedish Defence
University's Department of War Studies (IKV), and the co-convenor
of the Hybrid Threats Research Group (HTRG) and the
Land Warfare Research Group (LWRG)

'This book goes the extra mile by applying military design thinking to lead readers beyond simplistic solutions towards more nuanced and thoughtful ones. I highly recommend reading this book because provides two essential realizations: the Grey Zone is real and to explore it successfully, we have to use new skills and thinking.'

Sönke Marahrens, *Director of COI Strategy and Defense at the European Centre of Excellence for Countering Hybrid Threats*

'Competition between collectivities that perceive reality in different ways, especially in Grey Zones, is part of everyday life and small incidents can turn into armed conflicts. This book significantly contributes to a better understanding of the processes in Grey Zones in risk society, and therefore I highly recommend it to readers.'

Kire Sharlamanov, *Professor of Sociology at the International Balkan University in Skopje, North Macedonia. Editor of the Journal of Law and Politics and Sociology*

'Engaging, timely, and deeply important. This book is a wide-ranging, thought-provoking, and invaluable resource for analyzing war and conflict in the modern era. The Editors and Authors are an impressive, international array of scholars and practitioners who present a well-curated and coherent approach to many of the thorniest quandaries of international security.'

Richard A. Lacquement, Jr., *Colonel, United States Army (retired)*

'This book is a critical and timely contribution to the study of contemporary conflicts and hybrid threats. Marzena Żakowska and David Last offer a comprehensive and thoughtful approach to navigating the complexities of the grey zone for small states. The depth of analysis provided by the authors makes this an essential read for policymakers, military strategists, and academics alike.'

Omar Oscar Ashour, *Professor of Security & Military Studies, Doha Institute for Graduate Studies and Director of the Strategic Studies Unit at the Centre for Research and Policy Studies, Qatar*

'This book provides an overview of the contours of contemporary conflict in the Grey Zone. Its editors, Marzena Żakowska and David Last, bring together a cosmopolitan set of Authors to explore a range of Grey Zone threats and potential approaches to managing global war and conflict in order to sustain society, states, and the global commons they depend upon.'

John P. Sullivan, *instructor at the Safe Communities Institute at the University of Southern California and a research fellow at the Future Security Initiative at Arizona State University*

Modern War and Grey Zones

This book explores the challenges small states face in navigating the complexities of modern war, particularly within the ambiguous Grey Zones where the boundaries between peace and conflict blur.

This book addresses the multifaceted challenges of policy, strategy, operations, and tactics, providing valuable insights into how small states can deter and manage violence. It highlights the vital role of international organizations, alliance-building, cyber operations, information warfare, lawfare, and the balance of power. Additionally, it tackles the issue of strengthening state security and fostering social cohesion in the face of Grey Zone threats. A distinctive feature of this book is the inclusion of innovative design problems developed to assist small states in navigating complex security landscapes with both practical and strategic approaches.

This book will be essential reading for security practitioners and military professionals and of great interest to students of security studies, defense studies, and international relations.

Marzena Żakowska is an assistant professor and lecturer at the National Security Faculty of the War Studies University, Warsaw, Poland. She has published extensively on armed conflicts, hybrid threats, mediation, security in the Western Balkans, and social security issues.

David Last has been teaching at the Royal Military College of Canada and the Canadian Forces College since 1999. He has authored or edited eight books and over 80 articles and chapters on peacekeeping, conflict management, professional military education, and special operations.

Modern War and Grey Zones

Design for Small States

Edited by Marzena Żakowska
and David Last

Routledge
Taylor & Francis Group
LONDON AND NEW YORK

Designed cover image: Arkadiusz Kupiec, ARWAY Arkadiusz Merchant

First published 2025
by Routledge
4 Park Square, Milton Park, Abingdon, Oxon OX14 4RN

and by Routledge
605 Third Avenue, New York, NY 10158

Routledge is an imprint of the Taylor & Francis Group, an informa business

British Library Cataloguing-in-Publication Data
A catalogue record for this book is available from the British Library

ISBN: 9781032550374 (hbk)
ISBN: 9781032550367 (pbk)
ISBN: 9781003428701 (ebk)

DOI: 10.4324/9781003428701

Typeset in Times New Roman
by codeMantra

Contents

Figures

Tables

Contributors

Jowita Brudnicka is an assistant professor at the University of Social Sciences in Łódź, Poland. She graduated from the War Studies University in Warsaw, earning her PhD in security studies in 2018. She is currently pursuing certification as a clinical psychologist. Her research in security science focuses on state security, geopolitics, and fragile states, specifically emphasizing Afghanistan. In her psychological research, she explores the psychology of the battlefield, including the experiences of civilian participants in armed conflicts and cultural and psychosocial differences among migrants from non-European civilizations. She is a member of the Polish Geopolitical Society, where she served as a board member from 2020 to 2022, organizing the Annual Conference of Polish Geopolitics Experts.

Jeffrey V. Gardner is an Assistant Professor of Homeland Security Studies at the American Public University System (APUS). He is a US Air War College graduate and holds a PhD in Public Policy and Administration, specializing in Homeland Security Policy and Coordination. A retired US Army Lieutenant Colonel, he served over 21 years on active duty. As a Colonel in a state defense force, he commanded a brigade supporting the National Guard with command, control, communication, and incident response functions. Since 2013, he has collaborated with NATO commands and organizations on initiatives focused on counter-terrorism, hybrid threats, and comprehensive defense strategies. His research interests include advancing homeland security, enhancing defense strategies and capabilities, countering hybrid and asymmetric threats, improving intelligence sharing for counter-terrorism efforts, and fostering interagency and cross-ministry coordination.

Eryk Habowski holds a PhD from the War Studies University in Warsaw. He has collaborated with the Pilecki Institute, where he specializes in the study of totalitarianism and genocide. His work also involves research on UN diplomatic sources and archives. He served as the content coordinator for the multi-year international project *Lemkin 2018–2023* and was the Deputy Editor-in-Chief of the scientific yearbook *Totalitarian and 20th Century Studies*. As the Chief Expert for the Polish Section of the World Expo in Shanghai, he coordinated the Polish program and continued his work in the subsequent World Expo in Milan. He curated the 2019 exhibition on genocide and Raphael Lemkin's contribution to defining this "odious crime" at the UN Headquarters in New York. Habowski also holds a PhD in Law from the Polish Academy of Sciences. He is the author, co-author, and editor of numerous books on topics such as genocide, totalitarianism past and present, the history of diplomacy, conflict policy instruments, and international law. His research interests encompass information warfare, lawfare, the philosophy of law and politics, war crimes, genocide, the balance of power, and the threats posed by AI in social engineering, politics, and armed conflicts.

Duraid Jalili is Senior Lecturer in Environment and Security Education at King's College London and Senior Academic Advisor for Defense Futures at the UK Ministry of Defense. His research focuses on security sector responses to environmental insecurity and how to enhance strategic thinking and education for senior leaders. His teaching includes intermediate and advanced officer courses at the Defence Academy of the United Kingdom, continuous professional development and executive education courses across various government organizations, and civilian courses at King's College London. He is the founder and co-director of the Climate Security Expert Network and regularly contributes to policy, concepts, and doctrine development for governments and militaries worldwide.

Urszula Jaśkiewicz is a specialist at the European Project Implementation Center within the Ministry of the Interior and Administration in Warsaw, Poland. Previously, she served as a Communications Specialist at the Government Information Center in the Chancellery of the Prime Minister of Poland. She holds a master's in security studies from the War Studies University in Warsaw and is pursuing her doctorate. She contributed to organizing the 17th Common Security and Defense Policy (CSDP) High-Level Course 2021/2022, known as "Walter Hallstein," specifically Module 4: The Future of CSDP/CSDP in Neighboring Regions, in Warsaw in 2022. Additionally, she was involved in the Warsaw Security Forum in 2022 as the VIP relations-communication specialist. Her areas of interest include national security, the role of intelligence services in hybrid conflicts, conflict resolution, and security issues in the Western Balkans.

Mikkel Storm Jensen is a Major in the Danish Army and currently serves as an instructor at the Royal Danish Defense College's Institute for Strategy and War Studies. He completed his PhD in 2023, focusing on the role of offensive cyberspace operations capabilities and their impact on inter-alliance dynamics within NATO. His research interests span two main areas. From a defensive standpoint, he explores the state's role in developing and implementing resilience strategies to counter hybrid threats originating from the cyber domain. From an offensive perspective, his research examines the potential for offensive cyberspace operations, especially from the viewpoint of small, alliance-dependent states, within the context of strategic competition that remains below the threshold of armed conflict.

Arian Kadriu is Assistant Professor of Security Studies at the University of Business and Technology (UBT) in Pristina, Kosovo. He holds a PhD in National Security, with his dissertation focusing on radical terrorism and the security systems of the Western Balkans. At UBT, he also serves as the Vice Dean of the Department of Political Sciences and Security Studies, where he shapes the department's academic direction in the national security field. His research interests encompass national security, radical terrorism, hybrid threats, cybercrime, and organized crime. Since 2020, he has been the coordinator of the International Conference on Political Sciences and Security at UBT College, contributing significantly to the field's academic and professional discourse.

Danuta Kaźmierczak is an associate professor at the University of the National Education Commission in Krakow, Poland. Her research focuses on security, civil protection, and public management. She has participated in security and risk management projects, including the European Union's project *Science for Disaster Risk Management 2020: Acting Today, Protecting Tomorrow* and the Visegrad Fund project 2024, *Security Education on Procedures for Live Bombs and Other Military Explosive Materials*. Since 2020, she has been an editor of *SDirect24 Scientific Journal*. She is the director of the Publishing House at the Institute of Security and International Development. This international forum publishes original peer-reviewed research on defense, security, political studies, and international relations. Her recent research focuses on issues of cultural inclusion and the engagement of local communities in disaster risk management.

Waldemar Kitler is Professor of Security Science at War Studies University in Warsaw, Poland. As a leading authority in national security, he collaborates closely with the Ministry of Science and Higher Education, the Ministry of Climate and Environment, and the Supreme Audit Office. He has worked with the Department of Defense Systems in the Ministry of National Defense, the Crisis Management and Civil Protection Office, and the Main Headquarters of the State Fire Service. During his tenure at the National Defense University and War Studies University, he co-founded and led the Department of Law, served as a Vice Dean for Research at the Faculty of National Security and served as the Head of the Department of National Security Law and the Director of the Institute of Defense Law and Administration. From 2018 to October 2023, he headed the Institute of State Security. He currently serves as the President of the Board of the Defense Knowledge Society. He has served on the Strategic National Security Review Team (2010) and worked with the Ministry of National Defense, the National Security Bureau, and public administration bodies. He has led major research projects funded by the National Centre for Research and Development (NCBR), including the project *National Security System of the Republic of Poland*. He has authored over 270 publications in the field of security science. His research focuses on the national security system, national defense, crisis management, public administration, and the state's function under conditions of threat and wartime, as well as the role of public authorities in these scenarios.

Malwina Ewa Kołodziejczak holds a doctorate in security science and is an assistant professor at the War Studies University in Warsaw, Poland. She is currently pursuing a PhD in law at Cardinal Stefan Wyszyński University. She has successfully led numerous research projects in partnership with the Ministry of Justice, the Ministry of Science and Higher Education, and the Institute of Justice. She received the MOFA Taiwan Fellowship in 2018, during which she conducted research at National Chengchi University in Taipei. Her research interests include the national security system, extraordinary measures for state security, the role of the Commander-in-Chief, the concept of war, recognition in international public law, and constitutional law.

Suzana Lampreia is a Portuguese navy officer and military professor who currently serves as a lecturer in the Science and Technology Department at the Portuguese Naval Academy and the Portuguese Navy Research Center in Alfeite. She holds a degree in naval engineering from the Portuguese Naval Academy, a postgraduate degree in health and safety at work, and both a master's and a PhD in industrial engineering from the New University of Lisbon. Her research interests span cybersecurity, ship maintenance, reliability, condition monitoring, quality control, and data management. She has served as the chief of mechanical engineering and damage control on various Navy ships and at naval installations ashore. She has also held the role of director at the Ponta Delgada POL NATO Depot and served as a designated expert for pollution control in the Azores Islands, in addition to her work as a technical inspector for ships. She is the coordinator of both the naval engineering bachelor's degree program and the master's degree program in naval engineering, with a focus on the mechanical branch.

Mirosław Laskowski is a Lieutenant Colonel and an assistant professor at the War Studies University in Warsaw, Poland, where he serves as Head of the Military Didactics Facility within the Institute of Tactics at the Military Department. He graduated from the Military University of Technology in Warsaw, Maria Curie-Skłodowska University in Lublin, and the War Studies University. He is an active editorial committee member for the scientific journal *Annales Universitatis Mariae Curie-Skłodowska – Sectio J: Paedagogia-Psychologia*. He is a reviewer for the *SDirect24 Scientific Journal*, which operates under the auspices of the NATO Defense Education Enhancement Program. As an expert within the Polish Armed Forces, Lieutenant Colonel Laskowski contributes to developing doctrinal documents. His scientific interests primarily focus on military education, with a current emphasis on its role in building social cohesion and resilience in response to hybrid threats.

David Last is Associate Professor of Political Science at the Royal Military College of Canada, a Commonwealth Scholar and Fulbright Research Chair, Secretary of the International Society of Military Sciences, and co-chair of the Military and Police Relations Working Group of the European Research Group on Military and Society. He served in the Canadian Army for 30 years, including NATO and UN operations. He has taught political science and war studies at the Royal Military College of Canada and the Canadian Forces College since 1999. He has edited or co-edited eight books and published over 80 chapters, articles, and reports on peacekeeping, conflict, and higher education. He has served on program and institutional evaluations for higher military education in Finland, Botswana, Namibia, South Africa, and Portugal. Since 2009, his research has focused on conflict management, taking a global comparative perspective on the contribution of security education.

Victor Lobo is a professor at the Portuguese Naval Academy and a visiting professor at NOVA Information Management School (IMS) of the New University of Lisbon. He holds a master's degree in electrical and computer engineering from the Technical University of Lisbon, and an MSc and a PhD in Computer Science, specializing in Artificial Intelligence, from the Universidade Nova de Lisboa. Habilitation is postdoctoral degree, perhaps we can change to: He also earned a habilitation in Information Management, specializing in Decision Support Systems. He has served as a naval reserve officer and has held leadership roles, including the Director of the Centro de Estudos em Gestão de Informação (known as MAGIC, a research center for information management at NOVA IMS), Deputy Director and Director of the Portuguese Navy Research Center (CINAV), and President of AFCEA-Portugal for two years. He has participated in over 30 research projects funded by the Portuguese Science and Technology Foundation and the European Union (including FP7, H2020, PADR, EDF, PT2020) as a researcher and principal investigator. He has authored over 150 scientific papers in journals and conference proceedings. His primary research interests are machine learning, robotics, computer architecture, underwater acoustics, and cybersecurity.

Harri Ruoslahti is a principal lecturer at the ResLab of Laurea University of Applied Sciences in Finland. He holds a PhD in organizational communication, specializing in co-creation within EU-funded projects. His research and teaching interests are centered on security and business continuity management, societal and organizational resilience, and the social and societal aspects of cybersecurity, hybrid threats, and misinformation/disinformation. He has been engaged in several EU-funded projects, including ECHO and DYNAMO, which focus on enhancing cybersecurity, and INEGMA-E2, dedicated to developing evaluation methodologies for civil protection exercises.

Pedro Miguel da Silva Costa is the Head of the Portuguese Air Force's Space Operations Center and serves as a Defense Liaison Officer at the Portuguese Space Agency. In these roles, he has been actively involved in Space Domain Awareness, Space Intelligence (INTEL), and

Access to Space. He represents Portugal at the European Commission, the European Union's Space Strategy for Security and Defense, the European Defense Agency, NATO's Centre of Excellence for Space and is the National Contact Point for CapTech Space. He participates in the UN Open-Ended Working Group (OEWG) on reducing space threats by establishing norms, rules, and principles for responsible behavior in space. He holds a degree in military and aeronautical sciences from the Air Force Academy. He has completed various career development courses, including the Joint Staff Course (CEMC). He also holds a master's in military sciences, security, and defense from the Military University Institute (IUM). He is a PhD candidate in international relations at the Faculty of Social Sciences and Humanities of Universidade Nova Lisboa. He is the director of the research project "Space Power: A New Structuring Domain of Power," which is being conducted at the IUM Research Center.

André Simonyi is Assistant Professor of International Studies at the Royal Military College Saint-Jean in Quebec. He holds a PhD from the School of Political Studies at the University of Ottawa, Canada, and is a graduate of the War College in Paris, France. He served in international operations as a Canadian infantry officer and graduated from France's École de Guerre. He has been a visiting scholar at Harvard University's Davis Center for Russian and Eurasian Studies, the Harvard Ukrainian Research Institute, the Institute for Advanced Studies at the Central European University in Budapest, and the New School for Social Research in New York. His research focuses on complexity theories and systemic approaches to understanding the emergence of political violence and the rise of Grey Zone conflicts. As a globalist and ethnographer, his fieldwork includes Ukraine, Hungary, China, and the United States. He has co-directed books on the relationship between religion and politics, as well as the impact of the war in Ukraine on Canada's security posture. He has published articles on the emergence of the concept of Grey Zone conflicts and its relevance to political theory. His current work emphasizes incorporating complexity thinking into the theorization of International Relations through a relational perspective.

Per Skoglund is a senior lecturer at the Swedish Defense University (SEDU), serving as the Head of the Functions and Perspectives Division. He holds a master's degree in engineering and a PhD in business administration. Since joining SEDU in 2016, he has held military logistics and supply chain management teaching and research positions. In addition to his role at SEDU, he has been an associate professor at the Norwegian Defense University College since 2014. His research focuses on the evolution of military supply chains over time across contexts of peace, crisis, and war, with a particular emphasis on the perspective of small nations.

Tarik Solmaz holds a PhD in strategy and security from the University of Exeter. His doctoral thesis, *Rethinking the Concept of 'Hybrid Warfare': A Revisionist Perspective,* is a critical re-evaluation of hybrid warfare, examining both the concept and its practical applications. From 2014 to 2018, he served as a security analyst at the Undersecretariat of Public Order and Security in Turkey, responsible for developing counter-terrorism strategies. He participated in research projects on security threats conducted by the Turkish Ministry of Interior, including Hybrid Threats, *Terrorism and Biological Terrorism in the Context of CBRN Threats*. His primary research interests include hybrid warfare, insurgency, and terrorism. He works at as lecturer at the Turkish Military Academy.

Ilkka Tikanmäki is a retired Lieutenant Commander (engineering) and currently serves as Safety, Security, and Risk Management Project Specialist at Laurea University of Applied Sciences. He holds an MBA in information systems and a BSc in Information Technology.

He is pursuing a PhD in operational art and tactics at the Finnish National Defense Uni versity. His research focuses on situational awareness, understanding, decision-making, inter-authority cooperation, information sharing, and addressing hybrid threats. He contributes to projects funded by the European Union—AIRBEAM, EUCISE2020, MARISA, RANGER, ANDROMEDA, ECHO, and DYNAMO. Throughout his career, he has served in various roles within the Air Force, the Navy, and the Border Guard. His experience includes participation in NATO's SFOR operation in Bosnia-Herzegovina, focusing on civil-military cooperation, and involvement in the UN UNIFIL operation in Lebanon.

Valter Vairinhos is a senior lecturer at the Portuguese Naval Academy and a researcher at the Portuguese Navy Research Center (CINAV). He was an associate professor with ISLA Santarém and a researcher at the Intellectual Capital Association (ICLAB-ICAA) in Alfeite and Santarém. He holds a degree in naval engineering from the Portuguese Naval School, an degree in applied mathematics and a master's degree in statistics and operations research from the Faculdade de Ciências, University of Lisbon. He earned his doctoral degree in multivariate data analysis from the Statistics Department of Salamanca University. His research interests include multivariate data analysis, maintenance management, reliability, data mining, and intellectual capital.

Hajdi Xhixha is Assistant Professor of Political Science at the University for Business and Technology (UBT) in Pristina, Kosovo, where she currently serves as the Dean of the Faculty of Political Science and Security Studies. Her doctoral research focused on international relations and diplomacy, particularly Russian-Albanian relations. Her current research focuses on Russian policies and security dynamics in the Western Balkans to enhance state defense strategies for Kosovo and Albania and address broader implications for regional security. In addition to her administrative and research roles, she teaches Russian Foreign Policy in the UBT Master's program in political science, international relations, and governance (POSIG). She coordinates a partnership project between UBT and the Centre for Resilience and Sustainable Development at Cambridge University, the Cambridge Policy Boot Camp (CPBC) on Feeding Children Nutritiously for Kosovo.

Marzena Żakowska is Assistant Professor at War Studies University in Warsaw, Poland. She holds a PhD in security science from the National Defense University in Warsaw. She holds postgraduate diplomas in armed conflicts and european union studies. She has completed advanced courses in conflict resolution at the Academy for International Conflict Management and Peacebuilding, United States Institute of Peace. She served as Vice Dean for Student Affairs and is now Director of Global Affairs and Diplomacy Studies at the Faculty of National Security, War Studies University. Since 2018, she has chaired the War Studies Group of the International Society of Military Sciences. As a postdoctoral research fellow, she collaborated with the Royal Military College of Canada and the US Army War College. She contributed to the NATO-DEEP *Hybrid Threats and Warfare Reference Curriculum*. She collaborates with scientific journals such as the *Small Wars Journal (SWJ–El Centro)* and the *Journal of Law and Politics*. Her research emphasizes national security, war and armed conflicts, hybrid threats, security in the Western Balkans, and broader issues related to social security.

Archipelago of Design Team

Philippe Beaulieu-Brossard is Co-President of the Archipelago of Design (AoD) and a professor at the Canadian Forces College in Toronto. He holds a PhD in international relations from the University of St Andrews in the United Kingdom. Hc cmpowers national security leaders to shift mindsets in their teams to set conditions for evolution and address complex situations across NATO members and partners. He finds inspiration in transformative game design, design thinking, systems thinking, and foresight to propose active learning approaches to fulfill his mission.

Oliver Jones is a senior researcher at the Archipelago of Design, where he leads projects in serious game design and publishes on security and defense. He holds a PhD in english from York University and an MA in critical theory and cultural studies from McMaster University. His PhD dissertation, *Literary Friction: Poetics at the Operational Level of War*, explores literary and philosophical texts connected to the Israeli school of military design to understand how literary practices have been mobilized to support innovation in defense organizations from TE Lawrence to the present day.

Kassie Miedema is a design and foresight researcher at the Archipelago of Design. She holds a master's degree in strategic foresight and innovation from the Ontario College of Art and Design (OCAD) in Toronto and master's and bachelor's degrees from the School of Architecture at the University of Waterloo. At the Archipelago of Design, she applies her multidisciplinary expertise to tackle contemporary security and defense challenges through design thinking, foresight, gaming, workshop facilitation, and creative problem-solving methods. Her research interests include systems design, resilience, existential risks, urban agriculture, transformative cultural change, and strategic foresight.

Acknowledgments

This book is the culmination of the "Modern War and Grey Zones" project, initiated in 2021 by the War Studies Working Group (WG1) of the International Society of Military Science (ISMS). The editors are grateful to all ISMS WG1 contributors for their invaluable discussions, insightful viewpoints, and constructive comments, without which this book could not have been conceived and written. The consistent support of successive presidents of the ISMS is appreciated: Professor Pierre Jolicoeur, RMC (2021), Admiral João Paulo Ramalho Marreiros, Military University Institute of Portugal (2022), Professor Niels Bo Poulsen, Royal Danish Defence College (2023), and Professor Kjell Engelbrekt, Swedish Defence University (2024).

Many subject experts provided essential commentary and review throughout the development and editing process. It is a characteristic of work in the Grey Zone that some prefer not to be named. To you, to anonymous reviewers, and to any we have omitted, we are no less in your debt. The following people provided critical commentary, expertise, support, and reviews over the several years of preparing the manuscript: Susanna Bellander, Marina Caparini, Calvin Chrustie, Ian Ferrier, Cindy Gahunia, Ieva Galauskaite, John Gilmour, Dwayne Hobbes, Ken Holmes, Nancy Imbeault, William Jardine, Frederick Labarre, Carina Lamont, Yi Luo, Aiden McCaffrey, Peter Papler, Leah Sheriff-West, Kimio Tessier, Lotta Tillberg, Rob Williams, and Tyler Wentzel. The many different forms of expertise and wide range of employment—from banking and finance to law, policing, governance, intelligence, and the margins of society—illustrate the breadth of the Grey Zone problem.

In addition to the member institutions of the ISMS, conversations and events hosted by the National Guard University and the Kyiv Institute of Ukraine, the Slovenian Centre for Military Schools, the Faculty of Military Studies of Stellenbosch University and South African Military Academy, the School of Military Studies of the University of Namibia, the European Research Group on Military and Society (ERGOMAS), the Portuguese Higher Institute of Police Sciences and Internal Security, and Canadian Communications Security Establishment have made essential contributions to the body of knowledge on which this work draws.

We are grateful to the Archipelago of Design (AoD) team for their diligent work and continuing contributions to the valuable AoD website. Philippe Beaulieu-Brossard's leadership and the excellent work of Kassie Miedema and Oliver Jones have made design problems integral to this book. We sincerely thank Bartosz Mlyczyński, Maja Grochalska, and Jakub Krzyżanowski for their insightful discussions on the design problems.

The content does not necessarily reflect the views of any of the organizations to which the authors may belong. The editors are grateful to the Routledge team for their support and professionalism in producing this book and its electronic counterpart. Finally, we infinitely appreciate our families' support in completing this work.

Marzena Żakowska, Warsaw
David Last, Toronto

Foreword

Modern war in the Grey Zone is paradoxically more and less than past wars. It is more than traditional war because it engages political skullduggery, economic subterfuge, social impacts, new technologies, lawfare, and environmental vulnerability to a greater extent than military war plans of the past, even including the Cold War. It demands close cooperation from all branches of government and society in ways not even achieved in the Great Wars of the last century. However, it is less than those all-consuming, total wars-until-victory for which we once prepared. For small, democratic countries, our comparative advantage lies in pursuing all that peace offers our citizens. Good governance, prosperity, fair dealing, equity, rights, neighborly relations, free flows of information, and human development provide defenses against subversion and fragmentation that war in the Grey Zone threatens. These characteristics mostly support a cohesive society that can withstand and adapt to external threats and domestic challenges.

The "Grey Zone" concept has gained prominence as traditional distinctions between war and peace have broken down. It overlaps with ideas of hybrid war, political warfare, influence operations, and asymmetrical conflict. Indeed, overlapping and concentric Grey Zones have been defined for police, lawyers, special forces, bankers, forensic accountants, and others. For small states, navigating this complex environment requires innovative strategies and a sound understanding of global and local security issues. The authors in this volume have worked with a common understanding of the Grey Zone, which is arguably appropriate to most of the world's states, which are not the major powers for which much of the security literature is produced.

The ISMS—a society of defense universities in small democratic countries—furthers research and academic education in military arts and sciences in the broadest sense. It is a network for creating, developing, exchanging, and disseminating research and knowledge about war, conflict management, and peace support efforts. Its contributions include the open-source *Handbook of Military Sciences*, funded by the Norwegian Defence University College, and an annual conference hosted by its member institutions. Since the 2015 conference in Helsinki, its members have been seized with the problem of war below the threshold of war and the implications for peace and security.

This book is an important resource for security professionals, policymakers, and educators. It addresses the foundational elements of Grey Zone conflicts, including climate change, cybersecurity, space, and markets. It explores the strategic implications for small states, including policy development, crisis management, and international cooperation. A key strength of the volume is its emphasis on the social dimensions of modern conflict. Understanding how modern war in the Grey Zone affects social cohesion and state resilience is vital for today's security practitioners, whether serving in uniform or not. Complementing this book's four sections, the AoD has provided design problems as intellectual exercises to foster critical thinking and develop innovative solutions. The AoD's companion website expands on these exercises.

This volume and the Society that has supported its development are important resources for defense universities, policymakers, and scholars of the new security.

Professor Kjell Engelbrekt
Swedish Defence University
2024 President of the International Society
of Military Sciences, Stockholm

Abbreviations

ABW	Internal Security Agency (Poland)
ADF	Alternative für Deutschland
AI	Artificial Intelligence
AJP	Allied Joint Publication
AoD	Archipelago of Design
ASAT	Anti-satellite weapon
ASEAN	Association of Southeast Asian Nations
AU	African Union
AW	Foreign Intelligence Agency (Poland)
BCCI	Bank of Credit and Commerce International
BOA	Antiterrorism Operations Bureau (Poland)
BRICS	Brazil, Russia, India, China
CAF	Military Intelligence (Canada)
CBA	Central Anti-corruption Bureau (Poland)
CBRN	Chemical, Biological, Radiological, and Nuclear
CBSA	Canada Border Services Agency
CCDCOE	NATO Cooperative Cyber Defense Centre of Excellence
CN	China
CNCS	Portuguese National Cybersecurity Centre
CO	Cyberspace operations
COPUS	Committee on the Peaceful Uses of Outer Space
CSDP	Common Security and Defense Policy
CSE	Communications Security Establishment
CSIS	Canadian Security Intelligence Services
CSIS	Center for Strategic and International Studies
DIME	Diplomatic, Informational, Military, Economic
DIMEFIL	Diplomatic, Informational, Military, Economic, Financial, Intelligence, Law enforcement
ECOWAS	Economic Community of West African States
ECRE	European Council on Refugees and Exiles
EMP	Electromagnetic pulse
ERGOMAS	European Research Group on Military and Society
EU	European Union
FAO	Food and Agriculture Organization (of the United Nations)
FINTRAC	Financial Transactions and Reports Analysis Centre of Canada
FIRST	Forum of Incident Response and Security Teams

GAC	Global Affairs Canada
GDP	Gross Domestic Product
GEO	Geostationary Earth Orbit
GHG	Greenhouse Gases
GIIF	General Inspectorate of Financial Information (Poland)
GRU	Russian Military Intelligence Service
HADR	Humanitarian and Disaster Relief
HEO	High Elliptical Orbit
HNS	Host Nation Support
HUMINT	Human intelligence
IACHR	Inter-American Court of Human Rights
IASB	International Accounting Standards Board
IBM	International Business Machines Corporation
ICAIE	International Coalition against Illicit Economies
ICC	International Criminal Court
ICISS	International Commission on Intervention and State Sovereignty
ICJ	International Court of Justice
IEP	The Institute for Economic and Peace
ILO	International Labour Organization
IMF	International Monetary Fund
IOM	International Organization for Migration
IPBES	Intergovernmental Science-Policy Platform on Biodiversity and Ecosystem Services
IP	Intellectual Property
IPCC	Intergovernmental Panel on Climate Change
IR	International Relations
IRGC	Islamic Revolutionary Guard Corps
ISE	Institute for State Effectiveness
ISIL	Islamic State of Iraq and the Levant
ISMS	International Society of Military Sciences
ISR	Intelligence, surveillance, reconnaissance
ITAC	Integrated Terrorism Assessment Centre
IU	In-vehicle unit
JLSG	Joint Logistics Support Group
KSP	Warsaw Metropolitan Police
LEO	Low Earth orbit
MACA	Military assistance to civil authorities
MCDC	Multinational Capability Development Campaign
MCPP-N	Marine Corps Prepositioning Program—Norway
MDA	MacDonald Detwiler and Associates
MEO	Medium Earth Orbit
MI	Ministry of Infrastructure (Poland)
MPK	Krakow Municipal Transport Company
MTCR	Missile Technology Control Regime
NASA	National Aeronautical and Space Administration
NASDAQ	National Association of Securities Dealers Automated Quotations
NATO	North Atlantic Treaty Organization
NDC	Nationally Determined Contribution

NGO	Non-Governmental Organization
NORDEFCO	Nordic Defense Cooperation
NSA	National Security Agency
NSPA	NATO Support and Procurement
OAS	Organization of American States
OCHA	Office for the Coordination of Humanitarian Affairs
OCO	Offensive cyberspace operations
OECD	Organization for Economic Co-operation and Development
OFOP	National Federation of Polish NGOs
OIC	Organization of Islamic Cooperation
OSCE	Organization for Security and Cooperation in Europe
OSINT	Open-source intelligence
PESTLE	Political, Economic, Social, Technological, Legal, Environmental Factors
PGNiG	Polish Oil Mining and Gas Extraction
PMCs	Private military companies
PMESII	Political, Military, Economic, Social, Informational, Infrastructure
PME	Professional Military Education
PNT	Positioning, navigation, time
POLSA	Polish Space Agency
PPE	Personal protective equipment
PSCs	Private security companies
PwC	PricewaterhouseCoopers
REB	Radio-Electronic Warfare
RMA	Revolution in Military Affairs
RU	Russia
SC	Customs Service (Poland)
SCO	Shanghai Cooperation Organisation
SCS	Customs and Tax Service
SDGs	Sustainable Development Goals
SDI	Strategic Defense Initiative
SEC	Security Exchange Commission
SELEC	Southeast European Law Enforcement Center
SG	Border Guard (Poland)
SIGINT	Signals intelligence
SKW	Military Counterintelligence Service (Poland)
SOP	State Protection Service
SWIFT	Society for Worldwide Interbank Financial Telecommunication
SWW	Military Intelligence Service (Poland)
UAE	United Arab Emirates
UCDP	Uppsala Conflict Data Program
UK	United Kingdom
UN	United Nations
UNDP	United Nations Development Programme
UN GGE	United Nations Group of Governmental Experts
UNDRIP	UN Declaration on the Rights of Indigenous Peoples
UNFCCC	United Nations Framework Convention on Climate Change
UNHCR	United National High Commissioner for Refugees

UNRWA	United Nations Relief and Works Agency for Palestine Refugees in the Near East
USSOCOM	United States Special Operations Command
USSR	Union of Soviet Socialist Republics
UTK	Office of Rail Transport
VNSAs	Violent non-state actor
WB	World Bank
WB6	Western Balkans Six
WEF	World Economic Forum
WFP	World Food Programme
WG1	War Studies Working Group
WHO	World Health Organization
WOT	Territorial Defense Forces (Poland)
WS	Special Forces (Poland)
WTP	Warsaw Public Transport
WWII	World War II
ZW	Military Gendarmerie (Poland)

Introduction

Modern War, Grey Zones, and Security

Marzena Żakowska and David Last

War is not what it once was. Traditional war studies and histories of warfare drew a sharp distinction between war and peace. The line has always been blurred with labels like phony war or Cold War, but today, the war-peace distinction is less useful than ever. Soldiers may wait for a declaration of war, unaware that it is upon them already. Most countries need a new way to think about war, peace, and self-defense. Many governments are not seized with shaping the strategic environment or global balance of power but must find ways to protect national interests and mitigate threats to their preferred order. They find themselves increasingly subject to pressures short of war wielded by neighbors and larger states. Most of the world's military and security leaders spend their careers not fighting wars but preventing and avoiding them. Yet, this is not a benign environment. It is modern war in the Grey Zone—the subject of this book.

Modern War and the Grey Zone

Modern wars, variously called contemporary or 21st-century wars, blend recognized hostilities with non-military forms of aggression. Regular and irregular forces mix with civilians who fight remotely or conceal themselves in a target population. Advanced kinetic and non-kinetic weapons allow a palette of lethal and nonlethal attack modes. The source of any attack may be evident, ambiguous, or so obscure that a lengthy forensic investigation yields conclusive results.

Grey Zones are the conceptual spaces between positive peace and unequivocal war, illustrated in Figure IN 1. What makes it inextricably intertwined with modern war is that the many parties to a conflict will see the threshold of war—the very definition of war—in different terms. Sun Tzu's *Art of War* articulated the principle: "Therefore those who win every battle are not skillful—those who render others' armies helpless without fighting are the best of all" (Sun 1988:94). Intelligence, deception, information warfare, and tools that erode military capability without direct confrontation have long been employed. New technologies in space, the cyber domain, artificial intelligence, and autonomous weapons combine with perennial social dynamics like outgroups, criminals, and political entrepreneurs in new ways.

Figure IN 1 depicts well-established concepts of war—regular, irregular, and hybrid combinations—at the bottom (Bingöl 2017; Kaufman and Hartnett 2016:26). Asymmetrical conflict, defined as an asymmetry of forces, actors, or means, exists across the spectrum of modern war. When regular security forces confront secessionist irregulars, when mixed military and police units battle foreign-sponsored gangs, or when corporate raids seize defense assets, these are asymmetrical battles. They may be above or below a threshold of violence recognized as "war" by any party.

Throughout the Grey Zone (capitalized when used as we define it), hostile actors pursue kinetic and non-kinetic actions. Bombing infrastructure, assassinating leaders, and violent riots

DOI: 10.4324/9781003428701-1

are kinetic actions. However, infrastructure might be damaged or destroyed by a non-kinetic cyberattack. An assassinated leader may lead to non-kinetic political destabilization. Violent riots may be instigated by online manipulation and deception. These are not wholly new phenomena but combine with technologies new in the 21st century.

The actors responsible for hostile acts may be regular military forces whose identity might be obscured, irregular forces with foreign affiliations, or civilians without status as combatants. These actors may be inside a target country, cross boundaries to act, or engage targets from a distance, from bases in the hostile state, or anywhere else. The deliberate ambiguity of the Grey Zone allows for the denial of hostility while eroding the capability and will of the target country. Hostile states may wage information war, ally with disruptive elements, sponsor "fifth columns" and fellow travelers, use economic coercion, and manipulate the laws and freedoms in an open society to obscure any clear threshold of war.

The ambiguity of the Grey Zone may paralyze government decision-making and accentuate schisms in society. Allied countries may hesitate to commit to support. Deterrence or mutual aid may be based on clear hostilities and intent absent from the Grey Zone. There is no clear threshold between modern war and the Grey Zone of uncertain negative peace, in which multiple axes of competition compound violence and coercion from uncertain sources. The relative simplicity of East-West tensions in the first Cold War is replaced by multipolar contention. Small countries must navigate domestic unrest, and Grey Zone hostilities exacerbate many potential state and non-state instigators.

At the top of Figure IN 1 is positive peace. The Institute for Economics and Peace (IEP) offers a definition of positive peace resting on eight pillars: well-functioning government, sound business environment, equitable distribution of resources, acceptance of the rights of others, good relations with neighbors, free flow of information, high levels of human capital, and low levels of corruption. As communities and states slide from the "broad, sunlit uplands" (in Churchill's phrase) of positive peace, coercion, and fear can preserve the negative peace of contained violence, violence both within and between states may characterize periods of peace. The objective of security professionals in most countries is to contain and defeat violent threats

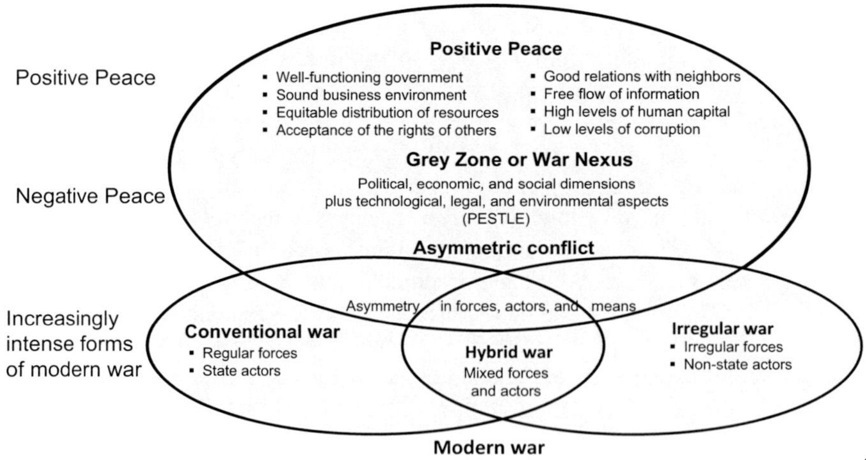

Figure IN 1 Modern War and Grey Zones.

Source: The depiction of war draws on Bingöl (2017) and Kaufman and Hartnett (2016:26); the depiction of peace draws on the IEP. Assembled by the Authors.

to national security, maximize the potential for human security, and contribute to operations that collectively enhance international security, usually through collective security (like the United Nations or Organization for Security and Cooperation in Europe (OSCE)) or collective defense (like NATO). The Grey Zone demands new knowledge and skills if military, police, paramilitary, and intelligence forces work effectively to support positive peace.

Thinking about War and Peace

We can begin with a taxonomy of military history and war studies to describe Grey Zones and modern war. A horizontal axis ranges from a narrow military focus to a broader view of political, economic, and social contexts for the use of intentional force and the pursuit of security. A vertical axis in Figure IN 2 describes more conventional disciplinary studies at the top and less conventional multidisciplinary and novel studies at the bottom. Work providing a conventional analysis of military conflict between states belongs in the upper left, and work examining broad implications of new threats and unconventional challenges belongs in the lower right. A discussion of forms of modern war must be informed by research and literature on the left, with a firm understanding of the implications of kinetic violence. In the Grey Zone, however, unconventional tools and political, economic, and social contexts are essential knowledge for security professionals.

Starting in the upper left of Figure IN 2, traditional military histories like the *Cambridge History of Warfare* describe conventional military strategy, operations, and tactics, usually for dominant powers of the age. Historians like John Keegan set their military histories in a broader social and political context, but they bifurcate war and peace, separating military and non-military functions. A.J.P. Taylor's description of *How Wars Begin* and Gideon Rose's book *How Wars End* illustrate this bifurcation of interstate wars. Licklider's work on the termination of civil wars does the same for intrastate violence. Isolating variables is a necessary social science tool, but doing so means omitting important developments. The bifurcation of war and peace is also evident in the US-centric collection by David Jordan and colleagues, *Understanding Modern Warfare*. The section on irregular warfare by James Kiras addresses guerrilla

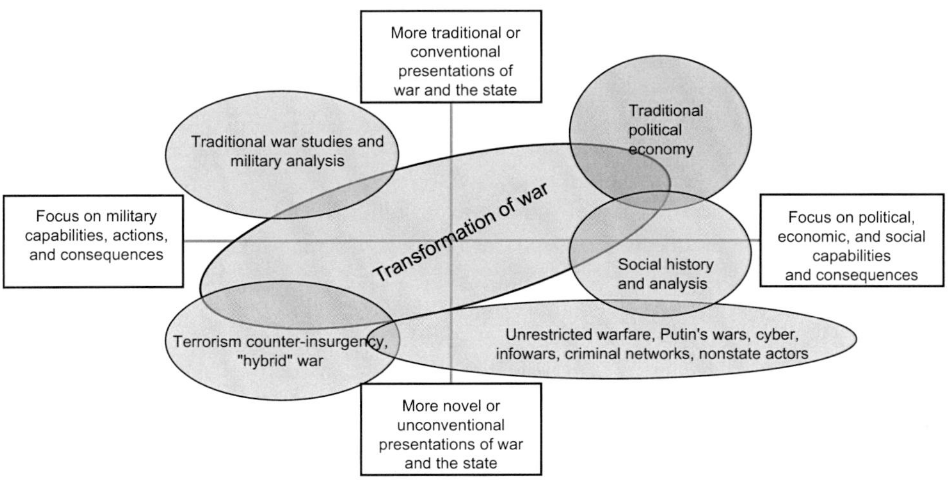

Figure IN 2 Relevant Knowledge for the Grey Zone.

Source: Compiled by the Authors from sources noted in the text.

war, insurgency and counterinsurgency, stability operations, foreign internal defense (an older concept), and coups, terrorism, revolution, and civil wars as forms of violence within the state (Jordan et al. 2016:305ff). Although political, economic, and social dimensions receive regular nods, the volume belongs in the upper left in Figure IN 2.

The lower left is dominated by studies of counterterrorism, insurgency, and military approaches to unconventional conflict, usually close to the intersection with conventional military thinking because that is where the market for these studies lies. The United States writing about the gray zone (US spelling) after Russia's 2014 seizure of Crimea also focuses on this lower left quadrant, concerned primarily with military responses to ambiguous circumstances. A US Special Operations Command white paper published in September 2015 makes it clear that US military thinkers are focused on strategic competition within the circles of war at the bottom of the Grey Zone in Figure IN 1 (USSOCOM 2015).

Writing about the Grey Zone from the US military-academic complex applies major-power strategic thinking, including unexpressed assumptions about the ascendancy of market capitalism, the legitimacy of US intervention, and strategic competition with Russia and China (Mazarr 2015; Freier 2016; Matisek 2017; Morris et al. 2019; Bolton 2021). These sources do not hint that the United States is a significant source of Grey Zone threats to many countries. For the security leaders of smaller states trying to manage the Grey Zone, neither the strategic perspective of major powers nor the focus on military responses to Grey Zone phenomena is appropriate. Political, economic, and social impacts, and the steps necessary to pursue positive peace are more important for most states.

Straddling the lower left and right are the 25 means of waging *Unrestricted Warfare*, described by two Colonels of the Chinese People's Liberation Army (Liang & Xiangsui 1999). Marc Galeotti's two books from 2022 also bridge the gap from lower left to lower right. *Putin's Wars* describes the blending of conventional and unconventional means in Russia's wars from Chechnya to Ukraine, including the February 2022 invasion. *The Weaponisation of Everything*, released before Putin's Wars, effectively articulates the breadth of Grey Zone conflicts. Galeotti addresses commercialized mercenaries, criminal business enterprise, the weaponization of life through hostages (something as old as warfare), lawfare and insurance pressures, the "deep fake" internet, cyberattacks and the release of hacked information, cultural influence, and permanent instability.

The upper right quadrant in Figure IN 2 includes descriptions of conventional conflict and competition extending to political, economic, and social means. Here, we find works on war and society, the transformation of war, and crossing into the lower left, work on information warfare, like *How to Lose the Information War* (Jankowicz 2020), and *Disinformation, Misinformation, and Fake News in Social Media* (Shu et al. 2020).

If war is an act of force to compel the enemy to our will, then the forms taken by force (including manipulation and deception) and the ability to interact directly with a target society rather than through the intermediary of violence have created new forms of Grey Zone conflict and modern war. Why is an actor motivated to engage in Grey Zone conflict? War above the acknowledgment threshold elicits an unpredictable response, but protracted instability and violence, without any clear victory, can be an attractively adjustable strategy for both state and non-state actors. Grey Zone attacks can be contained or used to justify extraordinary actions (the "state of exception" identified in Chapter 1). Chaos abroad can reduce the appeal of alternative governance systems, and overt actions can mobilize support for the government. Grey Zone operations can also serve non-state actor interests. Criminality thrives in the Grey Zone. Violent specialists (Tilly 2003) are in demand. Violence short of war can be profitable for legitimate businesses and high-paid specialists.

Framing Grey Zone conflict draws on recent books about the conditions of the early 21st century, which belong in the right half of Figure IN 2. Together, they explain why the Grey Zone seems more prominent now than during the Cold War or in the period of uncontested US hegemony immediately afterward. The Global War on terror and neoliberal excesses eroded trust and confidence in the US-led order. The Grey Zone has thrived on the unraveling of the neoliberal order. Gary Gerstle's book, *The Rise and Fall of the Neoliberal Order* (2022), is an appropriately US-centric description of this unraveling. Under rampant neoliberalism and financial capitalism, with no fear of communism as an alternative, markets again eroded society as they had in the first cycle of neoliberal globalization at the end of the 19th century (Polanyi 1957). The erosion of society has invited authoritarian solutions. Interpreting the end of the Cold War and the rise of authoritarianism, political historian Timothy Snyder's *Road to Unfreedom* (2018) describes the politics of inevitability and the politics of eternity. Here is a new motivation for eternal war in the Grey Zone. Disrupting neighbors and eroding truth helps preserve regimes like Putin's. Russia's leaders are confident that Russians can suffer more than the decadent West. The United States is confident that it can wage perpetual war if its citizens don't have to pay for it. Here are two significant players in the "war nexus" described by André Simonyi.

Who Needs This Book and Why

This book fills a gap in knowledge and thought processes for security leaders. As they prepare to advance from the specialized knowledge of their early appointments to the more general understanding demanded at mid-career, they will be expected to plan and conduct operations with other countries, agencies, and civilians. Many of this book's authors have military and government experience or teach in defense universities and staff colleges. Researchers and graduate students studying the phenomena will also find it valuable.

The International Society of Military Sciences (ISMS) War Studies Working Group initiated the project in 2021. ISMS is a consortium of defense universities serving the military community in small democratic countries. Through discussions, workshops, and conferences, it became clear that our colleagues and students needed new ways to think about conflict to develop resilient and adaptive security solutions. The linear operational planning process taught in staff colleges for generations is robust but inadequate to address the complex interaction of new actors, threats, and factors affecting security. Archipelago of Design (AoD)—another international consortium serving the security community—has been developing exercises, guidelines, and workbooks to challenge security leaders to think creatively and critically about confronting today's challenges. Philippe Beaulieu Brossard, Oliver Jones, and Kassie Miedema have worked with the editors to produce design exercises that introduce essential design tools such as divergent and convergent thinking in problem definition and framing of solutions. The design problems refer to resources on the AoD website. Readers are encouraged to explore the resources online.

Professional military education has reflected changes in warfare and domestic security. Major powers and coalition partners have typically dominated perspectives on interstate wars, while local views have influenced teaching on civil wars, insurgencies, and domestic assistance to governments. Modern war and Grey Zone conflicts have implications for scholars, researchers, mid-career professionals, and citizens, which must not be seen through the lens of major powers alone. The war nexus, challenges of climate change, dependence on space and the cyber domain, and the challenge of violent non-state actors (VNSAs) look different from Ottawa and Warsaw, Windhoek, and Ulaanbaatar than they do from Washington, Beijing, and Moscow.

Book Plan and Design Problems

The four sections address the characteristics of modern war in the Grey Zone, planning and decisions, deterring and managing violence, and society and people. An introduction to design thinking follows this introduction, and each section concludes with an exploration of one or two aspects of design, which can be applied to problems suggested by the chapters of the section. Educators and curriculum developers can use individual chapters or groups of chapters with the design exercises to encourage students and professionals to think creatively about challenges their predecessors have not encountered. The authors speak from experience directly to security professionals in states that are not superpowers, offering them practical insights and strategies relevant to their situations. A tabular summary of Grey Zone threats that serve to weaken the state and its institutions is included in the Appendix.

Section I addresses six novel aspects of modern war and the Grey Zone. Climate change, dependence on space assets, vulnerability to cyberattacks, globalized markets, and VNSAs affect small states in new and dangerous ways. In Chapter 1, André Simonyi analyzes the shift from total war, characterized by complete national mobilization, to the war nexus involving all domains of society. Simonyi argues that contemporary wars transcend traditional military engagements, incorporating economic, cyber, and social media dimensions. The war nexus represents a new ontology of war, where conflicts are ongoing, undeclared, and transcend borders. This paradigm shift challenges existing norms and necessitates a new approach to understanding and managing modern conflicts.

Hostile actors create the war nexus, but climate change affects all states and demands common action. In Chapter 2, Duraid Jalili links climate change to security. Global heating, pollution, and biodiversity loss exacerbate conflicts and generate new responsibilities for military forces—managing consequences and avoiding worsening the sources of climate change. Jalili discusses the implications for governance frameworks and strategic narratives, emphasizing the crucial role of security forces in managing disasters and complex emergencies. Declining habitability, environmental insecurity, rising temperatures, and extreme weather are reshaping the roles of security forces, prioritizing adaptation, resilience, and mitigation strategies.

Space access for communications, navigation, imagery, and intelligence is crucial in modern war. In Chapter 3, Pedro Miguel da Silva Costa, David Last, and Marzena Żakowska describe the strategic significance of space-based systems for communications, navigation, imagery, and intelligence. The United States, Russia, and China remain dominant, but the rise of commercial actors and new alliances gives small states options. The legal and financial frameworks for space management highlight geopolitical and security implications of space-based assets in the Grey Zone.

Chapter 4, by Suzana Lampreia, Victor Lobo, and Valter Vairinhos, examines cybersecurity challenges facing military systems. Cyberattacks are a major vulnerability in modern war and the Grey Zone. Any connected system is vulnerable to disruption and corrupted data. Risk mitigation strategies are essential. Blockchain technology offers promise but has limitations. Robust cybersecurity measures to protect networked systems are essential to modern military operations, but there is a balance between managing costs and maintaining performance in the face of cybersecurity risks.

In Chapter 5, David Last explores markets and Grey Zone conflicts. Market manipulation, commercial theft, and currency attacks impact national and human security. Small states must understand and leverage market and regulatory forces to manage their strategic environments. Commercial and financial power play a role in maintaining national capacity and security, even for small states.

Chapter 6, by Tarik Solmaz, examines the role of VNSAs in Grey Zone conflicts. Solmaz categorizes VNSAs as primarily political, criminal, or commercial, with overlapping motivations and characteristics. They are ready tools for hostile actors and transcend state boundaries easily. Both as instruments of modern war and on their own account, VNSAs can undermine the state's cohesion and capacity, demanding sophisticated responses by military, gendarmes, police, and intelligence.

Design Problem 1 focuses on choices about the nature of security and the balance of effort to achieve human security, national security, and international security, given the goal of positive peace and the constraints of the international system. Divergent thinking, horizon scanning, and identifying weak signals help frame the identification of security problems.

Section II explores the integration of ends, ways, and means. Most states have limited influence over the international factors affecting their security, but proactive policy choices and national strategies are still possible. Chapters on the policy-strategy link, crisis management, all-hazards intelligence, and logistics cooperation from the perspective of smaller states frame responses to modern war in the Grey Zone.

Chapter 7, by Jeffrey V. Gardner, Marzena Żakowska, and David Last, describes the connections between national security policies, strategic choices, and tactical actions within the Grey Zone. National security strategies for small states are developed through intergovernmental collaboration. Policy options translate into operational and tactical capabilities. The nature of the Grey Zone demands that whole-of-government policies be extended to whole-of-society actions in national defense. This chapter should be read with Section IV, which links society and people to national defense against Grey Zone threats.

In Chapter 8, Ilkka Tikanmäki and Harri Ruoslahti define crises in the context of hybrid threats and Grey Zone warfare. They present a three-part management strategy—detection, deterrence, and response—that guides strategic countermeasures. This chapter connects crisis management, leadership, resilience, and defense capabilities to Grey Zone attacks, underscoring the importance of alliances and networks of friendly states in managing crises and countering aggression.

Modern war and Grey Zone conflicts demand integrated intelligence for all hazards. In Chapter 9, David Last, Urszula Jaśkiewicz, and Marzena Żakowska examine how states with limited resources can draw on many sources of information and analysis to identify and track threats. Fusion centers and task forces can integrate intelligence beyond state security services to include whole-of-society assets supporting the management of Grey Zone conflicts.

In Chapter 10, Per Skoglund describes logistics cooperation for small states in Grey Zone conflicts and modern warfare. Planning for resilience and mobilization demands strategic choices about defense industries, sources of supply, and supply chain management. Alliance and alignment strategies can ensure supply continuity and mitigate risks or leave states vulnerable to abandonment.

Design Problem 2 introduces convergent thinking and the rich picture method to compress many ideas about a complex problem to arrive at an optimum solution. Design approaches often alternate between divergent and convergent thinking to explore options. The rich picture is a helpful visual technique for communicating across organizational and language barriers to build a common approach.

Section III deals with managing escalating violence in modern war—regular, irregular, hybrid, and asymmetric—as it bleeds into the Grey Zone. To deter and manage violence, security leaders must understand the use of international institutions, the challenges of deterrence strategies and alignment, the new potential for offensive cyber operations, and how information war and the use of law to wage war (lawfare) can affect the balance of power.

In Chapter 11, Marzena Żakowska and David Last explore how small states use international organizations to pursue security objectives that cannot be managed alone. Formal international organizations, supranational governance, and private transnational governance are options. Executive heads, secretariats, and assemblies can be managed. Decisions about caucusing, engagement, agenda-setting, and transactional contributions can be strategic. With international engagement, small states need not be alone in the Grey Zone.

In Chapter 12, Hajdi Xhixha, Arian Kadriu, David Last, and Marzena Żakowska discuss deterrence strategies for small states operating in complex geopolitical environments. Deterrence declines as uncertainty increases, making conventional attacks and Grey Zone operations difficult to deter. Foreign influence operations, diasporas, political entrepreneurs, and economic actors may increase violence. Alignment with major powers, regional alignment with peers, and application of domestic policy instruments might mitigate the potential for violence.

In Chapter 13, Mikkel Storm Jensen explores offensive cyber operations to deter cyberattacks. Cyberattacks are ambiguous, intended to fall below the response threshold and often below the attribution threshold. They are well-suited to Grey Zone operations but unlikely to serve as a deterrent. Allies may be reluctant to deploy cross-domain forces in response to cyber-only attacks. Generally, non-lethal and frequently without kinetic effects, cyberattacks are more than a nuisance but less than a superweapon. Investment in resilience and emergency responses may have better payoffs than reliance on offensive cyber operations to deter cyberattacks.

In Chapter 14, Eryk Habowski and David Last examine the intersection of power, law, and information. International and domestic laws are established to reflect power relations. Attacks on and through the legal system can upset international and domestic balances of power. Information warfare can exacerbate these attacks, eroding trust and confidence in government and state institutions.

Design Problem 3 asks readers to explore potential solutions to a complex conflict using divergent thinking and causal loops. Sketching causal loops helps to see the interaction of factors. These interactions help identify points for leverage and consequences.

Section IV moves to the frontiers of positive peace. Here, the challenge is preserving a cohesive and resilient society that combines national and human security. The essential functions of the state can be eroded in conditions short of war. Yet, states in regional condominiums and international agreements have less leeway to act—no state is an island. Culture and identity can be manipulated to weaken state cohesion, and group loyalties create exploitable fault lines.

In Chapter 15, Waldemar Kitler, Malwina Ewa Kołodziejczak, and David Last address the problem that modern war has no clear starting line, while national and international legal frameworks assume it does. Legal frameworks protecting normal liberties and economies do not respond well to states of exception that creep over the boundary to damage society by stealth. Basic functions like border security and public order can be attacked without triggering emergency legislation. Looking at examples from Poland, Finland, Estonia, and Canada, the authors suggest adaptable national legislation to address threats below the threshold of traditional international war. Even this may be challenged by lawfare and information war, or simply by due process in a liberal democracy.

In Chapter 16, Danuta Kaźmierczak, Mirosław Laskowski, David Last, and Marzena Żakowska analyze the impact of Grey Zone conflicts on social cohesion—a sense of belonging and effective relationships with others in society. Hostile states pursuing modern war in the Grey Zone may seek to destabilize rather than defeat a neighbor, or the attack on social cohesion may weaken a target state. Political, economic, social, technological, and legal threats are potent tools to weaken cohesion. Direct and indirect erosion of social cohesion can occur slowly or dramatically. Environmental stress and internal schisms can be as damaging as hostile action.

In Chapter 17, Jowita Brudnicka, Marzena Żakowska, and David Last offer prescriptions to strengthen the state. Putnam's concept of bridging and bonding capital and the sociological concepts of primary and secondary social organizations can be applied to reinforce social resilience in the face of attack or subversion. Civic nationalism, national narratives, and civil society can be attacked by information warfare and hostile "active measures." Comprehending the strength of social bonds and using effective policing to protect them is a vital defense against Grey Zone attacks.

Design Problem 4 introduces convergent thinking to develop a solution based on empathy mapping. Understanding and articulating what others say, think, do, and feel helps security leaders see their solutions from the perspective of stakeholders. An iterative process and "red teaming" to see alternative perspectives can help stress-test solutions.

The concluding chapter revisits key themes of mutual dependency, the need for creative solutions, society as an asset and target, and tools for responding to Grey Zone threats. There are implications for professional security education, which must be holistic and provide for cross-domain employment. Small states must make decisions about playing dirty, and policy and strategy must adapt to the new reality.

Conclusion

Modern war is not like the wars of the 20th century. It is embedded in the Grey Zone between unequivocal war and stable peace. The ontology of war has changed. Contemporary experience demonstrates that the Grey Zone is not separate from modern war but intimately connected to it. Much English-language writing about Grey Zone conflict and hybrid warfare emanates from the US military-academic complex, focusing on major-power perspectives and military responses, reflecting a scale and resource specialization beyond the means of most countries. For states with dangerous large neighbors, it is seductive to think that there is a clear threshold for war and that a strategic response with military means is possible, that old alliances and deterrence still work. That may be a dangerous illusion.

Traditional approaches to war and peace are not sufficient for small states. They contend with a spectrum of threats from regular military to irregular forces and civilian actors, from lethal violence to non-kinetic attacks, and from clearly attributable to ambiguous and unattributable threats. Even the smallest states must worry about climate change, space, the cyber domain, markets, and VNSAs. Their strategic responses must be innovative holistic approaches with whole-of-government crisis management, all-hazards intelligence, whole-of-society logistics, and international engagement. To harden themselves against information war and subversion, they must invest in social cohesion and the legal framework for national defense in the Grey Zone.

Design thinking will help generate the creative thinking needed to find new ways to survive modern wars in the Grey Zone and to get closer to positive peace. Ultimately, the success of small states in modern war will depend on their ability to anticipate and respond to threats in the Grey Zone, leveraging the strategic, economic, social, and technological tools at their disposal. This book is a step toward practical guidance and a better understanding of the challenges ahead.

References

Bingöl, O. (2017), Hybrid War and Its Strategic Implications to Turkey, *Gazi Akademik Bakış*, vol. 11, no. 21, pp. 107–132.

Bolton, M. (2021), US Civil-Military Relations in the Gray Zone, in A.J. Echevarria II (ed.), *The Gray Zone in Conflict: Dimensions, Definitions, and Implications*, Marine Corps University Press, Quantico, Virginia pp. 169–188.

Freier, N. et al. (2016), *Outplayed: Regaining Strategic Initiative in the Gray Zone, A Report Sponsored by the Army Capabilities Integration Center in Coordination with Joint Staff J-39/Strategic Multi-Layer Assessment Branch*, US Army War College Press, Carlisle, PA.

Galeotti, M. (2022a), *Putin's Wars from Chechnya to Ukraine*, Osprey Publishing, Oxford.

Galeotti, M. (2022b), *The Weaponization of Everything: A Field Guide to the New Way of War*, Yale University Press, New Haven, CT.

Gerstle, G. (2022), *The Rise and Fall of the Neoliberal Order: America and the World in the Free Market Era*, Oxford University Press, Oxford.

Jankowicz, N. (2020), *How to Lose the Information War: Russia, Fake News, and the Future of Conflict*, Bloomsbury Publishing, London.

Jordan, D., Kiras, J. D., Lonsdale, D. J., Speller, I., Tuck, C. and Walton, D. C. (Eds.) (2016), *Understanding Modern Warfare*, 2nd ed., Cambridge University Press, Cambridge.

Kaufman, A. A. and Hartnett, D. M. (2016), "Managing Conflict: Examining Recent PLA Writing on Escalation Control", Center for Naval Analyses.

Keegan, J. (2000), *The Book of War: 25 Centuries of Great War Writing*, Penguin Publishing Group, New York.

Liang, Q. and Xiangsui, W. (1999), *Unrestricted Warfare*, PLA Literature and Arts Publishing House Arts, Beijing.

Licklider, R. (Ed.). (1993), *Stopping the Killing: How Civil Wars End*, New York University Press, New York.

Matisek, J. W. (2017), Shades of Gray Deterrence: Issues of Fighting in the Gray Zone, *Journal of Strategic Security*, vol. 10, no. 3, pp. 1–26.

Mazarr, M. J. (2015), *Mastering the Gray Zone: Understanding a Changing Era of Conflict*, US Army War College Press, Carlisle, PA.

Morris, L. J., Mazarr, M. J., Hornung, J. W., Pezard, S., Binnendijk, A. and Kepe, M. (2019), *Gaining Competitive Advantage in the Gray Zone: Response Options for Coercive Aggression Below the Threshold of Major War*, Rand Corporation, Santa Monica, CA.

Polanyi, K. (1957), *The Great Transformation: The Political and Economic Origins of Our Time*, Beacon Press, Boston, MA.

Rose, G. (2010), *How Wars End: Why We Always Fight the Last Battle*, Simon & Schuster, New York.

Shu, K., Wang, S., Lee, D. and Liu, H. (2020), *Disinformation, Misinformation, and Fake News in Social Media*, Springer International Publishing, Cham.

Snyder, T. (2018), *The Road to Unfreedom: Russia,* Europe, America, Crown Publishing Group, New York.

Sun, T. (1988), *The Art of War: Complete Texts and Commentaries*, translated by Thomas Cleary, Shambhala Publications, Boston, MA and London.

Taylor, A. J. P. (2019), How Wars Begin, Lume Books, London.

Tilly, C. (2003), *The Politics of Collective Violence*, Cambridge University Press, Cambridge.

USSOCOM (2015), White Paper, 9 September. https://publicintelligence.net/ussocom-gray-zones/

Introduction to Design Thinking

Archipelago of Design (AoD)

Oliver Jones, Kassie Miedema, and Philippe Beaulieu-Brossard

Design thinking helps navigate the increasingly complex and interconnected challenges that characterize defense and security. Leaders and organizations practicing design aim to generate fundamentally different perspectives on complex, ambiguous, or unknown problem sets. Design thinking provides tools to collaborate and visualize when planning, problem-solving, or strategy-making. Positioned between positive peace and outright war, the ambiguous nature of modern war and Grey Zone conflicts requires small states to adopt strategies that can navigate deep social, political, military, economic, informational, and organizational complexities without defaulting to large states' security outlooks and solutions. The unique national security and defense contexts of small states, modern war, and Grey Zone conflicts are the organizational issues the design process addresses. Design thinking can also be used to strategize around emergent threats and to understand how changes in the larger security environment may challenge existing security capabilities and strategic outlooks. In short, design is a structured process for exploring changeable conditions that inhibit organizational creativity, effectiveness, and decision-making. Design's structured process is uniquely effective in enabling collaborating groups to define novel or unobserved problems better, scope their deeper implications, and target opportunities for interventions and solutions that would be otherwise overlooked.

Many design frameworks exist, but most shift between divergent and convergent phases. Divergent phases help cast a wide net, gathering information and brainstorming possibilities. Convergent phases help narrow down the most promising ideas, choose the most pertinent information, and focus on a few possibilities. Design Problems 1 and 3 in this book reflect divergent thinking, while Design Problems 2 and 4 use convergent thinking. Archipelago of Design (AoD) has developed *Breakaway: Reframing to Prevail* (2023) for the Canadian Armed Forces and *Safeguarding the Future* (2024), a foresight handbook for defense and security professionals with detailed guides for how to apply foresight tools and mindsets. Common design frameworks in the military sector include the US Army Design methodology (2015), and in the civilian sector, the Stanford d.school 5-Step Model (n.d.) and the UK Design Council's (2024) Double-Diamond model. References suggest additional resources.

Design is a series of activities that work by discovering and defining the problem(s) and developing and testing solutions. As an iterative sequence, these activities traverse divergent and convergent phases. Discovering and defining problems not only helps to understand them deeply but also ensures their root causes (not just symptoms) are addressed or, at least, considered. These activities also help understand interconnected problems and provide solutions to the most impactful problems. Developing and testing the solution ensures, it is fit for purpose and has the intended impact. The team tasked with implementing that solution can refine it in the process.

DOI: 10.4324/9781003428701-2

Sometimes, the design process is linear, moving in a single direction from problem-framing to solution design and implementation. However, it is more common to start working on a solution only to realize that the problem needs to be more clearly defined. The latter non-linear sequence maps to a pattern of iteration that some design researchers have characterized as the co-evolution between defining the problem and finding solutions (Beaulieu-Brossard and Dufort 2020). The critical point is that the design process is typically iterative and goes through multiple divergent and convergent thinking cycles. It might be tempting to jump straight to converging on what seems like the best solution. However, having a broad spectrum of initial ideas to refer to can enable a design team to redefine the problem or quickly pivot when the problem they are attempting to address mutates or changes in unexpected ways.

The AoD primer presents a variety of mindsets, approaches, methods, and tools for design thinking (AoD 2022). Spending time with these resources can enhance problem-solving in the complex space of modern war and of Grey Zone operations. Military design thinking is now a mature field, extensively reviewed (Wrigley, Mosely and Mosely 2021), and addressed by a growing number of encyclopedia entries in the *Handbook of Military Sciences* (Sookermany 2023).

The design exercises in each section will focus on the problem or solution and stimulate divergent or convergent thinking related to problems implied by the chapters in the section. The goal is to familiarize readers with the different phases of design and some common tools and techniques designers use at various stages. These exercises are not comprehensive. Rather, they provide concrete examples demonstrating the potential of design for addressing the challenging problems of modern war and Grey Zone conflicts. Further reading and resources are provided for those who would like to learn more about design thinking in a military context.

References

Archipelago of Design (2022), *Collaborative Innovative Thinking by Design for the Canadian Armed Forces*. Archipelago of Design. https://aodnetwork.ca/collaborative-innovative-thinking-by-design-for-the-canadian-armed-forces/

Archipelago of Design (2023), *Breakaway: Reframing to Prevail*. Archipelago of Design. https://aodnetwork.ca/breakaway-reframing-to-prevail/

Archipelago of Design, Kerr, H. and Miedema, K. (2024), *Safeguarding the Future: A Foresight Handbook for Defence and Security Professionals*. Archipelago of Design. https://aodnetwork.ca/safeguarding-the-future/

Beaulieu-Brossard, P. and Dufort, P. (2020), "Enhancing Challenge Framing in Defence Organisations: Toward Reflexive Methods", *Honvédségi Szemle*, vol. 148, no. SI 2, pp. 33–49.

Design Council (2024), *The Double Diamond: A Universally Accepted Depiction of the Design Process*. https://www.designcouncil.org.uk/our-resources/the-double-diamond/

Sookermany, A. (ed.) (2023), *Handbook of Military Sciences*. Springer, Cham.

Stanford University (n.d.), *An Introduction to Design Thinking: Process Guide*. https://web.stanford.edu/~mshanks/MichaelShanks/files/509554.pdf

US Department of the Army (2015), *Army Design Methodology (ATP* 5-0.1), Washington, D.C. https://rdl.train.army.mil/catalog-ws/view/ARIManagingComplexProblems/downloads/Army_Design_Methodology_ATP_5-0.1_July_2015.pdf

Wrigley, C., Mosely, G. and Mosely, M. (2021), "Defining Military Design Thinking: An Extensive, Critical Literature Review", *She Ji: The Journal of Design, Economics, and Innovation*, vol. 7, no. 1, pp. 104–143.

Further Readings

Barabrook-Johnson, P. and Penn, A. (2022), *Systems Mapping: How to Build and Use Causal Models of Systems*. Palgrave-Macmillan, Cham.

Braun, W. (2002), "The System Archetypes", *Systems*. https://www.academia.edu/28706184/The_System_Archetypes_The_System_Archetypes

Bury, P. and Murphy, D. (2023), "No Time to Hide: The Future of Irish Defence and Security", *War on the Rocks,* 3 March. https://warontherocks.com/2023/03/no-time-to-hide-the-future-for-irish-defense-and-security-and-how-our-partners-can-help/

Dupont, D. (Host) (2024), *Tension Podcast Ep. 7: The Bold and Messy Roller-Coaster Ride of Design Education for Military Officers* [Audio Podcast Episode]. In *Tension Podcast*. Archipelago of Design. https://aodnetwork.ca/tension-podcast-ep-7/

Ertan, A., Floyd K., Pernik P., and Stevens T. (eds.)/ NATO CCDCOE (2021), *Cyber Threats and NATO 2030: Horizon Scanning and Analysis*. NATO Cooperative Cyber Defence Centre of Excellence (CCDCOE). https://ccdcoe.org/library/publications/cyber-threats-and-nato-2030-horizon-scanning-and-analysis/

Gray, D. (2017), "Updated Empathy Map Canvas", *Medium,* 15 July. https://medium.com/@davegray/updated-empathy-map-canvas-46df22df3c8a

Meadows, D. H. (2008), *Thinking in Systems: A Primer*, edited by Donella Wright, Chelsea Green Publishing, White River Junction, Vermont.

Ryan, A. (2017), "Design in Civvies: The Promise of Creating Degrees of Freedom in Government", *Journal of Military and Strategic Studies*, vol. 17, no. 4, pp. 39–50.

Stroh, D. P. (2018), "A Systemic View of Israeli-Palestinian Conflict", *System Thinker*. https://thesystemsthinker.com/a-systemic-view-of-the-israeli-palestinian-conflict/

Section I

Modern War and Grey Zones

1 Conflicts in the Grey Zone

From Total War to War Nexus

André Simonyi

Introduction

Modern war reflects the complexity of the 21st-century global socio-political system. As these words are written, the world seems to have relapsed into a surreal repetition of history. The war in Ukraine reminds many of the invasions preceding World War II, with its unpredicted military action and apparent lack of response in kind by European and world political actors. Considering this eruption of military violence, the common misquote of Clausewitz's "war is the continuation of politics by other means" must be revisited. Contemporary wars are transitioning from the idea of total war—the absolute commitment to military victory—to a war nexus, the independent but connected commitment of all domains. In this emerging context, war is ongoing, through all means available, violent and non-violent, and is not defined by time and clear end-states. Spatially, war transcends the traditional borders defining sovereignty and the international system. From a closed battle space, we shift our representation of war and conflict to a complex environment, multilayered and multidimensional, in which victory is not defined by armistice but by the relative positioning of the actors involved.

War in the 21st century emerges as a phenomenon beyond the sole purview of military institutions. Sensing this change, belligerents adapt by separating the battle space into three ontologically different realms. The first, tactical level, maintains the traditional approach to operations—sensing, analyzing, planning, and acting within a generic environment, the battle space. The second realm engages with design thinking, expanding the battle space into a systemic perspective of uncertainty, feedback loops, and emergence (Blakesley et al. 2005). The third realm engages beyond and parallel to military deployments: other domains, such as the economic, cyber, and social media communication, among others, engage in conflict, not subordinated to, but in coordination with, the military domain. War in the 21st century cannot be reduced to victory in the battle space but emerges as a complex environment with all domains/systems engaged independently yet connected and open to each other.

Over the last decades, this ontological shift toward the recognition of a complex security environment was commented on in national strategies (The White House 2021), studied at command and staff colleges (Blakesley et al. 2005), and written about in books (Nordstrom 2004, 2007; Freedman 2019; Gray 1998; Hass 2019). Reframing contemporary conflict, Mary Kaldor's innovative concept of New Wars described an emerging environment "understood in the context of the process known as globalization" (Kaldor 2012:4), a world interconnected and networked in which differentiating actors and methods requires new approaches. But these new wars are different in other ways and evolving in real time. The concept of war nexus creates the necessary discursive and practical space to explore the emerging deep structural and ideational changes as an alternative to the traditional new versus old categorization of political violence.

DOI: 10.4324/9781003428701-4

The war nexus builds on Kaldor's New Wars theory (Kaldor 2012) by displacing the battlefield's dominance from exclusive military action and opening the possibility of a new complexity already expressed through deep studies of power (Nordstrom 2007), interests (Nordstrom 2004), identity (Hassner 2015), religion (Marsden 2019), ecology (Grove 2019), and others. The war nexus captures the ontological change in and of the phenomena of war and accommodates its fluidity, constant change, and emergence.

The emergence of the war nexus challenges the traditional approaches to warfare, its rules, norms, processes, and structures. Contemporary hybrid warfare evolved from tactical and operational "soft power," complementary with "hard power," becoming an independent strategic action domain. Enhanced by technology, it is an extremely efficient tool to wage war as it creates new capabilities to dislocate an opponent, using undetectable backdoor channels to penetrate technological structures (e.g., computer viruses) or the very fabric of societies within their domestic realm (e.g., election tampering), for example. These forms of hybrid warfare are undeclared and often hidden in anonymity through proxies.

By shifting from the military battlefield to the war nexus, we elevate hybrid warfare from technique to a new ontology of war. Wars of this form are undeclared and occur "below the threshold of war" across all domains. As these transcend borders, undeclared and timeless, they require a new name: Grey Zone conflicts.

But this is not the only extent of change: we are in a Grotian moment, a historical shift in the organization of global polity. Humankind's transition across time and space is a complex process involving all aspects of social life. War, as a human phenomenon, is one of the many reflections of these transitions. Semantically and analytically, it is not the conflicts that are gray but the zone that is gray.

Modernity and War

The history of modern war and warfare accompanies and informs the creation of the state system. From the late 18th until the early 20th century, modern strategic thinking emerged, differentiating the tactical concerns of warfare from campaigning and the integration of the politico-strategic level of war (Coutau-Bégarie 1999:173). From the writings of the Comte Jacques de Guibert (1977:53–84) prefacing the political integration of war by Clausewitz (Clausewitz 1989) to General Helmut Von Moltke's reflections on the total implication of nations in its undertaking (Moltke 1995:21–75), change in the art of war accompanied the creation of the European state system and the passage from a customary approach to warfare to the rational legal reframing of the concepts of war and peace. By the end of this period, war was a total engagement of the nation, mobilizing all its resources without exception to attain an ultimate objective: victory, represented by unconditional surrender rather than the negotiated settlement of limited political aims.

Its legal structure transitioned from a customary to a formal framework, adding to the emerging international system of agreements and functioning (Fazal 2018). This period of clarity was short-lived. In the 1930s Germany, Carl Schmitt- wrote:

> For several years now, bloody struggles have been carried out in the varied regions of the earth—struggles to which a more or less common understanding warily avoids attaching the term "war". It may be all too easy to scoff at this observation. In truth, however, it has become all too clear that old orders are unraveling just as new ones come to replace them. Indeed, the problems surrounding the concept of war reflect the disquietude of the current international situation.
>
> (Schmitt 2011:30–31)

Reflecting on the evolving international order regulated by the Treaty of Versailles and the Geneva League of Nations -failures, Carl Schmitt pointed at the legal gap in the concept of war, blurring its differentiation with the state of peace (Schmitt 2011:32). Post WWII, the ambivalence expressed by Schmitt was formalized. On the one hand, the creation of the United Nations asserted an internationalist legal framework denouncing illegal wars. On the other hand, the Cold War confrontation supported the realists' view that national security remains the focus of statesmanship at all costs (Butterfield and Wight 2019:42; Waltz 1979:134).

The Widening and Deepening of the Concept of Security

Between 1945 and 1991, the world confronted its first absolute existential threat, total annihilation, while it underwent profound changes: decolonization, the globalization of the economy, global technological connectivity, and global awareness through media-provoking social movements. It was an intense period of systemic adjustment and change during which diverse ideas about the world order and national security emerged. Strongly held in the grips of government agencies worldwide (Oren 2002), they were challenged by social movements and scholarly theorization. On the one hand, the survival of humanity trumped every aspect of security; on the other, challenges to the world order and alternative ideas on the objects of security were emerging.

The end of the Cold War released the possibility of a major global systemic readjustment. Themes of geopolitical reorganization, wars and revolutions, failing states, identity claims, environmental awareness, poverty, gender, racialization, and many more security issues challenged the idea of national security solely preoccupied with issues of high politics, namely, national and international security. This dramatic period also released a simmering world of ideas and theories about society and security. Theorized in Buzan's famous *People, States and Fear*, the concept of national security—oriented toward containing the nuclear threat—became wider and deeper, encompassing multiple sectors: military, environmental, economic, societal, and political (Buzan 1983). Challenging the realist levels-of-analysis approach to IR, individual–state–international (Singer 1961; Waltz 1979), the so-called Copenhagen School- shifted the conversation from the hierarchy of war to a systemic view of the world (Buzan, Waever, and Wilde 1997).

From this analytical point of view, states were subjected to military threats and an array of independent yet intersecting security sectors. Reflecting the emerging international system, this battle of ideas voiced debates on security between "traditionalists" and "wideners." The former saw security studies as "about the phenomenon of war and [argued] that it can be defined as 'the study of the threat, use, and control of military force'" (Walt in Buzan, Waever, and Wilde 1997:3). The latter promoted the expansion of the concept of security to embrace other sectors on equal footing (Buzan and Hansen 2009:187–191). These apparently conflicting ideas were the precursors of deeper questioning.

War and Security: The End of Exception?

Beyond these polarizing debates, the 1990s became the tipping point toward a still ongoing period of dramatic adjustments and changes at the international and global levels. The line separating domestic and international realms became blurred, symptomatic of the larger systemic readjustment at the center of which we find the cornerstone of the international system: sovereignty.

Declaring the state of exception differentiates between war and peace in the current international legal framework. It also defines the inside-outside realms, or domestic and international. But since the end of the Cold War, the line separating war as an extension of policy by other means from matters of national security became increasingly thin: climate change, pandemics, economy and trade, and international social movements, among others, transcend borders. Although these challenges blurred the domestic-international divide, they were not considered existential threats. The War on Terror of the 21st century changed that.

In 2005, Italian philosopher Giorgio Agamben addressed the effects of the "global civil war-" and denounced the normalization of the state of exception by the Bush government in 2001, and the passing of the Patriot Act. The legal decisions taken by the US government, followed subsequently by several democratic governments-, to suspend indefinitely the rights of individuals for reasons of national security created a new legal space, an ambivalence in which executive powers were concentrated while democracy remained in effect. For reasons of national security, the executive branch was empowered not only to wage an undeclared war but also to create a zone of exception within the territorial boundaries of the state. Initially intended to treat non-US citizens, the suspension of Habeas Corpus- was eventually extended to all US citizens-.

In the second decade of the 21st century, Russia's aggression against Ukraine exposes another variation of systemic change. In 2022, neither Russia's government nor its president had yet officially declared war on Ukraine. Moreover, four months after the beginning of full-scale hostilities, the Ukrainian government was warned that "any attempt to encroach on Crimea is a declaration of war on our state." These events do not represent separate incidents, as they are not the subjects of declarations of war. Instead, they are part of the same "war" for which the beginning and the end remain unknown. This means that not only for anyone involved, Ukrainians, Russians, but also the rest of the world, the legal difference between state of war and state of peace is irrelevant, both at the domestic and international levels.

War as the violent expression of politics in the 20th century emerged and transformed in the 21st century as a pernicious, omnipresent reality, not only blurring the traditional lines separating war and peace but also the international from the domestic. These are more than a simple adjustment to the traditional peace-war continuum and its rules. They announce a shift within the system itself, a profound rearrangement of the systemic equilibrium, a different order of things. The shift is incomplete, however, and our founding knowledge of laws, norms, and habits brings us back to our referent frameworks, namely, the international order. This is part of the motion of emergence: as the system changes, the old lingers in the form of discourse and tradition, while the new is being codified.

Differentiating War and Warfare

In many texts, the concepts of war and warfare are freely interchanged, as if they represented the same object of analysis. But by doing so, we lose the strength of two powerful ideas that, when placed in relation, offer an opportunity to observe difference and change. War and warfare reflect each other, revealing their similarities and differences, the distortions becoming clues for perceiving change.

What is war? Among the many definitions found in the literature on war, Hedley Bull provided a most elegant and simple one: "War is organized violence carried on by political units against each other" (Bull 2002:178). As with any working definition, Bull still raises analytical interrogations, such as the nature of violence and the definition of political units. First, is organized violence related exclusively to military power, or does it include other forms of aggression? Second, how do we define a "political unit" when considering recent discussions with so-called rogue groups or the penetration of politics by corporate or criminal interests? Our answer is that "violence" and "political units" are contingent on the periodization of observation and part of the analysis of the Grey Zone.

What is warfare? Warfare represents the conduct of the confrontation, and the doctrines, strategies, and tactics belligerents employ to achieve their objectives. Since the early legal modern codification of the international realm-, war and warfare have been generally consistent: both were engaged according to the legal framework. Since 1945, however, this general rule has begun to falter, and we have observed an increasing number of conflicts involving warfare without wars.

The term "hybrid warfare" gives us a clue in exploring this transformation. Historically, hybrid warfare refers to the use of non-conventional means—irregulars, such as guerillas, insurgents, and terrorists, or even propaganda, espionage, deception, and information operations. This technique of war-making has been part of the toolbox of warfare since ancient times (Mansoor 2012), engaged in conflict once the state of war has been confirmed.

A second, more recent, meaning of hybrid warfare relates to the standalone use of aggression engaged without a formal declaration of war. It blurs the lines between peace and war and between international and domestic. As a Canadian policy document describes:

> Below threshold tactics and hybrid warfare challenges our ability to identify, attribute, and categorize adversarial actions, thus making it increasingly difficult to distinguish between actions which, taken collectively undermine our national security and defense interests in a significant manner (including those that give rise to a use of force) and those that do not. The linkages between disparate spheres of activity are also difficult to understand and can mask broader strategic objectives.
>
> (Government of Canada 2021)

Hence, hybrid warfare, still a warfare technique, defies the rules regulating war. Now, war and warfare do not mirror each other: the first is regulated by international conventions and agreements; the second escapes those conventions, circumventing the clear identification of belligerents on the battlefield and escaping the attribution of responsibility for warlike actions.

Placed under tremendous pressure in the 20th century, the international system is changing its inner processes, expanding the concept of security and modifying de facto, not de jure, the processes regulating conflict. Hence, although the international rules for war remained relatively unchanged, warfare has become pervasive. Russian actions on Georgia (2008) and Crimea (2014); American aggressive actions defying sovereignty in Pakistan and Syria; cyber-attacks on Iran; the use of social media to destabilize democratic processes; and, at the time of writing, the Russian war against Ukraine are but a few examples of a new form of war, which, although regulated by international conventions de jure, is defied de facto.

This Grotian moment is not only a light adaptation of customary law: it fundamentally affects our most basic definitions. There is no longer a clear divide between war and peace, nor a war-peace continuum between states or groups of states. The normalization of the state of exception and the effacement of the line between war and warfare are the signs of an important systemic readjustment involving the whole global environment: we are in the Grey Zone itself.

Conflicts in the Grey Zone: The War Nexus

This evolving relationship between war and warfare illuminates a new ambivalence in the current systemic arrangement. On the one hand, we have a modus operandi regulating the international level, establishing rules and norms of behavior, and on the other, profound structural and procedural changes at the global level, allowing the emergence of new actors and behaviors. The changing relationship between war and warfare expresses this misalignment. To study and analyze this misalignment, we need analytical tools. The Grey Zone represents a time-space of systemic readjustment within which war is one phenomenon. The war nexus is the conceptual representation of war in the Grey Zone.

As war and warfare increasingly misalign and become incapable of providing coherence for the system to remain in relative equilibrium, new, unregulated behaviors emerge. In the war nexus, conflicts are reshaped and rules modified. Emergence within the war nexus occurs at different levels, the most apparent being the expanding spatiality of warfare. Operation Desert Storm in 1991 captures the essence of "traditional" warfare: reacting to an invasion, an allied coalition musters both soft and hard power to free Kuwait from Saddam Hussein's aggression. Spatially concentrated and legally framed (Goldman 1991), the war focused on military power aiming toward a clear objective.

In contrast, the war nexus does not privilege the military component. Contemporary warfare operates across domains—political, economic, social, technological, legal, environmental, and more—and the non-physical dimension of the virtual and cyber spectrum. This new formulation of hybrid warfare reframes the spatiality of war from a territorially oriented hierarchical perspective—all efforts directed to support the war effort—to a complex environment characterized by all domains of society acting independently but in relation to each other and their environment deterritorialized.

This spatial realignment of warfare from a defined battleground to a multi-domain battle space realigns the temporal perspective of contemporary conflicts. Since WWII, wars tend to stagnate and peter out without obvious closure. From Korea to Vietnam, Israel, and the Middle East, the "war on terror," and more, the idea of victory switched from an absolute meaning to one that is subjective and evolving (Strohn 2020). In the war nexus, wars and conflicts, including the attempts to impose peace, are not distinguished by achieving an end-state victory but rather represent a phase toward achieving strategic-political objectives. War and peace become confusingly entangled globally, conflicts emerge in different forms in an endless chess game, and relative positionality, rather than victory, is the decisive factor.

In the war nexus, we observe an expansion of the number of actors involved in contemporary conflicts. Challenging the states' sole legitimacy, today's geography of conflict is a complex mix not only of rogue groups, so-called terrorists, and proxies but also privateers and economically motivated parties, such as cartels and others. Although not new, these groups emerge differently and seek recognition as belligerents, creating an exception within the international framework of legitimate actors. Transcending boundaries and accumulating enough capabilities to engage in conflict with states and coalitions, they challenge the spatiality of war and create a new temporal framework. War is a never-ending process.

As systemic change is underway, we tend to apply the changing nature of war to its stylistic application. Viewed this way, "Grey Zone conflict" is simply a variation on an existing theme: hybrid warfare is part of the unchanging nature of war. However, fusing war and warfare only demonstrates the difficulty of incorporating the systemic change around us into our analytic vocabularies. While we need to keep war and warfare distinct concepts, we also need a new vocabulary that accommodates contemporary dissonances. The "war nexus" concept provides the discursive space to express, observe, and study the emergence of a new relationship between war and warfare.

Beyond New Wars: Into the Grey Zone

We are in a moment of change and transformation, a Grotian moment, part of a systemic process for which temporality is fluid and not set on a linear trajectory. These changes happen over time, often exceeding the life span of most observers, challenging our capacity to differentiate between "old" and "new" as events unfold. This, too, is a feature of the Grey Zone.

Observing the world and the war phenomena through the lens of complexity and systems helps break the paradigmatic and forceful repetition of past arrangements and capture the tremendous changes around us. These changes profoundly influence war and warfare. However, although we recognize that "…war antedates the state, diplomacy and strategy by many millennia" (Keegan 2000:3), we perpetuate its interpretation through the lens of past experience. Hence, the international system, in which sovereignty prevails, veils our capacity to look beyond and integrate change as it forces us to question the system we are bound to protect. We are locked in. International rules and norms become dissonant and fail to adapt. Divergence in our interpretation of war, perceived as one of the regulatory processes between states in the framework of the international system, announces either tipping into chaos or inflection into the emergence of a new paradigm-.

In the Grey Zone, the foundational premises of the international system are challenged. Essential to this coded agreement, sovereignty provides a clear separation between the international and domestic realms. And yet, challenged every day by increased globalization through technological advances, economic integration, industrialization, environmental concerns, and social movements, the boundary separating the realm of state-making and securing societies is blurred. National security and securing the world are entangled concepts to which rules and norms do not respond adequately.

These changes do not occur following a before-after model but rather happen concurrently, influencing each other and creating feedback loops and unexpected results. It is an ongoing movement, and our theoretical lenses must be constantly adjusted to reflect a world in constant change. The outcome is unpredictable. We are in the Grey Zone.

Conclusion

War is a powerful concept. It immediately instills images and feelings of violence and confrontation. Yet, it is one of the most difficult ideas to harness and define, and as we study the phenomena historically, we uncover its temporal malleability and adaptability. Today, Merriam-Webster defines war as "a state of usually open and declared armed hostile conflict between states or nations" and "a period of such armed conflict." In French, Larousse specifies that "war implies the application of special rules defining the relations between states; it begins with a declaration of war or an ultimatum and ends with an armistice, and in principle, a peace treaty ending the war." These definitions follow the intentions expressed in Article 2 of the Charter of the United Nations, founded on the principle of sovereignty. Could we apply these definitions to other historical times? Are these definitions applicable today?

In the 21st century, war is not total in the sense of the complete engagement of state assets to achieve a decisive victory; it is a nexus where all domains of society are engaged in conflict to maintain their system's coherence and survival. As an expression of the Grey Zone, the war nexus represents the emerging part of a bigger systemic movement. For the last several hundred years, the social world, enabled by an incredible scientific and technological explosion, moved toward the total integration we call globalization. This integration affects all domains; horizontally, we observe its effects on all systems, economic, legal, technical and technological, industrial, political, religious, etc.; vertically, we observe its effects at all systemic and sub-systemic levels, from the planetary to the individual.

From a systemic perspective, the 20th century does not represent a break from the 19th-century state-making process but rather its continuation through a series of changes. Placed under tremendous pressure, the international system of states globalized the 19th-century European way of war. To observe those changes, we have focused on four intersecting factors: first, war evolved from a time-bound phenomenon defined by victory to an endless strategic positioning process. Second, the physical boundaries of states defining the traditional physical settings for war transgressed outside the proper rules and norms of war. Third, the actors waging war are not solely states but encompass an array of other actors, from religiously or ideologically motivated actors to profiteers; fourth, the constant transgression of the rules, norms, and processes expressed through international law and other conventions demonstrates the distance between the concept of war and its enactment.

These four factors lead to the misunderstanding and mismanagement of conflict in the modern world and the creation of a discursive alternative: the Grey Zone. This term defines our inability to reconcile the conceptual and its enactment. In conformity to "war as a continuation of politics by other means," we usually extrapolate that it is through the technique of war—warfare—that we can explain and theorize contemporary conflicts. Hence, hybrid warfare, or the combined use of soft and hard power to pursue victory, explains the conflicts "below the threshold of war." But, consistent with the study of the four criteria of time, space, actors, and rules, war has expanded from the sole realm of military action. It is with the use of hybrid warfare that belligerents can challenge time, space, actors, and rules. Hence, hybrid warfare should be seen as an elaborate toolbox to wage war, not the conceptual explanation for the profound system changes underway. The Grey Zone is that conceptual space of change in the 21st century: there is a difference between hybrid warfare and conflicts in the Grey Zone.

We are witnessing the emergence of a new equilibrium. This process has been ongoing for over a century, progressively defying Hugo Grotius' *De Jure Belli ac Pacis* (*The law of War and Peace*) regulating conflicts between nations. War and warfare not only show that political violence extends well beyond the realm of military confrontation between states, but they also give us clues to understand the nature of the profound changes underway at the global level, where legacy and emergence meet.

References

Agamben, G. (2005), *State of Exception*, University of Chicago Press, Chicago.

Blakesley, P., Walk, J., Downing, G., Sorrells, W., Pendall, D. and Wallwork, R. (2005), *Systemic Operational Design: An Introduction*, School of Advanced Military Studies, Fort Leavenworth, Kansas.

Bull, H. (2002), *The Anarchical Society*, 3rd edition, Columbia University Press, New York.

Butterfield, H. and Wight, M. (eds.) (2019), *Diplomatic Investigations: Essays on the Theory of International Politics*, Oxford University Press, New York.

Buzan, B. (1983), *People, States and Fear: An Agenda for International Security Studies in the Post-Cold War Era*, Wheatsheaf Books, Brighton, Sussex.

Buzan, B. and Hansen, L. (2009), *The Evolution of International Security Studies*, Cambridge University Press, Cambridge.

Buzan, B., Waever, O. and De Wilde, J. (1997), *Security: A New Framework for Analysis*, UK edition, Lynne Rienner Publishers, Boulder, CO.

Clausewitz, C. (1989), *On War*, translated by Michael Eliot Howard and Peter Paret, Princeton University Press, Princeton, NJ.

Comte, J. (1977), *Guibert, Écrits militaires 1772–1790*, Nation armée, Copernic, Paris.

Coutau-Bégarie, H. (1999), *Traité de stratégie*, Economica, Copernic, Paris.

Fazal, M. (2018), *Wars of Law: Unintended Consequences in the Regulation of Armed Conflict*, Cornell University Press, Ithaca, NY.

Freedman, L. (2019), *The Future of War: A History*, PublicAffairs, New York.

Goldman, R. (1991), "The Legal Regime Governing the Conduct of Operation Desert Storm 1992 Symposium Issue: War and the Law," *University of Toledo Law Review,* vol. 23, no. 2, pp. 363–394.

Government of Canada (2021), *Mobilizing Insights in Defence and Security (MINDS)*. Annual Report 2021–2022, https://www.canada.ca/en/department-national-defence/corporate/reports-publications/minds-annual-reports-2021-2022.html

Gray, C. (1998), *Postmodern War: The New Politics of Conflict*, 1st edition, The Guilford Press, New York.

Grove, J. (2019), *Savage Ecology: War and Geopolitics at the End of the World,* illustrated edition, Duke University Press Books, Durham, NC.

Hass, C. (2019), *Aujourd'hui la guerre*, Fayard, Paris.

Hassner, P. (2015), *La revanche des passions*, Fayard, Paris.

Kaldor, M. (2012), *New and Old Wars: Organized Violence in a Global Era,* 3rd edition, Stanford University Press, Stanford, CA.

Keegan, J. (2000), *The Book of War: 25 Centuries of Great War Writing*, Penguin Books, London.

Mansoor, P. (2012), "Introduction: Hybrid Warfare in History", in P. R. Mansoor and Murray W. (eds.), *Hybrid Warfare: Fighting Complex Opponents from the Ancient World to the Present*, Cambridge University Press, Cambridge, pp. 1–17.

Marsden, L. (2019), *Religion Press and International Security*, 1st edition, Polity Press, Cambridge.

Moltke, H. (1995), *Moltke on the Art of War: Selected Writings*, edited by Daniel Hughes, Presidio Press, Novato.

Nordstrom, C. (2004), *Shadows of War: Violence, Power, and International Profiteering in the Twenty-First Century*, 1st edition, University of California Press, Berkeley.

Nordstrom, C. (2007), *Global Outlaws: Crime, Money, and Power in the Contemporary World*, 1st edition, University of California Press, Berkeley.

Oren, I. (2002), *Our Enemies and US: America's Rivalries and the Making of Political Science*, 1st edition, Cornell University Press, Ithaca, NY.

Schmitt, C. (2011), *Writings on War*, edited by Timothy Nunan, 1st edition, Polity Press, Cambridge, UK.

Singer, J. (1961), "The Level-of-Analysis Problem in International Relations", *World Politics,* vol. 14, no. 1, pp. 77–92.

Strohn, M. (2020), *Winning Wars: The Enduring Nature and Changing Character of Victory from Antiquity to the 21st Century*, Casemate Academic, Oxford, UK and Philadelphia, PA.

The White House (2021), Interim National Security Strategic Guidance. https://bidenwhitehouse.archives.gov/briefing-room/statements-releases/2021/03/03/interim-national-security-strategic-guidance/

Waltz, K. (1979), *Theory of International Politics,* 1st edition, McGraw-Hill, Boston, MA.

2 Environmental Insecurity and Modern Conflict

Duraid Jalili

Introduction

Global heating and biodiversity loss represent an existential challenge at all levels of security. The scale, pace, complexity, interdependence, and irreversibility of these environmental trends and their associated security implications are unprecedented and paradigm-changing. They are a constant background to Grey Zone threats and will increasingly be in the foreground. In this new environment, military forces face a dual obligation. First, they must respond to a rapidly growing number and range of operations, including non-kinetic roles such as humanitarian and disaster relief (HADR) operations. Second, they must enhance their sustainability to demonstrate commitment to changing environmental norms and maintain operational effectiveness in increasingly hazardous and resource-constrained environments. Achieving this dual obligation will require security practitioners who can operate in a scenario of permanent volatility and uncertainty and are comfortable in continually reappraising the validity of their organizations' objectives, methods, resources, and the security function itself.

This chapter provides insights from current research and debates to help security practitioners better understand this new environment. It focuses on four critical areas: first, the nature of current and future environmental shifts; second, governance frameworks and strategic "narratives" surrounding environmental security; third, the security implications of global warming and biodiversity loss; and fourth, the implications for armed forces.

An Overview of Environmental Trends

Since the Industrial Revolution, Greenhouse Gas (GHG) concentration in the atmosphere has increased exponentially due to human activity. This includes Carbon Dioxide (CO_2), Methane, and Nitrous Oxide. Most importantly, the concentration of CO_2 in the atmosphere is now higher than for at least 800,000 years. Over 35 billion metric tons of CO_2 alone are released into the atmosphere each year. As noted in the Working Group I report for the Intergovernmental Panel on Climate Change (IPCC)—which was drafted by 234 scientists from 66 countries using more than 14,000 scientific papers and approved by 195 governments—this scenario is "unequivocally caused by human activities" (IPCC 2021). Unfortunately, even if all GHG emissions were to suddenly cease, it would still take hundreds of years for enough CO_2 to be removed from the atmosphere to restore the planet's carbon cycle (Tans 2009). Why is this a problem? Simply put, these gases create a heat-trapping blanket around the Earth. They absorb and emit energy from the sun rather than allowing it to be reflected into space. IPCC data show the rapid rise in

DOI: 10.4324/9781003428701-5

average global temperature and the likelihood that this trend will continue into the foreseeable future (IPCC 2021:6, Figure SPM.1).

These rising temperatures significantly impact the earth's systems, from its soils, forests, and grasslands to its oceans, glaciers, and permafrost. These changes can result in more frequent and extreme versions of environmental phenomena, such as increased heat stress and heat-waves, altered rain patterns that lead to droughts, flooding, tropical cyclones, severe storms, increased seismic activity, and the acidification and oxygen depletion of aquatic environments. These events have a direct, negative impact on the health and survival of specific species, as well as undermining the natural resources and systems that support life in general. This is compounded by the increase of hazardous pollutants due to industrialization, which can have acute and chronic impacts on health (Kampa and Castanas 2008). Other forms of pollution—such as noise, light, plastics, and pesticides—can also compromise species' feeding, mating, and migration patterns, resulting in rapid losses to biodiversity, food crops, and carbon-capturing ecosystems (Richard et al. 2021). These impacts exacerbate the wider damage caused by the mass extraction of the earth's resources. As noted in the landmark IPBES-IPCC workshop on biodiversity and climate change, more than three-quarters of the world's surface has been modified by humans and their livestock, who now account for more than 95 percent of mammal biomass (Pörtner et al. 2021:14).

Humans have modified the natural landscape in vast and often detrimental ways to support economic and population growth, from urban expansion and deforestation to the unsustainable extraction of natural resources and groundwater. Modern methods for agricultural production have resulted in extensive GHG emissions and the creation of a monocultural global food system that increases the risk of mass crop failures (IPCC 2019). If global temperatures continue to rise, ecosystems will be unable to adapt naturally, and humans will be unable to restore them. This will lock in a vicious cycle in which higher levels of GHGs lead to extreme weather that erodes biodiversity, reducing the capacity for natural carbon capture and leading back to higher levels of GHGs.

These mutually reinforcing feedback loops create the potential for abrupt and irreversible changes to the ecosystems that support life. Indeed, various scientists argue that we have breached (or are about to breach) vital "planetary boundaries." These systems affect human survival, including freshwater use, ocean acidity levels, and biosphere integrity (Persson et al. 2022). Breaching these limits could set off negative tipping points across different ecological systems, including Arctic Sea ice, the Amazon rainforest, and the Atlantic Meridional Overturning Circulation system. These tipping points mean that such systems would start to generate self-perpetuating and cascading outputs with abrupt and catastrophic results (Lenton et al. 2019).

Prediction is difficult because environmental trends interact in seemingly chaotic ways, but impacts clearly generate significant near-term risks for civilization, and the pace and scale of risks are not equally distributed. Marginalized communities and developing nations in the Global South suffer the worst impacts, with fewer social, technological, and financial resources to prevent or adapt (Lenton et al. 2019). At the same time, the interconnected nature of these systems creates the potential for societal, economic, and technological interventions that act as circuit-breakers or catalysts for "positive" tipping points (e.g., rapid reductions in the cost of renewables). As shown in Table 2.1, these contexts highlight at least six key categories through which we can more easily understand the characteristics of environmental shifts.

Table 2.1 Categories for Understanding Environmental Change

Category	*Characteristics*
Scale	Already existential for various human groups and non-human species. Potentially existential for all.
Pace/Intensity	Risks are disproportionate to inputs and (often) exponential in nature. Significant potential for rapid concurrent and cascading effects leading to tipping points or derailment risks (i.e., in which a critical ecological or human infrastructure collapses or ceases to function in real terms). Direction of risk is predominantly negative but includes the potential for positive cascades and tipping points, depending on human activity.
Proximity	Already impacting all people/countries, but with disproportionately worse effects for developing nations and historically marginalized or disenfranchised groups (i.e., due to inequities facing different nationalities, races, religions, genders, physical abilities, neurological make-up, etc.).
Complexity	Impacts are highly context-dependent and manifest in complex and (often) difficult-to-predict ways. They connect with other complex variables within a "system-of-systems." This can make effects "appear" non-linear and chaotic, with connections between variables sometimes visible and sometimes obscured.
Interdependence	Climate variables are highly interdependent. No event occurs without almost infinite second-, third-, and nth-order effects. Events also have a compound effect, with climate risks accumulating over time until a significant tipping point is reached.
(Ir)reversibility	Some key climactic trends are already irreversible. Some might soon become irreversible, depending on human action. Even with rapid mitigation and adaptation progress, ecological threats will worsen significantly in the short- to medium-term. Each successive shift or tipping point would create a "new normal" baseline for insecurity, higher than the baseline preceding it.

Source: Author's elaboration.

Global Governance and Contested Environmental Narratives

Security forces operate in a global governance framework that has evolved over decades. Since the 1960s, United Nations environmental summits have helped to shape key features of current government responses, including the development of environmental ministries within governments, the growth of global networks of experts to support decision-making, the greater responsibility placed on northern hemisphere nations for climate action, and the awareness that some environmental challenges may take primacy over state sovereignty (Hough 2014). In December 2015, the UN Climate Change Conference (COP21) in Paris resulted in an international agreement on the need to substantially reduce GHG emissions, finance developing countries to mitigate change, and strengthen resilience and adaptation (UNFCCC 2015).

To meet their legal obligations, UN member countries must regularly outline the steps they will take to decrease GHGs and build resilience to the effects of climate change (known as "Nationally Determined Contributions" or NDCs). Unfortunately, there are significant roadblocks to progress. Most notably, current governance systems are dominated by stakeholders with the greatest political, economic, and military power who seek to protect their interests and maintain a level of status quo rather than engaging with more far-reaching solutions and marginalized communities. This status quo is influenced by large-scale fossil-fuel lobbying and reliance on economic and mitigation models that don't incorporate environmental realities (and that inherently produce technological and market-based solutions) (Stoddard et al. 2021).

In addition to these overarching governance contexts, security forces must operate in an environment of competing strategic narratives. In the public sphere, environmental discourse is heavily influenced by large-scale misinformation and disinformation campaigns (Treen et al. 2020), which often propagate "discourses of delay"—redirecting responsibility, advocating for non-transformative solutions, emphasizing the disruptive nature of change, or arguing that extinction is inevitable (Lamb et al. 2020). These can generate social and political polarization, skepticism, apathy, and mistrust toward environmental action and governmental leadership across civilian and military populations.

In the security sphere, there also exist a range of different narratives about what we should see as the most important victims, threats, agents, and responses to climate change. These are summarized in Table 2.2.

In general terms, nation-states tend to pursue "national" and "human" security discourses, while scientists tend to favor "international" and "ecological" security discourses. However, these discourses have led to different responses from politico-military leaders seeking to integrate environmental issues into their security strategies. This includes three common types of response:

Securitization. Stakeholders seek to gain support for "exceptional measures" by framing climate change as an existential security threat. Others may oppose these "securitizing" moves out of fear that they could support hidden agendas or counterproductive militarized solutions (Oels 2012).

Climatization/Environmentalization. Stakeholders seek to integrate security activities within wider environmental policies and norms while also incorporating existing environmental practices and concepts (such as "precaution" and "resilience") into the security field (Maertens 2019).

Riskification. Stakeholders seek to introduce the concept and criteria of "risk" into climate security policy and practice, increasing the number of options and strategies available for dealing with climate governance and action (von Lucke, Wellmann and Diez 2014).

Importantly, governments and armed forces have often pursued more than one of these responses simultaneously. Understanding these contexts can help when interpreting the potential objectives and degree of consistency in the environmental strategies of different state and non-state security actors.

Table 2.2 Discourses of Climate Security

Discourse	*Referent*	*Threat*	*Agent*	*Response*
National security	Nation-state	Conflict, sovereignty, and economic interests	State	Adaptation
Human security	People	Life and livelihood, core values and practices	States, NGOs, international community, and communities themselves	Mitigation
International security	International society	Conflict, global stability	International organizations	Mitigation and adaptation
Ecological security	Biosphere	Challenges to equilibrium associated with contemporary political, social and economic structures	People: changing political consciousness	Fundamental reorientation of societal patterns and behavior

Source: McDonald (2013:49).

The Security Implications of Environmental Shifts

Although there is significant certainty regarding the negative impact of environmental shifts on non-conventional threats (e.g., health, migration, economic instability), there is less certainty about their impact on conventional threats (e.g., armed conflict and violence). Furthermore, any impacts are highly dependent on the unique political, economic, socio-cultural, technological, and military contexts of a country or community. Environmental shifts tend to exacerbate or compound existing threats and insecurity, rather than creating conflict out of nowhere (Koubi 2019).

Non-conventional Threats

There is medium to high confidence that environmental shifts will have negative impacts on food, water and resource scarcity, mental and physical health, personal property, and critical national infrastructure in many regions (IPCC 2022:9 Figure SPM.2 Panel B). These impacts can compound wider issues such as direct or excess deaths, livelihood insecurity, resource competition, social polarization, displacement and migration, and threats to identity and heritage.

Death and Infrastructure Damage. The increased frequency and intensity of extreme environmental events—such as storm surges, hurricanes, cyclones, flooding, tsunamis, and forest fires—can result in both immediate deaths and longer term excess deaths (i.e., due to their detrimental impact on critical national infrastructures, such as sanitation, electricity and heating, health, water, and agriculture). Adaptation of infrastructure and processes can help reduce risk to life, e.g., early warning systems, shelters, evacuations, natural and constructed flood barriers, and fire breaks. However, debates exist on the value and viability of adaptation versus managed retreat for high-risk areas, e.g., low-lying coastal areas, river flood plains, and mountainous and forest-rich areas (Mach and Siders 2021).

Income and Food Insecurity. Climate extremes like flooding, drought, and heat stress, and slow-onset risks like gradual sea-level rise can reduce agricultural productivity, land and freshwater availability, soil fertility, and biodiversity. This can lead to income loss, food scarcity, declining health, and starvation. These impacts are damaging to agricultural economies and to communities dependent on food imports. There is particular risk for communities with reduced capacity to respond to famines, due to underlying issues such as poverty, corruption, economic mismanagement, armed conflict, or inadequate health, water, and sanitation infrastructures (IPCC 2019:441–466).

Health and Well-being. It has been estimated that climate change causes millions of deaths annually. Most of these deaths are the result of the negative health effects of fuel-based air pollution (e.g., fumes from transportation, smoke from indoor heating, or vegetation fires). However, global heating, environmental damage, and chemical pollution can also cause tens to hundreds of thousands of deaths through hunger, diarrheal infections, vector-borne diseases, and extreme working conditions, as well as having significant mental health impacts on individuals and communities (DARA 2012).

Habitability. Rising temperatures will make large swathes of the world uninhabitable for human life (Xu et al. 2020). In addition to heat stress, there is increasing risk of excessive "wet bulb" temperatures, in which the combination of high heat and high humidity reduces survival times to as little as a few hours (even with shade and unlimited water). This is already leading to thousands of deaths worldwide, with current GHG emission pathways placing at least one-fifth of the global human population at risk (Im, Pal and Eltahir 2017).

Migration and Land Loss. The factors discussed above have already resulted in the forced displacement and dispossession of millions of citizens, seeking temporary or permanent refuge from their homes. This is likely to increase, depending in part on the success of long-term adaptation policies, e.g., sea barriers, early warning systems, or the mass relocation of citizens, towns, cities, or entire nations (Martyr-Koller et al. 2021). The human security implications of these losses are significant, from the negative physical impact of food scarcity, dietary changes, and relocation, through to the mental health challenges that can result from the loss of community, heritage, and identity (Kelman et al. 2021).

Maladaptation. In seeking to mitigate and adapt to these challenges, governments, organizations, and communities may generate negative second- and third-order effects. The transition to low-carbon economies, for example, may result in policies that prioritize land ownership and resource extraction over the needs of local workers, ecosystems, and peace agreements (Mirumachi, Sawas and Workman 2020). Similarly, increasing pressure to mitigate environmental impacts could lead to the divestment of funding and impulsive deployment of novel energy solutions or geo-engineering initiatives—such as cloud brightening or stratospheric aerosol injection—which could have unforeseen consequences (Dabelko et al. 2013).

Conventional Threats

Environmental change may increase the risk of violent conflict but the connection is neither systematic nor unconditional (Bernauer, Böhmelt and Koubi 2012:1). Although there is a clear link between environmental shifts and non-conventional threats, the link between these trends and armed conflict or violence is less certain. Yet, as visualized in Figure 2.1, there is a broad understanding of how this could happen. Specifically, shifting environmental norms shape the type, scale, and pace of environmental hazards in specific locations and on earth more broadly. These environmental hazards impact the resilience and viability of critical ecological and human infrastructures, which in turn impacts on health, mental models, and behaviors of different species, including humans. In principle, these impacts could include or lead to violence. For example, flooding could precipitate a loss of income or higher cost of living, increasing the perceived cost-benefit of crime and increasing levels of direct and indirect violence

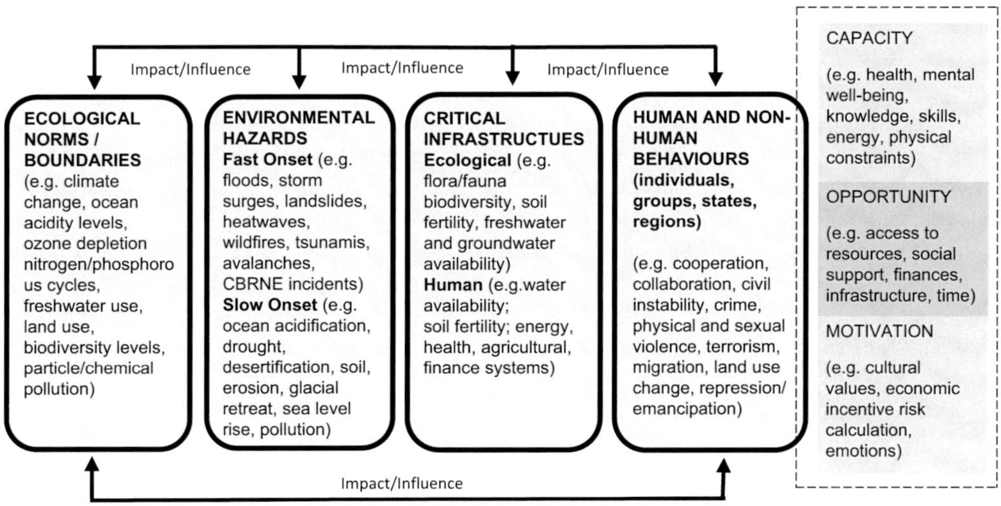

Figure 2.1 The Environmental (In)Security Cycle.

Source: Author's elaboration.

(Dell, Jones and Olken 2014). There are two critical additional components to this cycle, however. First, each step of this cycle impacts not only the step that follows it but also the step that precedes it. Second, although the current trajectory of this cycle is negative, the nature of the cycle means that governments, societies, and organizations can, in principle, implement circuit breakers at various points to help catalyze positive trajectories.

Although this conceptual "cycle" provides a useful macro-level framework for thinking, its manifestation in the real world is far more complex. The potential for environmental shifts to influence conflict and violence levels depends on many variables like geography, culture, political will, and resilience. The impacts of environmental shifts are also difficult to measure as they often transcend traditional borders and boundaries, occur over long periods, and interact with other inequities and drivers of insecurity in complex ways. Security officials must, therefore, seek to understand both macro-level trends and micro-level variables when creating strategies to mitigate or respond to environmental insecurity. One potential lens to achieve this is the nuanced links between specific environmental phenomena and acts of violence.

Temperature Shifts. Case studies demonstrate links between higher temperatures and criminal activity (e.g., rape, assault, burglary, murder), civil conflict (generally related to food scarcity during growing seasons), urban riots, and more frequent or less stable political transitions. The relationships are context-specific and have been questioned. Warm weather, for example, might catalyze violence not by reducing food supply but by increasing the attractiveness of looting.

Rainfall Levels. Drought may increase inter-community conflict in agricultural communities by reducing food production and livestock prices. Drought-induced scarcity can also trigger social unrest in countries dependent on low-cost food imports. Communal conflict can also increase during wetter periods due to resource abundance. Local variables influence these risks. Is it an election year? Is road infrastructure adequate? Do communities have mechanisms for inter-community dialogue or a history of land disputes? Are there traditional conflicts between herders and pastoralists? Shared rivers are a source of conflict but can stimulate diplomacy. Suitable water management infrastructures, flexible mechanisms for water allocation, and robust treaties between states can reduce conflict risks.

Extreme Weather. Extreme weather events such as flooding, storms, landslides, and tsunamis can cause economic losses and migration that might generate armed conflict, but there is weak evidence of links to armed conflict. It might fuel existing armed conflicts, exacerbating the frequency and severity of violent attacks, battle deaths, and casualties in insurgencies, increase the potential for ethnic fractionalization, demonstrations, strikes, and political unrest in a country, as well as the prevalence of transnational terrorism. In turn, extreme weather can lead to increases in government repression as a means of reducing the mobilization of dissidents and enhancing state control. Temporary migrants may be welcomed and supported, but longer term migration may increase social polarization, resource competition, risks of violence, and the potential for riots among a host population. Disaster-induced migration can increase the probability that existing civil conflicts will continue but has not been proven to ignite new conflicts. Local variables shape the risk of conflict. Conflict is less likely, for example, in scenarios where the host population can gain additional support from a central government.

Environmental changes can exacerbate existing violence and conflict, but their impact depends on other factors, such as demographic pressures, political and economic stability, the resilience of critical national infrastructures, and levels of corruption, social polarization, and ethnic fractionalization. The impact of environmental shifts on non-conventional threats is direct, with catastrophic risks to life, health, and well-being facing millions of citizens. While debates about environmental threats continue, there is broad agreement that increases in global heating and biodiversity loss will likely catalyze insecurity and conflict in far more significant, unpredictable, and devastating ways at the human, national, international, and planetary levels (Mach et al. 2019). Prioritizing mitigation, adaptation, and resilience will help reduce insecurity at all levels.

The Role of Defense

What does this mean for security forces? Defense leaders face a dual requirement. They must prepare their forces for continually increasing operational tempos and novel challenges. This may include expanding non-warfighting roles such as humanitarian assistance: and disaster relief (HARD), assisting civil authorities, border control, and diplomatic activities (to reduce environmentally exacerbated tensions and security dilemmas between nations). These non-warfighting activities will likely occur alongside increasing pressure to maintain operational effectiveness for high-intensity operations, in line with increasing inter-state tensions, resource competition, shifting migrant flows, and growing societal unrest and polarization. The second requirement is to mitigate their climate and biodiversity impact to avoid creating more problems and threats than they are solving.

Managing this dual requirement is not a zero-sum game. Focusing on environmental sustainability will not inevitably reduce military effectiveness. Rather, these actions can be mutually reinforcing. It is estimated, for example, that the world's militaries and the industries that provide their equipment create 5.5 percent of all global GHG emissions (Parkinson and Cottrell 2022). As a thought exercise, if we were to align this with the data from the Development Assistance Research Associates (DARA) report (2012), the mere existence of global military forces could be generating upward of 330,000 deaths per year. At the same time, violent conflict generates damaging environmental impacts that can lead to a vicious cycle of loss and damage (Weir 2020). This highlights the need for greater insight and reflection on the strategic trade-offs between the benefits of security forces for achieving deterrence and resolving conflicts and the negative impacts of militarization on environmental security.

By developing sustainable capabilities and strategies, it may be possible to generate greater resilience both for the military and the communities it serves. For example, a military with self-sustaining food, water, and fuel systems could maintain a forward operating presence for longer while reducing costly, risky, and ecologically damaging resupply missions. A national defense industry capable of recycling critical materials and repurposing them for new hardware could reduce a military's reliance on unstable and competitive global supply chains. Operational strategies that build adaptive capacity, protect ecological resources, and support local conflict resolution mechanisms could enhance prospects for resilient peace across vulnerable communities. Unfortunately, while some of these opportunities are recognized, there is still limited international or organizational agreement on the degree to which and the ways in which defense forces should adapt.

New Threats and Roles

A critical factor influencing this lack of agreement is the uncertainty about the role of defense forces in a future operating environment characterized by cascading, concurrent, and irreversible environmental threats. For some scholars, this environment will force militaries into a kind of permanent "secure and hold" position, in which they must try to deter inter-state conflict and prevent society from tearing itself apart for as long as possible either to buy time for political solutions or simply to ensure a more orderly process of civilizational decline. For other scholars, military forces—along with the police, gendarme, and coast guard—are at risk of becoming glorified border control agents tasked with controlling rapidly shifting internal and external climate migrant flows. There may be a significant increase in military support for HADR operations due to the military's role as a disciplined and biddable force with sizable medical, engineering, logistics, and communications expertise. Armed forces may be called upon for military assistance to civil authorities (MACA) operations, monitoring and protecting wildlife and critical ecosystems, supporting ecological regeneration efforts, or controlling the distribution of resources during food and energy crises. Such activities would require "proactive" and "reactive" responses, sometimes in support of other agencies and sometimes as the operational lead of last resort. The likelihood of these operations is difficult to determine. For example, the expansion of HADR activities will be affected by resource constraints, resilience-building efforts, and shifts in public support for overseas HADR deployments.

For military forces, is there a risk of securitizing problems that would be better addressed through developmental, diplomatic, or economic responses? How might states use traditional coercion or new domains (e.g., cyber and information operations) to gain monopolistic access or control of critical resources at the expense of smaller or weaker states? Is there a clear tipping point at which environmental threats require governments to significantly curtail social freedoms (enacted through security forces)? How far should military forces expand their doctrine and capabilities for these potential roles (Causevic and Al-Marashi 2020)? Should states expand civil defense forces like the UN White Helmets so that the military can maintain its core warfighting role (David 1997)? Might the overstretch of militaries result in irregular forces expanding for local security and stabilization? These are just a small sample of the issues security forces must reflect upon.

Adaptation and Mitigation

How can militaries best account for these challenges? Armed forces and inter-governmental alliances have published climate and security strategies. Central to many is an emphasis on reducing the military's negative environmental impact while making it more resilient to environmentally induced global shocks and operational challenges. Figure 2.2 illustrates these strategies in general terms. They present sustainability not as an "ethical" obligation but as a means of enhancing operational effectiveness, diplomatic opportunities, and personnel safety while reducing legal, financial, and reputational risks.

Figure 2.2 Military Framing of Risks and Opportunities from Adaptation and Sustainability Reforms.

Source: Author's elaboration.

Table 2.3 Potential Environmental Reform Priorities Across Defense Functions

Policy and Planning	• Integrate shifting operational requirements and implications of environmental trends in policy, strategy, doctrine, concepts, futures, battle orders, rules of engagement, etc. • Re-evaluate planning and strategy formulation processes to enhance systems thinking and reduce detrimental heuristics and biases (e.g., mini-maxing, short-termism, optimism bias, and protectionism). • Account for potential risks and liabilities created by future environmental laws and targets (national and international).
Organization	• Re-evaluate the balance of current military tasks, force postures, and the role and remit of the armed forces (e.g., creating specialized functions, reconsidering geographical distribution of forces, divesting capacity to local civilian resilience groups, etc.).
Education and Training	• Enhance whole-force education on environmental trends and risks. • Create core skills training to enable personnel—especially at the command level—to operate in increasingly uncertain and complex environments (e.g., critical and systems thinking, leadership skills). • Adapt training exercises (including wargaming) to account for large-scale disasters, cascading threats, and environmental extremes.
Personnel	• Show a commitment to climate issues to maintain attractiveness for recruitment of younger generations. • Recruit talent from relevant specializations (e.g., meteorology, engineering, search and rescue, diplomacy, construction). • Enhance physical and psychological security for personnel operating in climate-degraded conditions (e.g., PPE, immunizations). • Review personnel job specifications and replace less important duties with new duties for mitigation, adaptation, and resilience progress.
Infrastructure and equipment	• Enhance resilience and energy efficiency of infrastructure, capabilities. • Evaluate strategic benefits of maintaining or abandoning existing contracts, assets, capabilities, and bases (beyond pure financial" cost).
Information/ISR	• Enhance access to meteorological forecasting, build depth and breadth of localized human intelligence networks, and increase capacity to share intelligence with key stakeholders and enablers. • Provide transparency on sustainability and climate security targets, measures of effectiveness, and progress *across* various stakeholders (including internal decision-makers, external allies, and wider public).
Logistics	• Enhance resilience and energy efficiency of supply chains (national and international). • Ensure continued—ideally equitable—access to critical supplies (e.g., water, fuels, medicine, critical minerals).
Interoperability/ cooperation	• Enhance inter-state, inter-sectoral, inter-governmental, inter-agency, and inter-service cooperation and integration. • Enhance diplomacy and support for existing allies and potential partners in relevant areas of adaptation and mitigation.

Source: Adapted and expanded from Cox et al. (2020).

The relative importance of factors illustrated in Figure 2.2 varies across and within different security forces. Uncertainty clouds planning for military reform. Yet, as shown in Table 2.3, even a limited response to these risks would entail significant reforms across multiple defense functions.

For some, the reform priorities in Table 2.3 are unachievable and undesirable; for others, they must be revised. Simply retrofitting reforms to existing organizational models will fail to deal with inevitable military damage to the environment (Clark, Jorgenson and Kentor 2010). Military climate strategies generally emphasize solutions that pose limited risk to their own budgets and power rather than riskier, paradigm-changing strategies. They prefer efficiency and economy to shrinkage or divestment. These strategies may represent a pragmatic approach to environmental gains, but the goalposts for success are shifting. Armed forces must enhance mitigation, adaptation, and operational effectiveness. Serving and retired military leaders are calling for a change from compliance and liability to opportunity and strategic advantage.

Conclusion

Global heating and biodiversity loss are already exacerbating existing threats and creating new threats, from the human and local levels to the national, international, and planetary levels. Environmental modeling indicates that that the scale, pace and concurrency of these impacts are set to grow at an unprecedented rate over the coming decades. This will likely place security forces in a position where they must support a rapidly expanding range of traditional and non-traditional operations while simultaneously increasing their capacity to operate sustainably with strained financial and material resources. Although many predictions exist as to what this scenario might entail—from increasing HADR and MACA operations to regional conflict and total war—military personnel will certainly need to learn how to adapt and operate more flexibly in increasing complexity, ambiguity, and uncertainty.

References

Bernauer, T., Böhmelt, T. and Koubi, V. (2012), "Environmental Changes and Violent Conflict", *Environmental Research Letters*, vol. 7, no. 1, pp. 1–8.

Causevic, A. and Al-Marashi, I. (2020), "Can NATO Evolve into a Climate Alliance Treaty Organization in the Middle East?" *Bulletin of the Atomic Scientists*, vol. 76, no. 2, pp. 97–101.

Clark, B., Jorgenson, A. K. and Kentor, J. (2010), "Militarization and Energy Consumption", *International Journal of Sociology*, vol. 40, no. 2, pp. 23–43.

Cox, K., Knack, A., Robson, M., Adger, N., Paillé, P., Freeman, J., Black, J. and Harris, R. (2020), *A Changing Climate: Exploring the Implications of Climate Change for UK Defence and Security*, RAND Corporation, Santa Monica, CA.

Dabelko, G. D., Risi, L. H., Null, S., Parker, M. and Sticklor, R. (eds.) (2013), *Backdraft: The Conflict Potential of Climate Change Adaptation and Mitigation (Environmental Change & Security Program, Report vol. 14, no 2)*, Wilson Center, Washington, DC.

DARA (2012), *Climate Vulnerability Monitor 2nd Edition: A Guide to the Cold Calculus of a Hot Planet,* Fundación DARA International. https://daraint.org/climate-vulnerability-monitor/climate-vulnerability-monitor-2012/

David, C.-P. (1997), "After the Blue Helmets Come the White Helmets", *Peacekeeping & International Relations*, vol. 26, no. 1, pp. 3–7.

Dell, M., Jones B. F. and Olken, B. A. (2014), "What Do We Learn from the Weather? The New Climate-economy Literature", *Journal of Economic Literature*, vol. 52, no. 3, pp. 740–798.

Hough, P. (2014), *Environmental Security: An Introduction*, Routledge, London.

Im, E. S., Pal, J. S. and Eltahir, E. A. B. (2017), "Deadly Heat Waves Projected in the Densely Populated Agricultural Regions of South Asia", *Science Advances*, vol. 3, no. 8, pp. 1–7.

IPCC (2019), *Climate Change and Land: An IPCC Special Report on Climate Change, Desertification, Land Degradation, Sustainable Land Management, Food Security, and Greenhouse Gas Fluxes in Terrestrial Ecosystems*, Cambridge University Press, Cambridge, UK and New York, NY, US.

IPCC (2021), *Climate Change 2021: The Physical Science Basis. Contribution of Working Group I to the Sixth Assessment Report of the Intergovernmental Panel on Climate Change*, Cambridge University Press, Cambridge, UK and New York, NY, US.

IPCC (2022), *Climate Change 2022: Impacts, Adaptation, and Vulnerability.* Contribution of Working Group II to the Sixth Assessment Report of the Intergovernmental Panel on Climate Change, Cambridge University Press, Cambridge, UK and New York, NY, US.

Kampa, M. and Castanas, E. (2008), "Human Health Effects of Air Pollution", *Environmental Pollution*, vol. 151, no. 2, pp. 362–367.

Kelman, I., Ayeb-Karlsson, S., Rose-Clarke, K., Prost, A., Ronneberg, E., Wheeler, N. and Watts, N. (2021), "A Review of Mental Health and Wellbeing under Climate Change in Small Island Developing States (SIDS)", *Environmental Research Letters*, vol. 16, no. 3, pp. 1–13.

Koubi, V. (2019), "Climate Change and Conflict," *Annual Review of Political Science*, vol. 22, pp. 343–360, https://doi.org/10.1146/annurev-polisci-050317-070830.

Lamb, W. F., Mattioli, G., Levi, S., Roberts, J. T., Capstick, S., Creutzig, F., Minx, J. C., Müller-Hansen, F., Culhane, T. and Steinberger, J. K. (2020), "Discourses of Climate Delay", *Global Sustainability*, vol. 3, e17, pp. 1–5, https://doi.org/10.1017/sus.2020.13.

Lenton, T. M., Rockström, J., Gaffney, O., Rahmstorf, S., Richardson, K., Steffen, W. and Schellnhuber, H. J. (2019), "Climate Tipping Points - too Risky to Bet Against", *Nature*, vol. 575, pp. 592–595. https://doi.org/10.1038/d41586-019-03595-0

Lucke von, F., Wellmann, Z. and Diez, T. (2014), "What's at Stake in Securitising Climate Change? Towards a Differentiated Approach", *Geopolitics*, vol. 19, no. 4, pp. 857–884.

Mach, K. J., Kraan, C. M., Adger, W. N., Buhaug, H., Burke, M. Fearon, J. D., Field, C. B., Hendrix, C. S., Maystadt, J.F., O'Loughlin, J., Roessler, P., Scheffran, J., Schultz, K. A. and von Uexkull, N. (2019), "Climate as a Risk Factor for Armed Conflict", *Nature*, vol. 571, pp. 193–197. https://doi.org/10.1038/s41586-019-1300-6

Mach, K. J. and Siders, A. R. (2021), "Reframing Strategic, Managed Retreat for Transformative Climate Adaptation", *Science*, vol. 372, no. 6548, pp. 1294–1299.

Maertens, L. (2019), "From Blue to Green? Environmentalization and Securitization in UN Peacekeeping Practices", *International Peacekeeping*, vol. 26, no. 3, pp. 302–326.

Martyr-Koller, R., Thomas, A., Schleussner, C. F., Nauels, A. and Lissner, T. (2021), "Loss and Damage Implications of Sea-level Rise on Small Island Developing States", *Current Opinion in Environmental Sustainability*, vol. 50, pp. 245–59. https://doi.org/10.1016/j.cosust.2021.05.001

McDonald, M. (2013), "Discourses of Climate Security", *Political Geography*, vol. 33, no. 1, pp. 42–51.

Mirumachi, N., Sawas, A. and Workman, M. (2020), "Unveiling the Security Concerns of Low Carbon Development: Climate Security Analysis of the Undesirable and Unintended Effects of Mitigation and Adaptation", *Climate and Development*, vol. 12, no. 2, pp. 97–109.

Oels, A. (2012), "From 'Securitization' of Climate Change to 'Climatization' of the Security Field: Comparing Three Theoretical Perspectives", in J. Scheffran, M. Brzoska, H. G. Brauch, P. M. Link and J. Schilling (eds.), *Climate Change, Human Security and Violent Conflict: Challenges for Societal Stability*, pp. 185–205, Springer-Verlag, Berlin, Heidelberg.

Parkinson, S. and Cottrell, L. (2022), *Estimating the Military's Global Greenhouse Gas Emissions*, Scientists for Global Responsibility, and Conflict and Environment Observatory. https://www.sgr.org.uk/publications/estimating-military-s-global-greenhouse-gas-emissions

Persson, L., Carney Almroth, B. M., Collins, C. D., Cornell, S., de Wit, C. A., Diamond, M. L., Fantke, P., Hassellöv, M., MacLeod, M., Ryberg, M. W., Søgaard Jørgensen, P., Villarubbia-Gómez, P., Wang, Z. and Zwicky Hauschild, M. (2022), "Outside the Safe Operating Space of the Planetary Boundary for Novel Entities", *Environmental Science & Technology*, vol. 56, no. 3, pp. 1510–1521.

Pörtner, H. O., Scholes, R. J., Agard, J. and Ngo, H. T. (2021), "Scientific Outcome of the IPBES-IPCC Co-sponsored Workshop on Biodiversity and Climate Change. Intergovernmental Science-Policy Platform on Biodiversity and Ecosystem Services (IPBES) and Intergovernmental Panel on Climate Change (IPCC), Bonn, Germany.

Richard, F.-J., Southern, I., Gigauri, M., Bellini, G., Rojas, O. and Runde, A. (2021), "Warning on Nine Pollutants and Their Effects on Avian Communities", *Global Ecology and* Conservation, vol. 32, pp. 1–30. https://doi.org/10.1016/j.gecco.2021.e01898.

Stoddard, I., Anderson, K., Capstick, S., Carton, W., Depledge, J., Facer, K., Gough, C., Hache, F., Hoolohan, C., Hultman, M., Hällström, N., Kartha, S., Klinsky, S., Kuchler, M., Lövbrand, E., Nasiritousi, N., Newell, P., Peters, G. P., Sokona, Y., Stirling, A., Stilwell, M., Spash, C. L. and Williams, M. (2021), "Three Decades of Climate Mitigation: Why Haven't We Bent the Global Emissions Curve?", *Annual Review of Environment and Resources*, vol. 46, no. 1, pp. 653–689.

Tans, P. (2009), "An Accounting of the Observed Increase in Oceanic and Atmospheric CO2 and an Outlook for the Future", *Oceanography*, vol. 22, no. 4, pp. 26–35.

Treen, K. M. d.'l., Williams, H. T. and O'Neill, S. J. (2020), "Online Misinformation about Climate Change", *Wiley Interdisciplinary Reviews: Climate Change*, vol. 11, no. 5, pp. 1–20.

UNFCCC (2015), *The Paris Agreement*, TIAS no. 16–1104. https://unfccc.int/documents/184656?utm_source.

Weir, D. (2020), "How Does War Damage the Environment?", *Conflict and Environment Observatory*. https://ceobs.org/how-does-war-damage-the-environment/

Xu, C., Kohler, T. A., Lenton, T. M., Svenning, J.-C., and Scheffer, M. (2020), "Future of the Human Climate Niche", *Proceedings of the National Academy of Sciences*, vol. 117, no. 21, pp. 11350–11355.

3 Space and the Grey Zone

Pedro Miguel da Silva Costa, David Last, and Marzena Żakowska

Introduction

Anyone with a smartphone has recourse to satellite navigation, global positioning, synchronized timing, and satellite images. Aviation, shipping, and land transport rely on space-based navigation. Telecommunications, finance, and national security for almost every country intersect with space-based satellites at some point. In peacetime, most governments and citizens can take for granted these links to space. In the Grey Zone between peace and emerging war, security planners should think more strategically about dependence on space. The operating environment is complex, marked by regional conflicts, dynamic threats from states and non-state actors, instability, and insecurity on a global scale. Technology plays a balancing role between major and minor actors but simultaneously threatens to destabilize relationships and revolutionize the means of waging war (Roland 2009).

The technology that enables military operations in or supported from space overlaps with military and civilian technology developed for other purposes. All the rockets that propelled the first wave of satellites into orbit were developed for surface-to-surface military use (Chun 2006:15). Sensors and electronic communications have been developed to support civilian science and economies. The products, services, and capabilities of space-based assets multiply security and defense capabilities for even small countries. The NATO Space Handbook and Allied Joint Doctrine for Air and Space Operations (AJP-3.3) define space capabilities in three segments: space, ground, and links between (NATO 2013:6; 2016:2-5).

This chapter describes the geography of space, actors in space, geopolitical thinking about space, implications for Grey Zone conflict, and strategic options for small countries. Understanding the development of space technology and considering theories about power dynamics help to illuminate choices small states can make to manage dependence on space in the information age.

Geography of Space

Space may be framed as a Global Commons, defined as areas outside the political control or sovereignty of any one state. States vie for access and control without the legal framework of boundaries and sovereignty. Space, like international water before it, has allowed states and commerce to operate in a globalized economy, fostering development, strategic partnerships, and alliances, competing for value by using access to the Commons wisely. Like access to the oceans, space is a domain that generates influence in the international system. We learn to think differently about geography, as Portuguese navigators did in the age of Vasco Da

DOI: 10.4324/9781003428701-6

Gama—Oceans became highways, not barriers. A better understanding of space will enhance security and prosperity.

Space power consists of more than rockets and satellites; it includes everything that can be achieved through access to space. Michael Krepon, a co-founder of the think tank, the Stimson Centre, describes "space power" as "the sum of all the skills that contribute to a nation's ability to benefit from the use of space" (Krepon, Hitchens and Katz-Hyman 2011:119). Space belongs to all, for the benefit of all, according to the Outer Space Treaty (1967), to which 172 states are parties or signatories (UN Office of Disarmament Affairs 2021). There are no constraints of sovereignty in space, making its use a power multiplier for economic, informational, political, and military purposes (Boniface 2003:124).

Space power rests on a system composed of elements in space, terrestrial elements, and the links between them (Tabarcia and Stroescu 2010:21). The space element includes satellites themselves, equipment carried, and vehicles that operate in space. The terrestrial element includes support and exploitation: launch systems, industrial installations, production facilities, command and control networks, communication nodes, information analysis, and distribution centers. The links between space and terrestrial elements are the frequencies through which not only satellites communicate but also the people who operate the systems enabling space. The signals can be disrupted by solar winds, terrestrial weather, or hostile action, and the ground stations receiving the signals are generally considered more vulnerable than the satellites, both to hacking and to physical attack (Parliamentary Office of Science and Technology 2006).

The physical characteristics of space should be understood to appreciate different kinds of space assets, threats, and regulations. The Kármán line represents the point at which conventional aircraft can no longer fly because their wings will not provide lift in the thinner atmosphere. It is about 100 kilometers above sea level, and the Fédération Aéronautique Internationale identifies it as the beginning of space. This is not universally accepted. The Outer Space Treaty (1967) does not define outer space, and the Convention on International Civil Aviation (1944) states that "every State has complete and exclusive sovereignty over the airspace above its territory." Modern jet fighters can fly to about 17 kilometers, and the record for jet flight is about 37 kilometers, although the theoretical limit for air-breathing engines is about 75 kilometers, so the Kármán line at 100 kilometers would seem a reasonable limit for sovereign airspace.

Table 3.1 presents the orbits. All elliptical orbits must cross the equatorial plane and have the earth's center at one focus of the orbital ellipse (France and Sellers 2011:64). A satellite's weight influences its orbit (Ribeiro 2007).

Low Earth Orbit (LEO) is mainly used for earth observation and reconnaissance technologies that do not require real-time communication and for manned space missions. The advantages of a satellite constellation in LEO are lower launch costs per satellite and relatively low power signal transmission. The commercial operator SpaceX began deploying Starlink satellites in LEO in 2019. The Starlink network now consists of over 4,000 small, mass-produced communication

Table 3.1 Usual Orbits of Artificial Satellites

Name of the orbit	*Abbreviation*	*Altitude (min-max)*
Low Earth Orbit	LEO	500–2,000 km
Mid-Earth Orbit	MEO	8,000–20,000 km
Geostationary Earth Orbit	GEO	36,000 km
High Earth Orbit	HEO	36,000–100,000 km
Polar Orbit	PO	200–1,000 km

Source: Adopted from Hall (2006:3).

satellites providing continuous global coverage. The strategy of mass compensates for orbital disturbances and disruption from atmospheric friction. LEOs between 500 and 2,000 kilometers above sea level encompass the van Allen radiation belt (about 1,000–1,200 kilometers above sea level), the charged particles of which interfere with communications. Most satellites are in LEO, and the space is increasingly congested.

Medium Earth Orbit (MEO) is 8,000 and 20,000 kilometers above sea level. Satellites in MEO transit roughly every 12 hours and can maintain visibility for several hours over a point. A constellation of 24–36 satellites can provide continuous navigation coverage of the entire Earth (Cepik 2015:12). With less atmospheric friction, the orbit is more stable than LEO, but cosmic radiation is greater, reducing the useful life of satellite components.

Geostationary Earth Orbit (GEO) is characterized by an orbital period the same as one complete Earth rotation, about 36,000 kilometers above sea level at a fixed latitude and longitude. Only three to five satellites are needed to cover the entire surface of the earth, and satellites tend to be quite large (over 600 kilograms). GEO satellites dominate global communications and house most meteorological equipment (Cepik 2015:12). The dominance of GEO satellites in telecommunications is now being challenged by Starlink's constellations of smaller LEO satellites (Frackiewicz 2023).

Finally, satellites in High Elliptical Orbit (HEO) are committed to supporting Arctic and Antarctic zones. They move slowest at their apogee, up to 50,000 kilometers above the surface, and fastest at their perigee, about 500 kilometers above the surface.

Space geography provides other strategic points. The van Allen radiation belt is an obstacle that constrains low-Earth orbiting. There are pathways for slingshot maneuvers that help spacecraft escape the gravitational pull, and there are five Lagrange or libration points near the earth where the gravitational pull of the earth and moon cancels each other, allowing objects to remain in place with minimal fuel use (Marshall 2021:250).

Larger satellites generally orbit farther from Earth, and smaller satellites orbit closer. Weight classifications have had to expand to encompass new and smaller designs. The US National Aeronautical and Space Administration (NASA) began in 1999 a project to enable broad research access to space through smaller satellites based on a standard design (Chin et al. 2008). Pico satellites, weighing about a kilogram and measuring about 10 cm cubed, may be the future of satellite-based research, with lower costs, faster development, and wider access. Large satellites are more than 1,000 kilograms and may require a dedicated launch vehicle. Medium satellites of 500–1,000 kilograms are common in MEO and might be launched in clusters. Minisatellites (100–500 kilograms), microsatellites (10–100 kilograms), nanosatellites (1–10 kilograms), and the newest picosatellites (less than 1 kilogram) are all limited to LEO. Although small satellites are lighter, cheaper, and faster to develop and launch, they have a shorter life span due to low orbit, mass production, and susceptibility to terrestrial gravity. Falling out of orbit and interfering with other orbiting objects is a risk (RT.com 2014)

As of September 2021, more than 4,550 satellites were orbiting Earth. The majority (3,790) were in LEO. More than 40 percent of these are owned and operated by companies rather than countries, with SpaceX the uncontested leader, operating 1,655 satellites in 2021 and claiming more than 3,500 satellites by January 2023. The US Federal Communications Commission has approved the launch of 7,500 more Starlink satellites, acknowledging SpaceX's plan to launch 30,000 internet satellites to provide continuous global coverage (Wolfe 2023). This makes LEO especially congested and potentially contested. Space debris is a growing problem. While collision incidents are rare, the US Department of Defense's Space Surveillance Network tracks over 27,000 pieces of space debris. As orbits become more crowded, there is the potential for cascades of space junk to remove swathes of satellites indiscriminately. The potential for accidental damage is a good cover for malicious attacks by actors with the capability.

The threat of space debris may force international cooperation to maintain the potential to operate safely in LEO. The UN Fourth Committee (Special Political and Decolonization) devoted a session in 2022 to the work of the UN Committee on the Peaceful Uses of Outer Space (COPUS), focusing on the need to close the technological gap, manage space debris, and develop sustainable norms for the peaceful use of space. In the Fourth Committee, more than a dozen countries spoke about the need for cooperation, norms, and international law. The United Arab Emirates (UAE) is chair of COPUS and leader of an innovative international space program. Argentina is actively advancing international law on the regulation of space. Pakistan, Indonesia, and Colombia discussed the need to mitigate space debris (UN General Assembly, Fourth Committee 2022).

Evolving Actors in Space

Four space races have shaped the actors in space. The initial race during the Cold War between the United States and the USSR produced rocket technology, ballistic missile navigation, and satellite communications. The second race saw the entry of new state actors, including China, India, Brazil, and the European Space Agency (formed in 1975). The third race entailed the rise of commercial actors beginning with the NASA-backed Space Services launch in 1989, which has accelerated since 2013 with the rise of commercial clients for private launch services like SpaceX, Virgin Galactic, and Blue Origin. The fourth race followed China's entry into competition with the United States, with growing risks of militarization of space.

Actors in space are differentiated by their access to technology. The basic sciences of rocketry, astrophysics, and telecommunications are open, but national security and commercial secrecy restrict access to specific knowledge needed to exploit space. Gaps in access to specialized science and technology for space have consequences for international power structures. This goes beyond military technology; all space science is considered dual use, with military and civilian applications, and this has driven investment strategies at the national level and by commercial actors. Space investment contributes to strategic autonomy in military and commercial development (European Parliament 2019:paragraph 10). State-based commercial space exploration and development has accelerated the use of space and made space assets accessible to small countries.

Emerging space powers include the many states that can indigenously develop, deploy, and exploit satellites but lack the launch capability to do so autonomously. Medium space powers can launch, develop, and control satellites. This includes all the members of and contributors to the European Space Agency, Japan, India, Israel, Iran, and Ukraine. Major space powers have all these capabilities and the capacity for crewed space flight.

The top 50 commercial owner-operators in 2022 accounted for 3,751 satellites, but this number includes only 1,655 for SpaceX, which is launching satellites faster than any state. The top national hosts of these 50 commercial space ventures are the United States (21/50), China (10/50), Japan, and Russia (3/50 each) (Rome 2023). Four companies are listed as multinational (SES Space Agency, European Space Agency, EUTELSAT Space Agency, and European Organization for the Exploitation of Meteorological Satellites). All four operate primarily from the European Union, demonstrating the impact of the scientific ecosystem established by the European Space Agency, from which smaller countries can benefit.

The distinction between state-owned and commercial space assets may be irrelevant. Commercially operated Chinese, Russian, or American assets would not operate against those countries' interests. It should be assumed that their host countries can access information traveling through those assets should governments wish to do so. This assumption can be relaxed only with multinational supervision (as for European agencies) and effective encryption. Clients of commercial and foreign-owned space assets should assume that someone is listening in and that access may be interrupted. National efforts to establish autonomous communications build on this assumption (Doicariu 2022).

The top space power is undoubtedly the United States by any measure. According to one industry source, the United States accounts for 60 percent of the small satellites launched between 2012 and 2019, with China accounting for 10 percent, Japan and Russia for 4 percent each, Germany for 3 percent, with the remaining 19 percent distributed among 62 countries (Bryce Space and Technology 2023). Rome (2023) compares ownership of more than 4,400 satellites giving the United States a larger share, at 64 percent. The United States also accounts for about 60 percent of space budgets, but the total is probably much higher when the US defense expenditure is included. China, Russia, Japan, and India are no surprises. However, the explosion of commercial competition and the renewed space races for technological and national advantage have expanded the number of space players, and most are not major powers. Table 3.2 lists the 30 countries with the most nationally owned satellites in 2022 (Rome 2023). The number of satellites correlates strongly to space budget (.98) and to GDP (.88) but not to population (.20).

Small, wealthy, scientifically adept countries with educated populations will likely perform well in the new technological race for benefits from space. The shaded countries in Table 3.2 represent contributors to the European Space Agency. The surprising number of satellites attributed to tiny Luxembourg reflects the collective European effort and Luxembourg's favorable tax regime. One policy deduction is that there is a broad international constituency for equitable space management as a shared resource. Disruption of space as a Global Commons will not be taken lightly by states that have invested in it so heavily.

What do satellites do? In 2022, 63 percent of the 4,550 satellites in orbit supported communications. Earth observation, including satellite imagery, accounted for 22 percent. About 8 percent supported technological development. Surprisingly, only 3.6 percent (about 164 satellites) supported navigation or global positioning. Earth science, space observation, and space science accounted for less than 3 percent (Rome 2023). Looking at the growth of satellite traffic over time, commercial and non-profit satellites have accounted for the largest growth since 2012 (Bryce Space and Technology 2023), with the former heavily focused on communications.

Another measure of the growth of engagement with space can be found in the number of states with space launching capacity. In 1966, this group was limited to the United States, USSR, and France. By 2020, it included all the members, associate members, and cooperating states of the European Space Agency—30 states. China, India, Iran, Japan, North Korea, and Ukraine also demonstrated launch capability by 2020, according to the Union of Concerned Scientists (UCS 2023).

Commercial options and multinational collaboration have increased launch capacity, and smaller and lighter satellites can be launched into LEO faster and more easily. Military and security forces are increasingly dependent upon satellites for positioning, navigation, and time (PNT), making it important for most countries to have reliable satellite arrays. It is now conceivable that more states can launch independent small satellite arrays to replace dependence on allied or commercial assets for intelligence, surveillance, and reconnaissance (ISR). The acquisition could be as short as two years, with individual satellites costing €1 million, or €36 million for the array, and launch costs as little as €40 million (Bowman 2012:19; Tyvak Nano-Satellite Systems 2018; Hanlon 2018).

Small satellites might respond quickly in theatres with low satellite coverage (Chun 2006:25). They might replace damaged satellite networks from anti-satellite (ASAT) attacks or space

Table 3.2 Countries in Space 2022

Country	*Satellites*	*GDP (billion USD)*	*Population*	*Space budget*
The United States	2,804	25,463	333,287,557	61.97
China	467	17,963	1,412,175,000	11.94
The United Kingdom	349	3,707	66,971,411	1.15
Russia	168	2,204	143,555,736	3.42
Japan	93	423	125,124,989	4.9
India	61	3,385	1,417,173,173	1.93
Canada	57	2,139	38,007,166	0.54
Germany	47	4,072	84,079,811	2.53
Luxembourg	40	82	650,774	0.28
Argentina	34	632	46,234,830	0.81
France	31	2,789	67,935,660	4.2
Spain	24	1,397	47,615,034	0.76
Italy	21	2,010	58,856,847	1.74
Israel	19	522	9,550,600	0.16
South Korea	18	1,665	51,628,117	0.72
Netherlands	16	991	17,703,090	0.23
Brazil	16	1,920	5,457,127	0.45
Finland	15	280	5,556,880	0.17
Australia	14	1,675	25,978,935	0.11
UAE	13	507	9,441,129	0.8
Taiwan	13	761	23,945,553	0.32
Switzerland	13	807	8,769,741	0.16
Saudi Arabia	13	1,108	36,408,820	2.1
Singapore	11	466	5,637,022	0.15
Turkey	9	905	85,341,241	0.06
Norway	8	579	5,457,127	0.14
Mexico	8	1,414	127,504,125	0.47
Indonesia	8	1,319	275,501,339	0.55
Thailand	7	495	71,697,030	0.09
Kazakhstan	6	220	19,621,972	0.06

Source: Table compiled by Authors from World Bank data (2022), Rome (2023), and statistical sources. Space budgets are from Statista; those shaded (right column) are from diverse sources and may not be comparable.

debris (France and Sellers 2011). To the extent that small satellites can be guided before orbit, they can be used as anti-satellite weapons. A Chinese satellite cube 40 centimeters a side weighing 40 kilograms was deployed from the Shenzhou-7 capsule and shown to be effective as an ASAT (Saxena 2016:7). Small states are generally not involved in testing space weapons. Their security leaders, however, need to think strategically about the use of space.

Geopolitical Thinking about Space

It is natural to see geopolitical thinking about space in terms of earlier theories of spatial pressure. Halford Mackinder's World Island, Alfred Thayer Mahan and Julian Corbett's theories about sea control, and Giulio Douhet and Alexander de Seversky's air power theories have all been referenced in geopolitical thinking about the control and use of space. Recent American thinkers Everett Dolman, David Lupton, and James Oberg are associated with American space programs.

In his 1904 lecture, the *Geographical Pivot of History*, the British geographer Halford Mackinder described a global ocean covering three-quarters of the world and a great continent encompassing Europe, Asia, and Africa, with some outlying islands and land masses. His prescription for world domination was uncontested supremacy on the world island, with the ability to mobilize resources and direct them to any outer area, using the efficiency of interior lines. The flaw he pointed to is that uncontested supremacy in such a large and diverse area is difficult. The dominance of space may be a similarly quixotic quest despite SpaceX's best efforts.

Colonel John Collins, working for the US Congressional Research Service, sought to transpose Mackinder's theory to space, producing the report *Military Space Forces: The Next 50 Years*, in 1989. Collins described space up to 85,000 kilometers as the next key military domain and marked the five Lagrange equilibrium points (L-1–L-5) for attention: "Who rules circumterrestrial space commands Planet Earth; Who rules the moon commands circumterrestrial space; Who rules L-4 and L-5 commands the Earth-Moon System" (Collins 1989:1). Unlike the other Lagrange points, a spacecraft at L-4 or L-5 is stable, like a ball in a bowl. This makes them ideal places for durable space stations. Vehicles and equipment going into orbit must pass through these points, making them effective strategic choke points and raising the possibility of a race to control them.

US Admiral Alfred Thayer Mahan (1840–1914) developed theories of sea power for a rising United States. Today, many see space as essential to structuring power between states, an immense, continuous area subjected to exploitation with rapidly changing technology. The congestion of orbits, presence of physical choke points, and commercial benefits from the exploitation of space make Mahan's thinking relevant today not only for space powers but also for smaller states with a stake in space. Destroying space-based assets may be a strategic option. Nuclear weapons generate an electromagnetic pulse (EMP) that can destroy circuitry, and US Homeland Security researchers have identified credible evidence of Chinese capability and intent to deploy high-altitude EMP to defeat technologically superior US forces (Pry 2020). For most countries, this suggests the need for alternatives to satellite dependence and contingency plans for a rapid launch of LEO satellite arrays.

Italian General Giulio Douhet was one of the early proponents of victory through aerial attacks on civilian populations with explosives, incendiaries, and poison gas. In *Command of the Air* (1927), he describes the wholesale destruction of cities. Russian-American aviator and engineer Alexander de Seversky was an enthusiastic advocate. In *Victory through Air Power* (1942), Seversky extolled American capacity to end World War II through direct and independent action by its air force. The lesson of the earlier era, from the London Blitz to the US devastation of Vietnam, seems to be that destruction by itself cannot guarantee victory. The space-based equivalent of destruction by strategic bombers might be directed-energy weapons from satellites targeting terrestrial assets with pinpoint accuracy. They were seen as anti-satellite and anti-missile weapons for the US Strategic Defense Initiative in the 1980s (Bloembergen and Patel 1987) and remain in development by the United States, China, and Russia as next-generation systems (Geis 2003), but they are unlikely to progress to mainstream use without tests in battle (Horkovich 2006).

US Air Force Officer David Lupton describes four "belief structures" (Lupton 1998:20–22) about the military utility of space. The *sanctuary school* sees space as a means of seeing beyond terrestrial boundaries and as a confidence-building medium, possibly making nuclear war less likely. Space should, therefore, be preserved as a demilitarized sanctuary. *Survivability doctrine* rests on the relative vulnerability of space-based assets, and its proponents resist reliance on space. An interesting corollary is the worry that space assets might be destroyed without initiating war, foreshadowing ideas of Grey Zone or below-threshold conflict. Lupton labels the third school *high-ground doctrine*, in which space becomes a platform for missile defense, altering the balance of nuclear terror in favor of the defense. This school of thought demands the militarization of space, while the first two suggest demilitarization. The final school of thought, *space control*, draws direct analogies with air and sea control. Control entails both the ability to defend assets and deny use to the enemy. This requires militarization of space, point defenses of key space assets, control of choke points, and the ability to move in space to conduct operations, thus constraining space control to major space powers, those with the capacity for crewed space flight.

Everett Dolman, professor of strategy at the US Air Command and Staff College, is a prolific writer about space strategy and theory. He follows the Control school of thought and advocates space-based assets to exploit space potential and deny it to enemies. *Astropolitik: Classical Geopolitics in the Space Age* (2002) takes a realpolitik view of using space for military purposes. *Pure Strategy: Power and Principle in the Space and Information Age* (2005) attempts to derive rules for air and space strategy, combining the emerging characteristics of the information age with space requirements. In the space and information age, strategists must discard the idea of victory—appropriate only in tactics (Dolman 2005:5–6)—and settle for effective management of interests. Writing from the perspective of US power, most of Dolman's writings advocate militarization of space, control to exploit it, and deny its use to the enemy.

James Oberg, a US defense scientist, former NASA engineer, and earlier contributor to ARPANET (precursor to the internet), also builds on Lupton's Control school of thought, presenting elements of space power to control space as a domain contributing to state power. State sponsorship of technological development yields national access to information, situational knowledge, and intrusive intelligence without violating sovereignty. Oberg's emphasis is not on warfare in or from space but on control of the medium as a vehicle for state power, inevitably leading to militarization (Wilkerson 2008:6).

One more American contribution to strategic thinking about space hearkens back to British maritime strategist Julian Corbett. It falls in Lupton's Control school of thinking but goes beyond it to consider less powerful actors. Consultant and professor John Klein (2006) is concerned with war in and from space. He offers a strategic framework for waging space warfare, drawing on history and maritime strategic thinking. Klein (2019) spells out space strategy for great, medium, and emerging powers as a business domain. Klein's 2019 prescriptions draw on the distinctions developed by Paikowsky (2007) cited above. Medium and emerging powers have options for space strategy, including non-military and diplomatic approaches, buying military power, waging limited war, concentration and dispersal, blockades, and commercial raiding. An important insight is that an attack on a ground station on an enemy's territory may be a *casus belli*, but a very disruptive non-lethal attack in international space may not (Klein 2019:74).

Space and Grey Zone Conflict

The Grey Zone is the space between stable, positive peace and overt interstate war. Actions that are non-attributable and occur in international waters, disputed territory, cyberspace, or outer space are probably Grey Zone phenomena. This is a problem for the many small and medium powers relying on a US-led world order for their prosperity and security. US technological supremacy and its concept of network-centric warfare may jeopardize interoperability with US forces, leaving allies open to manipulation or subservience (Vicente 2008). For states without full-spectrum information and access to space-based positioning, ISR, the fog of war is denser and the friction of uncertainty greater, whether in war, peace, or the Grey Zone between. This is not to say that space-based technology confers omniscience nor that technological supremacy leads to an easy victory. There is, therefore, room to maneuver for small states or what Paikowsky calls medium and emerging space powers.

Information age society is dependent upon space technology. Global positioning, navigation, communications, meteorology, fishing, agriculture, and commerce rely on space as an accessible Global Commons, taking on greater strategic value. The ability of great space powers to dominate space is undermined by increasing activity by medium and emerging space powers and commercial actors. Space is one of 25 domains in which Chinese PLA Colonels Liang and Xiangsui describe unrestricted warfare (Liang and Xiangsui 1999). Unacknowledged attacks in space are part of the new Grey Zone of space power competition (Luke 2012). Identifying sources of information or advantage for a rival and finding ways to degrade that advantage without going to war will confer significant advantages (Luke 2012:13).

Preserving space as a demilitarized sanctuary is a priority for most medium and emerging space powers. The erosion of the sanctuary is a characteristic of Grey Zone conflict. Resisting militarization and maintaining trust and confidence in universal access to space are characteristics of positive peace. Faster decision-making, greater global connectivity, and ready access to pertinent information must be preserved for common use. Non-attributable or evasive attacks on common assets may be attacks on satellites and ground stations, commercial embargoes, boycotts, cyberattacks, or malware. These actions would erode peace and prosperity for states without crossing the threshold of war.

As actors approach the threshold of regular, irregular, and asymmetric conflict, Grey Zone offenses in and from space may accelerate, specifically targeting military capabilities. The Gulf War of 1991 was the first to use space technology for military purposes (Johnson-Freese 2017:68). US Secretary of Defense William Cohen declared a Revolution in Military Affairs (RMA)—an opportunity to transform its strategy, doctrine, training, equipment, and tactics to achieve decisive results in a new way (Cohen 1999). The Rumsfeld Commission (2001) identified space technology as pivotal in the new way of making war, replacing reactive responses with action based on continuous updates to situational awareness (Moon 2017). Not just the United States but its NATO and coalition allies relied heavily on space-based assets by the time of the Afghan operations of 2001, the Iraq invasion of 2003, and deterrence operations in the Baltic states in 2008. In Afghanistan and Iraq, Western dominance was complete, but in NATO deployments in the Baltic states, the electromagnetic spectrum is more contested. Major General Yuriy Lastochkin, head of the Russian Defense Ministry's Radio-Electronic Warfare force, advocates targeting Western weakness—its "heavy reliance on continuous, high-bandwidth networks—and in particular space-based assets—for almost every facet of warfare" (Thomas 2020:2).

Five factors were conducive to the RMA among the United States and its allies and have implications for small countries facing new Grey Zone threats. *Structural changes* in Western market economies shifted thinking from centralized to decentralized control, enhancing resource access. *Societal changes* allowed governments to take advantage of widely educated and highly educated populations. Ukraine's response to Russia's attacks since 2014 illustrates the value of this mobilization. *Organizational changes* reduced the rigid hierarchies that were still typical of large industrial and military organizations in the 1980s, preparing the way for smaller, more agile organizations. *Cultural change* created more open, accepting, curious, and technically competent young soldiers and employees with whom new organizations could engage. Finally, the RMA would not have occurred without the *technological changes* that created new capacity to develop, produce, and employ technologies hitherto unimagined. The rapid spread of smartphones to a mass consumer market is both a driver and a symptom of these changes. All these changes are ambiguous; each opens new risks in the Grey Zone.

Reliance on space for PNT has coincided with an evolution in the command-and-control principle of warfare, from its foundation on terrestrial radios and satellite communications dating back to the 1970s to the introduction of computers and the integration of global communication networks. Satellite specialist Jules McNeff describes this evolution as breaching the C2 divide between reactive response and continuous situational awareness (McNeff 2010). The massive force structure investment of the 1980s and 1990s that brought deployed GPS, improved information throughput, and global digital data communications also brought the vulnerability identified by Lastochkin, not only for the United States but also for all the states that had begun to adopt the new way of working.

The Grey Zone blurs the distinction between peace and war. Not only military capabilities are now vulnerable to reliance on space and high bandwidth. National economies and societies addicted to smartphones, instant knowledge, commerce, and finance, driven by data and connectivity, are similarly vulnerable to disruption or attack. Starlink's promise to provide global broadband access may be a mixed blessing.

Options for Small Countries

In *Maritime Strategy for Medium Powers* (1986), British Admiral J. Richard Hill described the medium power as falling between the self-sufficient and the insufficient, forever confronting the question, "How much should one be able to do on one's own before being forced to call on help from an ally?" (Hill 2000:3). Notwithstanding a desire for national control, resources are constrained. The sphere of autonomous action will likely be small for medium and emerging space powers.

The largest group of medium space powers comprises the states collaborating through the European Space Agency, and this offers a model for the ascent of others, given sufficient trust and cooperation. The UAE is poised to lead a consortium of like-minded small states in new space ventures through generous funding, contracting, an attractive regulatory framework, and commercial incentives (Al Rashedi, Al Shamsi and Al Hosani 2020). Long-term collaborations and institutional arrangements like the European Space Agency offer the best hope for emerging space powers.

Commercial contracting or "buying power" is another option. The decision to buy power comes with caveats. Reuters reported in February 2023 that "SpaceX had taken steps to prevent Ukraine's military from using the company's Starlink satellite internet service for controlling drones in the region during the country's war with Russia" (Roulette 2023). Machiavelli's caution about the use of mercenaries is equally relevant for space services.

Alliance arrangements offer secure supply of expertise, services, and information if it suits the ally's interests. Security, nationalism, and commercial interests can clash. Canada relies on Radarsat2 to monitor its Arctic territories. The advanced technology is owned by MacDonald Dettwiler and Associates (MDA). In 2008, the Canadian government blocked the sale of MDA's satellite unit to US space company Alliant Techsystems Inc. because it did not want to lose control of sensitive technology, even to a close ally. In 2017, the United States rebuked the Canadian Government for allowing a Chinese company to purchase Norsat International, a high-tech company that supplied the US Department of Defense with satellite communications equipment. Both incidents damaged Canadian defense relations with the United States.

Active engagement in international organizations can help mobilize common interests and shape the space environment. With more actors in space, there is a growing need for communication frequencies. COPUS and the International Telecommunications Union have been active venues for discussing terrestrial orbits, frequency allocation, and geopolitical consequences. These are venues for horse trading for medium and emerging space powers.

Small states lacking significant assets may enhance their position in the international system through state policies to focus on strategic industries. Taiwan's successful domination of the silicon chip market is an envied example. Aerospace industries are seen as strategic assets. Space industries have growth potential and development opportunities and serve as a source of national pride hitherto reserved for superpowers. Specialization in space technology is inherently dual-use, driven by both commercial and military interests that shape state expenditures. Several countries have become space incubators for small companies. Young and innovative companies have developed new business models and technologies, leveraging state investments attracted to space. Luxembourg has become a significant space center by creating an attractive legal framework and financial structure to attract investments in the space sector. Singapore has positioned itself as one of the world leaders in the miniaturized satellite industry. The UAE has gone further, leading space exploration projects. Monaco has two space companies in the tiny Principality (population 38 thousand). Orbital Solutions Monaco launched its first nanosatellite just a year after it started operations in 2020.

In addition to incubating start-ups, emerging space powers may seek partner arrangements with more advanced states to develop their own programs. The first UAE astronaut on the International Space Station represents scores of agreements and technical partnerships between the European Space Agency and the UAE. In many cases, expertise rests with private companies, which can make their own decisions about the investment opportunities presented by the UAE, as they have about the tax regime in Luxembourg. China's growing economic and technological strength has led European states and the United States to step back from cooperation. In contrast, smaller countries like Israel, UAE, and Singapore are not a threat and may be attractive partners. If rivalry between major space powers intensifies, there may be growing concern about the leakage of sensitive technical knowledge. During the Cold War, the Missile Technology Control Regime was established by 35 countries to limit the spread of technology that undermined arms control agreements (Ozga 1994). Similar arrangements might be developed for space technology.

Conclusion

Is it true that dominating space enables world domination? Major space powers are betting that it does, but for most of the world's states, space remains an increasingly accessible asset for peaceful and military uses. Numerous cheap, small satellites in LEO now dominate communications and have changed command, control, surveillance, and navigation even for small states. Even emerging state powers have space options with increased commercialization; anyone with a laptop and a credit card can access satellite information. New thinking about space among the major powers includes American thinking about militarization and control (Dolman 2002; Klein 2006), Russian focus on disruption, and Chinese thinking about unrestricted warfare (Liang and Xiangsui 1999). For smaller states managing the Grey Zone and seeking to preserve peaceful access, cooperation in multinational space agencies, building international norms, and commercial ventures seem the most promising avenues. If space is to be dominated by a major space power, most countries' interests would seem to be served best by an open regime, and the United States is most likely (for now) to sponsor an open and democratic space regime. Admiral Hall's question remains central: how much should a state be able to do on its own? His pragmatic if unhelpful answer was, "as much as necessary under the circumstances."

References

Al Rashedi, N., Al Shamsi, F. and Al Hosani, H. (2020), "UAE Approach to Space and Security", in J. Pelton, S. Madry, and S. Camacho-Lara (eds.), *Handbook of Space Security: Policies, Applications and Programs*, Springer International Publishing, Cham, pp. 621–652.

Bloembergen, N. and Patel, C. K. N. (1987), "Review of the American Physical Society Report on Directed Energy Weapons", in T. Hansch, D. Grischkowsky, D. Pritchard and C. Shank (eds.), *International Quantum Electronics Conference*, vol. 21, OSA Technical Digest (Optica Publishing Group, 1987), paper THGG28.

Boniface, P. (2003), "*Guerras do Amanha (Wars of Tomorrow)", Editorial Inquérito*, Mem Martins, Portugal.

Bowman, E. C. (2012), *Intelligence, Surveillance, and Reconnaissance Processing*, Exploitation, and Dissemination System in Support of Global Strike in 2035, (Master's Thesis), Air War College, Maxwell Air Force Base, Alabama.

Bryce Tech (2023), *Smallsats by the Numbers*, updated July 2023. https://brycetech.com/reports/report-documents/Bryce_Smallsats_2023.pdf

Cepik, M. (2015), *Espaço e Relações Internacionais*, Universidade Federal do Rio Grande do Sul, Porto Alegre, Rio Grande do Sul.

Chin, A., Coelho, R., Nugent, R., Munakata, R. and Puig-Suari, J. (2008), "CubeSat: the Pico-Satellite Standard for Research and Education", in *Proceedings of the AIAA Space 2008 Conference & Exposition,* American Institute of Aeronautics and Astronautics, 9-11 September 2008, pp. 19–23, San Diego, California.

Chun, C. (2006), *Defending Space - US Anti-Satellite Warfare and Space Weaponry*, Osprey Publishing, Oxford.

Cohen, W. (1999), *Annual Report to the President and the Congress*, US Government Printing Office, Washington, DC.

Collins, J. (1989), *Military Space Forces: The Next Fifty Years*, Congressional Research Service Report for Congress, no 89–578. https://docslib.org/doc/11167552/lf-military-space-forces-the-next-fifty-years

Doicariu, D. (2022), "Satellite Communications System-A Perspective on National Security". *Presented at the Scientific Research & Education in the Air Force-AFASES, 26-28 May 2022,* pp. 19–23. "Henri Coandă" Air Force Academy, Brașov, Romania.

Dolman, E. C. (2002), *Astropolitik: Classical Geopolitics in the Space Age*, Routledge, London, UK and New York, NY, US.

Dolman, E. C. (2005), *Pure Strategy: Power and Principle in the Space and Information Age*, Routledge, London.

Douhet, G. (1927), *The Command of the Air*, Air University Press, Maxwell Air Force Base, Alabama.

European Parliament (2019), European Parliament Legislative Resolution of 17 April 2019 on the Proposal for a Regulation of the European Parliament and of the Council Establishing the Space Programme of the Union and the European Union Agency for the Space Programme and Repealing Regulations (EU) No 912/2010, (EU) No 1285/2013, (EU) No 377/2014 and Decision 541/2014/EU (COM(2018)0447 — C8–0258/2018 — 2018/0236(COD)).

Frackiewicz, M. (2023), "How Starlink is Disrupting the Telecom Industry", TS2 Space.

France, M. E. B. and Sellers, J. J. (2011), "Real Constraints on Spacepower", in C. D. Lutes and P. L. Hays (eds.), *Toward a Theory of Spacepower: Selected Essays*, Occasional Paper No 32, Institute for National Strategic Studies, National Defence University.

Geis, J. P. (2003), *Directed Energy Weapons on the Battlefield: A New Vision for 2025, Center for Strategy and Technology*, Report No. 32, Air War College, Air University, Maxwell Air Force Base, Alabama.

Hall, K. (2006), *Near Space - Should Air Force Space Command Take Control of Its Shore?*, Air War College, Air University Press, Maxwell Air Force Base, Alabama.

Hanlon, E. (2018), "Survivability Analysis of a Small Satellite Constellation". *Presented at the 34th Space Symposium Technical Track, 16 April 2018,* Colorado Springs, Colorado, United States,

Hill, J. R. (1986), *Maritime Strategy for Medium Powers*, Routledge, New York.

Hill, J. R. (2000), "Medium Power Strategy Revisited", Royal Australian Navy, Sea Power Centre, Working Paper No. 3 (March 2000), pp. 1–27.

Horkovich, J. (2006), "Directed Energy Weapons: Promise & Reality". *Presented at the 37th American Institute of Aeronautics and Astronautics (AIAA) Plasmadynamics and Lasers Conference, 5–8 June 2006*, San Francisco, California.

Johnson-Freese, J. (2017), *Space Warfare in the 21st Century. Arming the Heavens*, Routledge, New York.

Klein, J. J. (2006), *Space Warfare: Strategy, Principles and Policy*, Routledge, New York.

Klein, J. J. (2019), *Understanding Space Strategy: The Art of War in Space* Routledge, New York.

Krepon, M., Hitchens, T. and Katz-Hyman, M. (2011), "Preserving Freedom of Action in Space: Realizing the Potential and Limits of US Spacepower", in C. D. Lutes and P. L. Hays (eds.), *Toward a Theory of Spacepower Selected Essays*, National Defense University Press, Washington, DC, pp. 201–216.

Liang, Q. and Xiangsui, W. (1999), "Unrestricted Warfare: China's Master Plan to Destroy America", PLA Literature and Arts Publishing House Arts, Beijing.

Luke, B. K. (2012), *Recognizing and Adapting to Unrestricted Warfare Practices by China*, Air War College, Air University Press, Maxwell Air Force Base, Alabama.

Lupton, D. (1998), *On Space Warfare: A Space Power Doctrine*, Airpower Research, Air University Press, Maxwell Air Force Base, Alabama.

Marshall, T. (2021), *The Power of Geography: Ten Maps That Reveal the Future of Our World*, Scribner, New York.

McNeff, J. (2010), "Changing the Game Changer: The Way Ahead for Military PNT", *GNSS Inside,* November/December 2010, pp. 44–51. https://insidegnss.com/military-pnt-the-way-ahead/

Moon, M. (2017), "The Space Domain and Allied Defence - DSCFC REPORT", NATO Parliamentary Assembly, Defence and Security Committee. https://www.nato-pa.int/download-file?filename=/sites/default/files/2017-11/2017%20-%20162%20DSCFC%2017%20E%20rev%201%20fin%20-%20SPACE%20-%20MOON%20REPORT.pdf

NATO (2013), *NATO Space Handbook: Practical Considerations for Space Support to NATO Operations*, SACT - Norfolk.

NATO (2016), AJP-3.3- Allied Joint Doctrine for Air and Space Operations, NATO Standardization Office (NSO).

Ozga, D. A. (1994), "A Chronology of the Missile Technology Control Regime", *The Nonproliferation Review*, vol. 1, no. 2, pp. 66–93.

Paikowsky, D. (2007), *Space Technology, Patterns of Warfare and Force Build-up: Between a Power and a Small State,* Working Paper Series - Economics of National Security, vol. 3. Samuel Neaman Institute for Advanced Studies in Science and Technology; Technion, Israel Institute of Technology. https://www.neaman.org.il/wp-content/uploads/2024/02/Space-Technology-Patterns-of-Warfare-and-Force-Build-up-Between-a-Power-and-a-Small-State.pdf

Parliamentary Office of Science and Technology (2006), "MILITARY USES OF SPACE", Postnote No 273. https://post.parliament.uk/research-briefings/post-pn-273/

Pry, P. V. (2020), "China: EMP Threat. The People's Republic of China Military Doctrine, Plans, and Capabilities for Electromagnetic Pulse (EMP) Attack", *EMP Task Force on National and Homeland Security*, pp. 1–14. https://info.publicintelligence.net/US-ChinaEMPThreat.pdf

Ribeiro, L. D. (2007), "Avaliação do Sistema Nacional de Desenvolvimento das Atividades Espaciais" (Dissertação de Mestrado em Administração Pública), Escola Brasileira de Administração Pública e de Empresas, Fundação Getúlio Vargas, Rio de Janeiro.

Roland, A. (2009), "War and Technology". Foreign Policy Research Institute. https://www.fpri.org/article/2009/02/war-and-technology/

Rome, P. (2023), "Every Satellite Orbiting Earth and Who Owns Them", DEWESoft, 9 February. https://dewesoft.com/blog/every-satellite-orbiting-earth-and-who-owns-them

Roulette, J. (2023), "SpaceX Curbed Ukraine's Use of Starlink Internet for Drones - company President", *Reuters*, 9 February. https://www.reuters.com/business/aerospace-defense/spacex-curbed-ukraines-use-starlink-internet-drones-company-president-2023-02-09/?utm

RT.com (2014), "US Military Satellites Vulnerable in Future Space War – Space Command Chief", *RT News*, 9 January. https://www.rt.com/usa/us-satellites-vulnerable-shelton-381/

Rumsfeld Commission (2001), *Report to the Commission to Assess United States National Security Space Management and Organization*, Defense Technical Information Center, Fort Belvoir. https://spp.fas.org/military/commission/report.htm

Saxena, V. (2016), "Anti Satellite Weapons: A Likely Future Trajectory", Vivekananda International Foundation, New Delhi. https://www.vifindia.org/sites/default/files/anti-satellite-weapons-a-likely-future-trajectory.pdf

Seversky, A. P. (1942), *Victory Through Air Power*, Simon and Schuster; Book Club Edition, New York.

Tabarcia, N. and Stroescu D. (2010), *Space - an Area of Strategic Importance to NATO*, Strategic Impact, National Defence University "Carol I" Printing House, Bucharest, Romania, vol. 2, no. 35, pp. 19–29.

Thomas, T. (2020), "Russia's Electronic Warfare Force. Blending Concepts with Capabilities", MITRE Center for Technology and National Security. MITRE Corporation, McLean, VA. https://www.mitre.org/sites/default/files/2021-11/prs-19-2714-russias-electronic-warfare-force-blending-concepts-with-capabilities.pdf

Tyvak Nano-Satellite Systems (2018), "Application for Authority for Ground Testing, Launch, and Operation of Experimental Non-Geostationary Low Earth Orbit Satellites", Federal Communications Commission, Washington DC.

UCS (2023), USC Satellite Database. https://www.ucs.org/resources/satellite-database?utm

UN General Assembly, Fourth Committee (2022), "Fourth Committee Shifts Focus to Peaceful Uses of Outer Space with Speakers Emphasizing Need to Close Technological Gap", 77th session, 14th meeting GA/SPD/760, 26 October 2022.

UNODA (2021), *Member States and Non-member States*, United Nations Office of Disarmament Affairs. UN Office for Outer Space Affairs, Treaty on Principles Governing the Activities of States in the Exploration and Use of Outer Space, including the Moon and Other Celestial Bodies (1966).

Vicente, J. (2008), "Operações em rede: da promessa à realidade" (*Network-Centric Operations: From Promise to Reality*), Nação e Defesa, No 120, 3a, Série, pp. 51–76.

Wilkerson, D. (2008), *Space Power Theory: Controlling the Medium without Weapons in Space, US.*, Army War College, Carlisle, PA.

Wolfe, D. (2023), "Starling Satellite Trains: Is this the Future of the Night Sky?", *Washington Post*, 6 January. https://www.washingtonpost.com/business/interactive/2023/starlink-satellite-train-spacex-visibility/

4 Cybersecurity in Military Systems

Risk and Mitigation

Suzana Lampreia,[1] *Victor Lobo, and Valter Vairinhos*

Introduction

In military environments, as in other organizations, people and systems are increasingly connected by information networks. These networks improve efficiency but result in cyber risks. All countries face rising cybersecurity risks because cyberattacks are the ideal Grey Zone weapon, eroding security and diminishing capability in peace while leaving a defender less capable if the aggressor chooses war. Erosion of capability in the Grey Zone can be a prelude to war, as it was for Ukraine between 2014 and 2022. This chapter considers assessing and mitigating cyber risks, comparing several national approaches, and offering risk assessment and mitigation matrices. Cybersecurity analysis should be a basic skill within every security force. National approaches, however, will be constrained by resources, and cybersecurity will compete with other force enablers. Blockchain technology offers some promise but is not a panacea because of the energy use and computing power that it requires. Finally, effective analysis can help select appropriate risk mitigation strategies.

Cyber Risks in the Grey Zone between War and Peace

We face rising cybersecurity risks because cyberattacks are the ideal Grey Zone weapon. Wired and wireless links connect land, sea, air, and space facilities. These elements of cyber systems are all at risk (Wilson 2019). The sources of cyber threat may be solitary hackers, even youth looking for a challenge, organized teams of criminal hackers seeking ransom, sabotage groups, terrorists, foreign countries, or dissatisfied factions or individuals within an organization. Maintaining security awareness and defensive postures is difficult when the object at risk is not physically tangible. Cyberattacks are unlike armed attacks with identifiable perpetrators, so recourse under international law is often impossible (Waxman 2011). Yet the sequence of a cyberattack may be like a physical attack, beginning with identifying a target, reconnaissance and data collection about the target, the attack itself, and then an electronic battle-damage assessment (Li and Liu 2021).

Consider some of the many attacks that have occurred in recent memory. Examples of cyberattacks are provided by the Center for Strategic and International Studies (CSIS):

- A distributed denial of service (DDoS) attack targeted the Port of London Authority, forcing its website to go offline. A group linked to Iran took responsibility for the hack.
- A phishing campaign targeted the Jordanian Ministry of Foreign Affairs. Researchers attributed the attack to an Iranian cyber espionage actor.

DOI: 10.4324/9781003428701-7

- The Ethiopian Information Network Security Agency stated that hackers targeted the Grand Ethiopian Renaissance Dam. Ethiopia's communication security agency thwarted the attacks before hackers could access the networks.
- Hackers targeted Greenland's healthcare system, causing networks to crash throughout the island. While the attack did not damage or expose citizens' data, it limited health services.
- Since 2019, a Chinese hacking group has stolen intellectual property assets from US and European companies largely undetected. Researchers believe that the Chinese government backs the group.
- Russian hackers hit Italian websites with a DDoS attack, including the Senate, the Ministry of Defense, and the National Health Institute. The group stated that its goal was to target NATO countries and Ukraine.

Leadership and coordination are significant challenges in addressing cyber risks. The pillars of cybersecurity readiness are education and awareness, planning and preparation, detection and recovery, sharing and collaboration, and ethics and certification (ACS 2016:51–52). While open sharing and collaboration are important for defense, open access to defensive tactics can also be used by cybercriminals or hackers to hone their attacks. Information about impacts can help hackers, so attacks like those listed above are seldom acknowledged.

A cyberattack on military facilities or equipment may cause financial or operational loss (Myung and Dorantes 2006). The impact of cyberattacks qualifies them as a nonlethal weapon system, although they can have lethal consequences (Votel, Julazadeh and Lin 2018). Cyberattacks have been called translational because they threaten many systems (Domingues 2015). The June 2017 ransom attack on Maersk shipping affected global shipping and supply chains (Capano 2021). Cyberattacks on civilian infrastructure can degrade operational readiness, undermine social cohesion, or reduce support for authorities unable to protect the public. If social order breaks down or cyberattacks compromise essential services, the military may be called upon to respond as a last resort to restore essential infrastructure (Solar 2020).

Cybersecurity Analysis for Military Contexts

The rapid growth of this research on cybersecurity and blockchain technology leaves security leaders scrambling to keep up. They live with the certainty that there are aspects of cyber risk and cybersecurity that they do not understand. Security planners can draw on well-established analytical processes to understand their environment despite this challenge. A 2021 NATO technical report suggested four steps for cybersecurity analysis. The four steps are: *collecting inputs*, *preliminary risk assessment*, *test plan productions*, and *full risk assessment*. This is a good process when starting a new position. Gathering all the available information about the system at risk will kick-start an understanding of how the system works and drives subsequent analysis. What documentation is available? What is the architecture of the system, or how its parts connect? What is the scope of the system's capabilities and responsibilities, and how does it fulfill its mission? How is the system controlled, and what parts are critical to the mission? For example, considering a ship as a system, a new commander might review all the sources of information and controls available on the ship, as well as those integrating the ship with other systems at sea, in space, and onshore.

A preliminary risk assessment proceeds on two tracks. The first track is the threat event and source, and the second is the set of security controls and the plan to manage a threat. These two tracks converge on a threat scenario, which can be assessed for likelihood and consequences. For example, if software controls a ship's bilge pumps, what would happen if the software becomes corrupted and fails to turn on automatically? How would the crew know that something was amiss? Would they be able to fix it or bypass the problem? Working through this thought process and conducting non-destructive experiments would result in a test plan. A series of test plans to address different scenarios would result in a full risk assessment, compiling the likelihood and impact of multiple vulnerabilities to yield a robust assessment of how badly things could go wrong. A basic cybersecurity analysis is not the preserve of specialized personnel who work with computer programming; anyone who uses a smartphone should have a plan for when it stops working. Every security leader has a responsibility for cybersecurity.

National Approaches to Cybersecurity

Most states work with imperfect cyber defenses. There is a sense that the hackers will always be able to penetrate somewhere. Russia, China, and the United States have the advantage of scale. The United Kingdom, Portugal, and South Africa provide examples of national approaches to defensive cybersecurity. Some states have invested in offensive cyber capabilities (see Jensen, Chapter 13 in this volume). Large countries can afford specialized units and formations to wage offensive and defensive cyber operations. Russia has actively courted hackers, setting up the Internet Research Agency in 2013 before major cyberattacks on Ukraine (Greenberg 2019). The United States has a long history of working with the private sector and specialized agencies for offensive and defensive cyber operations (Buchanan 2020). Cyber offense often engages black hat hackers in the criminal world. Public-private partnerships may bring the advantages of large multinational corporations to small government agencies. It is common to find government departments and agencies using Gmail accounts, Google Drive storage, and shared Google documents; a company like Alphabet, with 150,000 IT employees, may offer better security than a government like Portugal, with only 27,000 defense department employees.

China is self-conscious about the West's commercial and technical dominance and has worked hard to diminish the gap. National control of Chinese networks is established as a strategic defense against cyberattacks and a national asset for future information dominance. Military and civilian information management and defense against encroachment in the cyber domain stem from the belief that any loss of national control constitutes a threat to the state (Raud 2023).

In March 2023, the US Government released a National Cybersecurity Strategy (WEFORUM 2023). A fundamental aim was to prevent cyberattacks on critical infrastructure like hospitals and energy facilities. The second pillar of the strategy is an international coalition against cyber threats to preserve a healthy digital ecosystem. The United States argues that a national cybersecurity strategy is essential because of technological dependence; failure results in economic impact and serious consequences for human lives. The Strategy articulates a widespread understanding: all organized societies must trust technology, and trust demands an effective cybersecurity strategy. Big powers like Russia, China, the United States, and India may operate at scales beyond the reach of smaller states, but all states need strategies to secure the technology on which both military and civilian assets rely.

Portugal has had a cybersecurity strategy since 2015, revised and presented as the National Strategy for Cyberspace Security, 2019–2023. The strategy aims to create a safer, more prosperous country through innovative, inclusive, and resilient action that preserves the fundamental values of the democratic rule of law and guarantees the regular functioning of institutions in the face of society's digital evolution (Portugal Digital 2022). To pursue these objectives, the Portuguese National Cybersecurity Centre (CNCS) serves as the national authority for the strategy. The CNCS engages citizens, public administration, commercial and private organizations, and IT professionals.

The South African National Cybersecurity Policy Framework, also from 2015, refers to the rising threats of cyberattacks and cybercrime. It includes a Cybersecurity Response Committee within the State Security Agency responsible for Justice, Crime Prevention, and Security. It is responsible for both public and commercial engagement and coordination of government responses (South Africa 2015). However, the focus within national borders is only one aspect of cybersecurity.

Philosopher of technology Benjamin Bratton has described the stack as a physical infrastructure of satellites, undersea cables, server farms, power sources, and networked computation that makes the cyber domain resilient and open to manipulation and destruction (Bratton 2016). The stack transcends national boundaries and makes states dependent on interconnected physical assets. Security groupings, like NATO, are interdependent in cyberspace. There have been proposals for common databases to help identify the components and vulnerabilities of the system. Common definitions, shared information about threats and incursions, and an operational framework for common responses to attacks would help to improve national responses (Efthymiopoulos 2019). Even with regional cooperation, states must provide the legal and technical framework for cyberspace management and resources to secure infrastructure.

Cyber operations are inherently non-national. The UK Active Cyber Defense program has been developed to protect the public sector and prevent cybercrime. A cyber defense system can be compared to a vaccination program. Outside the framework of national law, one jurisdiction cannot compel another or deflect threats to other jurisdictions (Stevens et al. 2019). International cooperation is, therefore, an inherent part of the design of cyber defense. National strategies for cybersecurity have been increasingly common in Europe since Russia's attacks in 2014. Ransomware and infrastructure attacks have broadened cyber defense from a military concern to a whole-of-society problem. We are all dependent on the digital ecosystem.

The Potential for Blockchain Security

The fragility of the digital ecosystem has led to thinking about codes to secure the digital transactions on which modern economies and management systems depend. Blockchain technology offers the promise of a more robust digital ecosystem in which data remain secure and processes cannot be interrupted by hostile actors. Cryptography is not a panacea because it depends upon a lot of high-speed computing power and the high energy consumption that entails.

Blockchain is a cryptographic system that promotes the decentralized security of data and software with multiple steps validated by users (Angin 2020). Blockchain technology has been applied to complex industries where equipment and processes are managed; secure encryption allows information to be stored and processes to be verified without fear of interruption. The US Defense Advanced Projects Agency began experimenting with blockchain in 2019, and the US Department of Defense's Digital Modernization strategy highlights the benefits of cybersecurity.

Data or information can be tokenized within a blockchain (Korsak and Fuqua 2021). Battlefield operations management, logistics and supply chain management, commands to autonomous swarms of drones, and border protection have different profiles for storing, managing, and exchanging information. They might have different blockchain patterns (Ahmad et al. 2021). Blockchains are structured around blocks with a unique digital signature (Woodfield 2019). These signatures are renewed every time a change of contents or format is introduced.

The computer company IBM describes four types of blockchain security based on who can participate and who has access to the data. A *public blockchain* network allows anyone to participate, and participants can remain anonymous. Transactions are validated by internet-connected computers, which undertake complex calculations to achieve consensus or validate work. Bitcoin is the most widely known public blockchain network, in which anyone can participate, and anyone can access data anonymously (or pseudonymously) when provided with a key. *Private blockchains* require a proven identity and allow only authorized organizations or individuals to participate. This becomes a private business-only network secured by encryption. Networks for battlefield management, drone control, or supply management would probably be based on private blockchains. *Permissionless blockchains* have no restrictions on servers—any server can participate in the network—while *permissioned blockchains* require a proven server identity validated by certificates (IBM 2023).

Most military applications would probably follow the practice of high-security banking and finance, requiring authorization of individuals and servers. Blockchain encryption might ensure higher confidence in data transmission through defense networks, but they would have to work in real time under the stress of operational conditions. Because blockchain encryption is costly in energy and computing power, it is not always possible. Considering military intelligence applications, McAbee, Tummala, and McEachen (2019) suggest four circumstances when collaboration could benefit from blockchain encryption: if trust between participants is uncertain, if dispersed locations are necessary to prevent service disruption, if there are problems related to traceability or transparency, or if dispersion requires secure command methods. Digital signatures alone, however sophisticated, cannot resolve all issues of trust, nor do blockchain signatures solve all problems of privacy or data access, as in medical records (Woodfield 2019).

Blockchain technology is now being tested for application to autonomous ships and to minimize human error by enhancing ship controller security (Petkovic, Mihanovic and Vujovic 2019). While secure digital transactions confer great advantages, they are not without cost. The blockchain grows as new blocks attach to the chain. Decentralizing blockchain ledgers means that more calculations are required. With more transactions, blockchains become longer, and the cost of each transaction in computing power and energy grows. Using modeling and simulation, programmers Golosova and Romanovs estimate that average transaction costs are \$75–\$160, mostly accounted for by the energy cost associated with computing (Golosova and Romanovs 2018). The advantages of blockchain for operations are security, data integrity, decentralization, and organizational confidence (Mohril et al. 2021; Green et al. 2020). However, high-speed computing is not free; time and energy costs may mean that the blockchain's advantages are largely theoretical.

Cost is not the only potential disadvantage to the widespread use of blockchain technology. Information storage requirements would increase for every node to which encryption is decentralized. Erasing data would be difficult. Power consumption of communication devices and networked machinery like drones would increase. Finally, blockchain security cannot preclude hacking if users confer access (Islam et al. 2020; Blockchain 2022). Blockchain should be seen as a sophisticated encryption tool with various options but potentially high costs.

Cybersecurity Assessment and Risk Mitigation

Blockchain technology cannot secure defensive dominance in cyberspace. So, what do cyberattacks look like, and what makes cyber systems vulnerable? We should consider hostile intentions, inherent problems of defense, opportunities for better data security, and frameworks for risk mitigation.

In the Idaho National Laboratory experiment of 2007, a \$300,000 diesel generator self-destroyed on the command of 140 kilobytes of code (Greenberg 2019:67). Software attacks on energy systems, electrical grids, pipelines, and refineries have been in the public lexicon of cyberattacks since then (Kott and Colbert 2016). The US-Israeli attack on Iran's Nuclear centrifuges using the Stuxnet worm represents an early case of strategic attack using computer networks (Buchanan 2020:Chapter 6). Russia's absorption of Crimea in 2014 was accompanied by cyberattacks on electrical and telephone infrastructure, government information systems, fire, police, ambulance, taxation, and banking services (Greenberg 2019:238). North Korean hackers attacked Sony Pictures in December 2014 in retaliation for The Interview (2014)—a comedy film that depicted the assassination of Kim Jong Un. Proof that the Russian and North Korean operations were state-sponsored required sophisticated detective work. False-flag operations are the norm in cyberattacks and a common dimension of Grey Zone conflict.

The motivation for state-led attacks can overlap with criminal and commercial motivations. Ransomware attacks that shut down businesses or commercial networks can generate income to support future operations. They can undermine governments and confidence in the ability of government to provide security and services to the population. However, not all cyberattacks are strategic or centrally directed. Hackers and online criminals may be individuals, organized groups, loose confederacies, transient alliances, or contracted employees of the state (Hald and Pedersen 2012). Attacks can originate from almost anywhere, with diverse motivations and capabilities.

Governments and military forces may not be serious enough about cybersecurity. Senior leaders must be generalists, and specialized knowledge of current cyber issues may reside in more junior leaders with little access to policy-making. Resources are always constrained, and defending everything everywhere is impossible. Government, defense, and security establishments must compete with high-paying commercial employers. When governments habitually outsource key tasks to the private sector, they often lack the expertise to supervise software development and may end up taking completion on faith or unwittingly introducing vulnerabilities.

Operational deployments introduce additional challenges for cyber-defenders. Vulnerabilities can be managed in a stable environment. A deployment forces units and headquarters to work in new teams, in austere conditions, with temporary networks, and without standard operating procedures. The Italian army identifies information leakage on operations attributable to these problems (Mulazzani and Sarcià 2011).

Blockchain encryption is one prescription for cybersecurity, but cyber defense requires layers of different tools for different circumstances. Other encryption requires less computing power and energy. Meta's commercial messaging service WhatsApp provides effective end-to-end encryption and Tools for network monitoring that will help to identify incursions. Web vulnerability scanning tools can identify weaknesses that permit intrusion or data manipulation through official websites. Packet sniffers log and analyze network traffic. Penetration testing simulates an attack on a website or database. Firewalls are the norm for official networks but vary widely in their effectiveness. One source of frustration is that greater security usually means degraded performance. People get used to good performance on their personal systems and then must adjust to less connectivity and slower processing speed when they are working on official networks that might be targets of attack. We all use information, and we all depend on cyber networks and technology. Whether trained cyber specialists or simply system users, leaders need to assess the risk of cyberattacks and take action to mitigate them.

Drawing on research in the *Journal of Cyber Security*, a cause-effect analysis can help to align risk perceptions with reality. Table 4.1 lists 12 cause-effect combinations. Each of these events can be assessed for likelihood of occurrence, exposure, impact on safety, security impact, and overall risk. Working through this exercise for any unit or system provides a way to check perceptions of risk against the more granular elements comprising threats. Table 4.1 lists only five causes, but different effects make these distinct events.

Table 4.2 summarizes risk assessment scales for a naval example. How would a ship's captain or naval task force commander assess these risks? The likelihood of an event can be estimated by its prior occurrence, though anything can happen first time. If an event has not occurred in the last three years, it might score 0, while if it has occurred more than twice in the last year, it might score 4. Exposure or impact on the environment considers the extent to which electronic systems affect the operation of the ship and safety of its crew. Safety impact is similar but with a focus on maintenance software. What is the risk that an attack on the ship's databases will

Table 4.1 Cyber Security Threats—Cause-Effect Analysis

Event	*Cause*	*Effect*
1	Misuse by personnel in the Organization	Information available to uncertified persons from organization
2	Misuse by personnel in the Organization	Information available to other persons outside the organization
3	Misuse by personnel in the Organization—virus infection	Information loss or inaccessible
4	Misuse by personnel in the Organization—virus infection	Infection of other organizations
5	Misuse by personnel in the Organization—virus infection	Impossible to control autonomous vehicles
6	Hackers/criminals gain access to information system	Information available to uncertified persons from organization
7	Hackers/criminals gain access to information system	Information available to other persons out of the organization
8	Hackers/criminals gain access to information system	Available information that can prejudice the organization actions
9	Hackers/criminals disrupt internet access by the organization	Impossible to online sharing information (in and out)
10	Hackers/criminals disrupt internet access by the organization	Impossible to control autonomous vehicles
11	Destruction of command facilities	Local security structure destroyed
12	Destruction of command facilities	Inaccessible information

Source: Adapted by Authors from Goh, Hou and Cho (2022).

Table 4.2 Scales for Risk Assessment (Naval Example)

Factor	*0*	*1*	*2*	*3*
Likelihood (L)	No occurrence in last three years	May have occurred in last three years	Occurred at least once in last three years	Occurred more than twice in last three years
Exposure (E)	No impact	Minor software anomalies	Serious software anomalies	Ship inoperable, may cause personal injury
Safety impact (Safety)	No safety impact	Minor software anomalies, some safety impacts	Serious anomaly in safety software	Fatal breach in ship safety, loss of data, and possible injuries
Security impact (Security)	No security impact	Minor software anomalies	Serious anomalies in security software	Fatal breach in ship security, loss of data, and possible injuries

Source: Adapted by Authors from Goh, Hou and Cho (2022).

cause inaccurate assessment of a situation or inappropriate maintenance action? For example, automatic bilge pumps might fail to trigger, or essential maintenance might not be flagged. A security impact affects the ship's weapon systems and ammunition.

The risk arising from a cyberattack can be calculated as the product of likelihood, exposure, and the sum of safety and security impacts:

Risk = Likelihood x Exposure x (Safety + Security)

The formula and scoring can be adapted to circumstances and make a good basis for discussion about risk and mitigation. The numbers in the example here were generated from discussion of news stories about cyberattacks, filtered through the professional expertise of naval officers who understand their reliance on technology and information systems to maintain and serve their ships and the ports that their ships rely upon. Incidents publicly recorded like those compiled by CSIS (2022) summarizing cyber incidents are a good place to start, but most defense and security organizations have their own lists of incidents affecting them. These are often classified but are a better indicator of the threat environment than generic lists.

Applying the formula to the 12 events listed in Table 4.1 gives a range of scores from 0 to 71. Scores over 35 might be accepted as significant, and scores over 50 as very significant. Event 11, destruction of command facilities, is clearly unacceptable. Disciplinary action and user education (like those annoying messages on external emails) can mitigate only the least serious events arising from misuse. The need for backups and less automated systems in case of catastrophic

Table 4.3 Risk Analysis Ranking of Events and Mitigation Strategies

Event	*Description*	*Exposure*	*Likelihood*	*Safety*	*Security*	*Risk*	*Mitigation strategies*
11	Destruction of command facilities destroys local security structure	3	3	3	3	71	Could blockchain mitigate this?
10	Disrupted access makes control of autonomous vehicles impossible	3	3	3	3	54	Cease autonomous vehicle operations
9	Disrupted access makes information sharing impossible	2	2	3	3	24	Rely only on saved data and local systems
12	Destruction of command facilities makes information inaccessible	2	3	1	2	23	Counterattack or deterrence strategies
7	Hackers make organization information available to outsiders	3	1	3	3	18	Time-expiry of information (e.g. ID cards) or changing sensitive data, locations
3	Virus causes information loss	2	2	2	2	16	Backup data, hard copies of essential material

(*Continued*)

Table 4.3 (Continued)

Event	*Description*	*Exposure*	*Likelihood*	*Safety*	*Security*	*Risk*	*Mitigation strategies*
5	Virus causes loss of control of autonomous vehicles	2	3	0	2	12	Cease operations and recall for reprogramming
8	Hackers gain access that compromises organization's activities	2	1	3	3	12	Revise operational plans, cease operations at risk
6	Hackers make information available to unauthorized personnel	2	2	1	1	8	Time-expiry of information (e.g. ID cards) or changing sensitive data, locations
2	User indiscretion makes information available outside organization	1	1	2	1	3	Disciplinary action, education programs
1	User indiscretion makes information available to unauthorized personnel	0	0	1	1	0	Disciplinary action, education programs
4	Misuse causes infection of another organization	2	2	0	0	0	Disciplinary action, education programs

Source: Author's elaboration.

failure is a reminder that small countries have some advantages over countries with large budgets to buy all the latest automated systems.

Table 4.3 is an example of risk analysis and mitigation strategies. Every leader is in the cybersecurity business and every organization from coastal patrol vessels to pension administration offices and border posts can be affected by cyberattack or simple internet failure. A resilience exercise might consist of a thought experiment: How would we operate if our information systems ceased to function? If we could not use the internet, how would we communicate? Can we perform checks to validate the information that our screens are giving us?

Conclusions

With technological advances, all military, political, and social systems are now cyber systems. Beyond national boundaries, states and companies rely on the connected "stack" of undersea cables, satellites, overland towers and cables, server farms, and technical support. As it has with traditional military hardware, technology confers advantages, sometimes on offensive and sometimes on defensive operations. A fully functioning blockchain architecture might make networked systems more impervious to cyberattacks but would not preclude physical disruption of satellites, cutting of undersea cables, or electromagnetic pulse destruction of printed circuits. Nor would it help when the batteries die, when the memory banks are full, or when the enemy is inside the gate. Thus, the advantage today lies with the attackers, and every security leader must think through the evolving problems of cybersecurity.

Note

1 The Authors acknowledge the contributions of the Editors and anonymous Reviewers to this chapter.

References

ACS (2016), Cybersecurity - Threats, Challenges, Opportunities. https://www.acs.org.au/insightsandpublications/reports-publications/cybersecurity-threats-challenges-opportunities.html

Ahmad, R., Hasan, H., Yaqoob, I., Salah, K. and Jayaraman, R. (2021), "Blockchain for Aerospace and Defense: Opportunities and Open Research Challenges", *Computers & Industrial Engineering*, vol. 151, no. 12, pp. 1–14.

Angin, P. (2020), "Blockchain-Based Data Security in Military Autonomous Systems", *European Journal of Science and Technology* (Special Issue), pp. 362–368. http://dx.doi.org/10.31590/ejosat.824196.

Blockchain (2022), Blockchain - Hot or Not?. https://softengi.com/blog/what-is-blockchain/

Bratton, B. (2016), *The Stack: On Software and Sovereignty*, MIT Press, Cambridge, MA.

Buchanan, B. (2020), *The Hacker and the State: Cyber Attacks and the New Normal of Geopolitics*, Chapter 6, Harvard University Press, Cambridge, MA.

Capano, D. (2021), "Throwback Attack: How NotPetya Accidentally Took Down Global Shipping Giant Maersk", *Control Engineering*, September 30. https://www.controleng.com/throwback-attack-how-notpetya-accidentally-took-down-global-shipping-giant-maersk/

CSIS (2022), Significant Cyber Incidents. Strategic Technologies Program. https://www.csis.org/programs/strategic-technologies-program/significant-cyber-incidents

Domingues, E. (2015), "Os Ciberataques como um Novo Desafio para a Segurança: o Hacktivismo" (Dissertação de Mestrado em Ciências Policiais). Instituto Superior de Ciências Policiais e Segurança Interna.

Efthymiopoulos, M. (2019), "A Cyber-Security Framework for Development, Defense and Innovation at NATO", *Journal of Innovation and Entrepreneurship*, vol. 8, no. 1, pp. 1–26.

Goh, Z., Hou, M. and Cho, H. (2022), "The Impact of a Cause–effect Elaboration Procedure on Information Security Risk Perceptions: A Construal Fit Perspective", *Journal of Cybersecurity*, vol. 8, no. 1, pp. 1–11.

Golosova, J. and Romanovs, A. (2018), "The Advantages and Disadvantages of the Blockchain Technology", in *Proceedings of the IEEE 6th Workshop on Advances in Information, Electronic and Electrical Engineering (AIEEE)*, 8-10 November 2018, pp. 1–6, Vilnius, Lithuania.

Green, E., Carr, E., Winebrake, J. and Corbett, J. (2020), *Blockchain Technology and Maritime Shipping*. US Maritime Administration. https://www.maritime.dot.gov/sites/marad.dot.gov/files/2020-07/MARAD%20Blockchain%20Final%20Primer%20%2820200622%29.pdf

Greenberg, A. (2019), *Sandworm: A New Era of Cyberwar and the Hunt for the Kremlin's Most Dangerous Hackers*, Knopf Doubleday Publishing Group, New York.

Hald, S. and Pedersen, J. (2012), "An Updated Taxonomy for Characterizing Hackers According to Their Threat Properties", in *Proceedings of the 14th International Conference on Advanced Communication Technology (ICACT)*, 19-22 February 2012, pp. 81–86, Pyeongchang, South Korea.

IBM (2023), What Is Blockchain Security?. https://www.ibm.com/think/topics/blockchain-security

Islam, I., Munim, K., Oishwee, S., Islam, A. and Islam, M. (2020), "A Critical Review of Concepts, Benefits, and Pitfalls of Blockchain Technology Using Concept Map", *IEEE Access*, vol. 8, pp. 68333–68341. https://ieeexplore.ieee.org/document/9056816

Korsak, D. and Fuqua, E. (2021), "Decrypting Bitcoin and Blockchain for Military Lawyers", *JAG Reporter Post*, September 23, pp. 1–11. https://www.jagreporter.af.mil/Post/Article-View-Post/Article/2753657/decrypting-bitcoin-and-blockchain-for-military-lawyers/

Kott, A. and Colbert, J. (2016), *Cyber-Security of SCADA and Other Industrial Control Systems*, Springer International Publishing, Cham.

Li, Y. and Liu, Q. (2021), "A Comprehensive Review Study of Cyber-attacks and Cyber Security; Emerging Trends and Recent Developments", *Energy Reports*, vol. 7, no. 8, pp. 8176–8186.

McAbee, A., Tummala, M. and McEachen, J. (2019), "Military Intelligence Applications for Blockchain Technology", in *Proceedings of the 52nd Hawaii International Conference on System Sciences*, 8-11 January 2019, pp. 6031–6040, Maui, Hawaii.

Mohril, R. S., Solanki, B. S., Lad, B. K. and Kulkarni, M. S. (2021), "Blockchain Enabled Maintenance Management Framework for Military Equipment", *IEEE Transactions on Engineering Management*, vol. 69, no. 6, pp. 1–14.

Mulazzani, F. and Sarcià, S. (2011), "Cyber Security on Military Deployed Networks - A Case Study on Real Information Leakage", in Em C. Czosseck, E. Tyugu and T. Wingfield (eds.), *Proceedings of the 3rd International Conference on Cyber Conflict*, 7-11 June 2011, pp. 1–15, CCD COE Publications, Tallinn, Estonia.

Myung, K. and Dorantes, C. (2006), "The Impact of Information Security Breaches on Financial Performance of the Breached Firms: An Empirical Investigation", *Journal of Information Technology Management*, vol. 17, no. 2, pp. 13–22.

NATO (2021), *Cyber Security Risk Assessment Process for Military Systems. NATO, Science and Technology Organization*. STO Technical Report. http://dx.doi:10.14339/STO-TR-IST-151.

Petkovic, M., Mihanovic, V. and Vujovic, I. (2019), "Blockchain Security of Autonomous Maritime Transport", *Journal of Applied Engineering Science*, vol. 17, no. 3, pp. 333–337.

Portugal Digital (2022), National Strategy for Cyberspace Security. https://portugaldigital.gov.pt/en/accelerating-digital-transition-in-portugal/get-to-know-the-digital-transition-strategies/national-strategy-for-cyberspace-security/

Raud, M. (2023), *China and Cyber: Attitudes, Strategies, Organization*, The NATO Cooperative Cyber Defence Centre of Excellence, Tallinn, Estonia.

Solar, C. (2020), "Cybersecurity and Cyber Defence in the Emerging Democracies", *Journal of Cyber Policy*, vol. 5, no. 3, pp. 392–412.

South Africa (2015), "The National Cybersecurity Policy Framework (NCPF)", State Security Agency, Government Gazette, no. 39475, pp. 39–95. https://www.gov.za/sites/default/files/gcis_document/201512/39475gon609.pdf

Stevens, T., O'Brien, K., Overill, R., Wilkinson, B., Pildegovics, T. and Hill, S. (2019), *UK Active Cyber Defence: A Public Good for the Private Sector*, Policy Institute, King's College, London.

Votel, J., Julazadeh, D. and Lin, W. (2018), "Operationalizing the Information Environment: Lessons Learned from Cyber Integration in the USCENTCOM AOR", *The Cyber Defense Review*, vol. 3, no. 3, pp. 15–20.

Waxman, M. (2011), "Cyber-Attacks and the Use of Force: Back to the Future of Article 2(4)", *The Yale Journal of International Law Future of Article*, vol. 36, pp. 421–459. http://dx.doi.org/10.2139/ssrn.1674565.

WEFORUM (2023), *The US Has Announced Its National Cybersecurity Strategy: Here's What You Need to Know*, World Economic Forum, March 9. https://www.weforum.org/stories/2023/03/us-national-cybersecurity-strategy/#:~:text=1%20The%20White%20House%20released%20the%20new%20US,partnerships%20to%20counter%20threats%20to%20the%20digital%20ecosystem.

Wilson, J. (2019), *Military Cyber Security: Threats and Solutions*. Military and Aerospace Electronics, vol.30, no.12. https://www.militaryaerospace.com/magazine/61187

Woodfield, D. (2019), "The Emerging Impacts of Blockchain Technology on DOD Asset Cyber Security", Air Command and Staff College, Air University, Maxwell Air Force Base, Alabama.

5 Markets and Grey Zone Wars

David Last

Introduction

Security leaders in small countries should understand how markets affect human and national security: they will be tasked to address the problems created by markets and the security problems exacerbated by how markets work. Whatever their trade or specialty, security professionals are embedded in markets. Beyond the violent non-state actors addressed in Chapter 6, the global market system affects all aspects of security. Markets create prosperity and freedom for some and exploitation and penury for others. Markets can support international security through cooperative regulation or disrupt and marginalize state priorities. This chapter describes connections between markets, Grey Zone threats, and the use of markets in modern war.

A market is a gathering for trading or exchange. Markets require cooperation as well as competition. They entail communication, learning, cheating, stealing, and coercion. Security and regulation are essential to prevent theft and exploitation in any exchange. Liberalism connotes individual freedoms, the protection of property, and the pursuit of prosperity through markets. But liberalism also has a long association with slavery, inequality, and oppression while it generates wealth for some (Losurdo 2014:35). Capitalism is the now-universal system of production for profit in markets, and free markets are the most efficient means of generating wealth (Milanovic 2019). However, capitalism's free markets lead to widening income disparities (Piketty 2014). Economist James Galbraith (2021) links inequality to market instability, and polymath Peter Turchin (2023) links inequality to political collapse. The Washington consensus articulated an open field for global investment with minimal state intervention. It reflected a belief in the offshoring of production, growth of non-national corporations, free flow of capital, tax havens, and new accounting standards that benefited transnational finance (Stiglitz 2002; Palan, Murphy and Chavagneux 2013).

Economic globalization was not slowed by the 2008 financial crisis (Mirowski 2013), but the inequities of late 20th-century neoliberalism spurred renewed authoritarianism. China's shift from "hiding capabilities and biding time" to "actively accomplishing something" dates to the 2008 crisis (Doshi 2021). Russia's attack on Georgia during the crisis and subsequent protracted campaign in Crimea and eastern Ukraine from 2014 to the 2022 invasion has been linked to the West's economic distraction and Russia's need for distraction from economic woes (Gerstel 2016). The rise of authoritarian populism in the United States and Europe is linked to economics and demographics (Brown 2022). The rise of Grey Zone conflict below the threshold of war, modern wars that span the threshold, and the problems all states experience in seeking positive peace are fundamentally related to markets. This chapter links markets and security, explores the Grey Zone vulnerabilities of small states, and commercial and financial power in modern war.

DOI: 10.4324/9781003428701-8

Markets and Security

No state today is isolated from international markets. Markets for resources, technology, and finance shape state power. The prosperity of individuals hinges on market access. National security requires access to resources and technology abroad, often financed by international capital. The stability of interstate relations is grounded on a functioning system of exchange, finance, and the mobility of specialized knowledge. Ideas about the value of market freedoms and the primacy of capitalist production have dominated international relations since the end of the Cold War.

Political economy shapes the duties of security professionals. Military forces are deployed abroad to secure resources against interstate rivalry. Police and paramilitary forces secure borders against smuggling and resource raiding to preserve state incomes. Access to advanced weapon systems, space-based communications, and surveillance are purchased or provided by states advancing their commercial interests. International finance and loans may have conditions that constrain the public sector, affecting security salaries and operations. Grey Zone threats and modern wars spanning the threshold of open violence entail market manipulation and economic tools to erode the security of enemies.

Four economies are relevant to security (Silwal et al. 2021:4–5). The *exchange economy* is two-sided and mutually beneficial. The *grants economy* is one-sided, consisting of the voluntary donation of resources. The *appropriation economy* is a one-sided coercive taking-of-resources, which may be violent or non-violent. The *mutual threat economy* includes mutually harmful reciprocal threats and violence, such as war. The first two describe peace economics, and the second two describe conflict economics (Brauer and Anderton 2020:444). These ideas, however, are socially constructed and situational. Paying taxes in a liberal democracy might be constructed as an exchange economy by citizens who benefit, but an oppressed minority might construct taxation as appropriation. Misinformation and manipulation can drive economic wedges.

These four economies exist both domestically and internationally. International loans from states or financial institutions can be seen as tools for growth, for the defense of stability and order, or as offensive instruments to gain control of economies and resources. They combine peace and conflict economics across every economic sector. Economic sectors may be defined by expenditures (public, corporate, personal), by groupings of industries (primary, secondary, etc.), or for specific purposes (organized, sensitive, etc.) (Black, Hashimzade and Myles 2017:990). Any major sector of the economy can be manipulated to move between the peaceful exchange economy and the coercive mutual threat or expropriation economy.

In the *primary sector*—including mining, agriculture, fishing, and forestry—extractive mines are more profitable if environmental rules can be set aside, and host nations are left with clean-up costs. Foreign ownership of plantations and export-led growth can benefit foreign owners and consumers at the expense of local food security. Corporate paramilitary forces prevent union activity and expropriate peasant land to expand internationally funded plantations (Hristov 2014).

The *secondary sector* of processing, manufacturing, and construction includes industries considered vital for national security: shipbuilding, aerospace, automotive, and heavy industries. In a recent special report, *The Economist* magazine highlights a turn toward "homeland economics," which privileges domestic production and secure supply chains, arguing that this trend will leave the world poorer without changing the fundamental characteristics of a global economy with inherently transnational markets (Economist 2023b). Shipbuilding, textiles, paper, and military industries were the original sources of industrialization and international trade rivalry

(Sen 1984). Aerospace, telecommunications, and advanced electronics can be added to this list, and all are now sources of potential manipulation and coercion. All are essential for national security.

The *tertiary sector* of services includes finance, legal, and security services and may be the sector most vulnerable to Grey Zone threats. Financial consultant John Perkins has popularized the concept of the "economic hit man," manipulating sovereign debt, obligations, and conditions of exchange for the benefit of powerful commercial interests and foreign powers (Perkins 2004, 2007). Financial power can confer ownership and control of raw materials, industrial capacity, technology through patents, and human skills through employment. In a capitalist world order, states do not always get these benefits. American foreign policy across Democratic and Republican administrations is a function of the influence of corporate elite networks seeking an open door for access to capital. Benefits have accrued to corporations (Van Apeldoorn and de Graaff 2016).

Legal services in the tertiary sector are a strength of liberal democratic economies because they operate independently from political authority and are vehicles for resolving disputes. They are a reason for peaceful exchange economies to function effectively. Grey Zone threats can erode or weaponize this effectiveness by mobilizing legal services to attack market economies from within. They can do this by masking actors and strategies, seeking disclosure of government information, manipulating legal actors through intimidation, false witnesses, or *kompromat* tactics, lawsuits, and manipulative industry tribunals under international agreements (see also Chapter 14 and the Appendix).

The market for security services is global. Beyond logistics, security guards, and private military companies, the secure management of information and control of surveillance systems has become a major private sector adjunct to the security state. The proliferation of security cameras managed by private companies in transit systems, airports, and cities and reliance on private companies for data management of passports, visas, and support to police services leaves sensitive private information in the hands of transnational corporations (Abrahamsen and Williams 2011). Most of the world's states do not have the technical capacity to manage the tasks they contract multinational private security companies (PSCs) like G4S, Securitas, GardaWorld, and DynCorp, particularly as these tasks extend into cyber-security and management of national data systems.

The fourth sector is the *intellectual economy* of research and development, education, consultancy, and publishing. The public sector has difficulty competing with the salaries offered by major corporations for the latest technical expertise, "The huge salaries of the tech firms have lured so many professionals that there is no one left to teach the next generation of students" (Zuboff 2019:189). The United States may act for corporate interests in the fastest-growing areas of financial capitalism and technology. Since 2016, however, China's increasingly assertive use of its own financial and economic power, the infrastructure of the Belt and Road Initiative, and the muscle of state-owned enterprises offer the prospect of commercial power in the service of the state (Doshi 2021:261ff).

Does it make a difference to small states whether the market power of major players benefits corporations or state interests? "Rules-based international order" has become a catchphrase for the 1989 Washington Consensus, emphasizing free markets, capital movement, and investor protections. It permitted the national champions of small states, like Finland's Nokia or Canada's Nortel and Blackberry, to thrive (for a while) in competitive international markets, and it gave small states access to technology and investment finance which could be trusted, it was thought, to be politically neutral. States like Sweden are now facing lawsuits resulting from their efforts to exclude China's telecommunications giant Huawei from their national infrastructure. There are obvious security implications if a potentially hostile power allied to neighboring Russia is to have control over 5G domestic communications. The effort to exclude it has triggered legal action under a 1984 investor protection agreement (UNCTAD 2023).

The Swedish case illustrates market vulnerability to hostile intent. Operating in any market permits moving between mutually beneficial exchange and coercive or manipulative extraction. Exporters can establish commodity cartels like the Organization of Petroleum Exporting Countries to manipulate supply and price on international markets. Consumers can manipulate demand, substitute alternatives, or stockpile reserves to offset the power of suppliers or cartels. With more power and wealth, national and economic actors may manipulate governments, establish extraction regimes in foreign countries, or purchase land to establish plantations for food supply (Hairong and Sautman 2010). How do security leaders know what is going on?

Economic Intelligence

Economic intelligence is a growing field for small countries faced with globalization, trade dependency, risks of economic espionage, and intellectual property theft. It is not a new field. France's Louis XIV sent agents to neighboring states to steal industrial secrets (Denecé 2016:135). British defense economic intelligence activities and analyses were widely distributed through British Ministries and services, even before World War I (Davies 2019). British-based companies, banks, business leaders, and foreign partners were enlisted to collect data about markets, actors, and activities. Smaller countries recognize the need for economic intelligence to understand the impact of markets and foreign economic activities on their security. As an American-led rule-based hegemony shifts toward rivalry, economic vulnerabilities have become more evident.

Each of the five areas of intelligence activity identified by the Oxford Handbook of National Security Intelligence is identifiable in efforts to understand and manage economic pressures. These include *collection and processing*, *analysis and production*, *dissemination* of intelligence products, *counterintelligence*, and *covert action* (Johnson 2010). All these activities are habitually conducted below the threshold of war. While these activities came naturally to states with imperial ambitions and international obligations, they had to be learned by smaller states and applied to their new circumstances (see Chapter 9). At the end of the 20th century, Canada was the eighth-largest trading nation in the world and a leader in several fields of technology, but its international competitiveness had been slipping since it joined the North American Free Trade Area and its share of world trade dropped by more than 25 percent (Potter 1998:5–6). In this environment, economic intelligence was framed first as a tool to enhance competitiveness in collaborative exchange. Open-source intelligence, national economic coordination, and competitive intelligence programs in federal and provincial ministries were the keys to improved economic performance (Calof 2017).

The Economic Espionage Act passed by the United States in 1996 was a wake-up call for America's allies, revealing details of widespread commercial spying by state and commercial actors in the United States and against America's allies (Jones 2008; Reid 2015). Awareness of hostile economic intelligence increased the urgency of economic analysis, production, and counterintelligence. Canada seconded analysts from its Security Intelligence Service to its Department for Industry to monitor the behavior of foreign investors. Business analysis, forensic accounting, and Security Exchange Commission disclosures were the primary sources for this oversight. More than a decade later, Canada was still trying to organize the essential resources to monitor foreign ownership, control, and influence in key sectors (Holmes 2012). Response to foreign ownership and control of critical industries, theft of intellectual property and technology, and subversion of trade and domestic industry begins with awareness, often only after damaging leaks or irreversible transitions.

Small democratic countries are challenged to mobilize and coordinate resources to understand and respond to market-related Grey Zone threats and hostile activities. Domestic interests may raise alarms about foreign influence or hostile action to resist foreign competition. Domestic interests may be actively, if unwittingly, enrolled to support hostile actions. The following section maps the extent of modern Grey Zone vulnerabilities and some tools and institutions available for small countries to counter them.

Grey Zone Vulnerabilities

The tabular summary of Grey Zone threats in the Appendix lists 20 specifically economic threats, but several threats in other categories are directly related to markets. Here, the market-related threats are treated in five groups: finance, supply chains, industry-related tactics, criminal enterprises, and other vulnerabilities that support market-based threats.

Finance. Attacks on national currencies, credit ratings, or foreign reserves can weaken a state and constrain its options. Manipulation of banking arrangements or national finance and cascading bank failures erode confidence and may cause governments to fall. Small countries can be more vulnerable to these threats than larger states. US wargames demonstrated that the financial crisis of 2008 could be engineered and targeted at countries, currencies, markets, or commodities (Rickards 2012). Sovereign Wealth Funds are government-owned investment funds that may operate openly or through parties that conceal their identity (cut-outs). Hedge funds are private investment funds that pool resources to seek investment profit. Very large public and private funds can be wielded to have a destabilizing effect on markets. When major investors sell assets, they may cause a loss of confidence in banks, currencies, commodity markets, or a country's economy. A government may exhaust its reserves and International Monetary Fund (IMF) borrowing limits to prop up its currency value. Loss of confidence in a currency can be followed by depression and political crisis. Currency crises affected the Mexican Peso in 1994, the Thai Baht and other Asian currencies in 1997, the Russian Ruble and the Argentine Peso in 1998, and the currencies and economies of Venezuela and Turkey in 2016. In each of these cases, IMF reports found structural weaknesses and regulatory errors compounding the crises, which generated profits for some actors. Rickards makes the case that these problems can be engineered. His focus is US vulnerability. How much more vulnerable are small countries?

In 2014, about a billion dollars, or more than 10 percent of Moldova's GDP, was stolen from three banks in Moldova through fraudulent loans engineered by insiders. The fraud was related to Russian money laundering, and pro-European political leaders were blamed for corruption. Support for joining the EU declined, although the accusations were not proven, and Russia was implicated in disinformation and influence operations from pro-Russian Transdniestria (Barojan 2021:92). Are there security sector implications? On April 6, 2001, Members of the NATO Stabilization Force supporting the Office of the High Representative in Bosnia-Herzegovina raided the Hercegovacka Bank in Mostar "to assist the Bosnian government in the fight against organized crime and to stamp out illegal financial activities…" (OCHA Relief Web 2001).

Supply chains. Domination of supply chains creates dependencies, and disruption of supply chains can impose economic costs, possibly being wielded as a deterrent. Technology, raw materials, and shipping illustrate three aspects of security problems related to supply chains.

Technology supply is critical to military effectiveness. Three-dimensional (3D) printing of spare parts, components, and complete drones has enhanced Ukraine's military capability (Stewart 2023). Ukraine's ingenuity in producing drones has run into supply chain problems because China dominates the commercial drone industry (Mozur and Hopkins 2023). Taiwan has a dominant position in the world's supply of advanced semiconductors. Semiconductors have growing military significance (Miller 2022). The United Kingdom's efforts to revive a strategic semiconductor industry illustrate the disadvantages of smaller countries and the challenges facing states trying to establish secure supply (Economist 2023a). Stockpiling generic processor chips is not a solution, so strategic partnerships with major producers are the best option for preserving access to technology.

Uninterrupted access to strategic materials is critical for small-country capability, particularly if they pursue a just-in-time delivery strategy of the latest technology. Resins, plastics, metals, and carbon fiber in the right forms for 3D printers are highly specialized products that must be sourced from a limited number of US, Chinese, and German producers. Leaving weaponry aside, food and fuel are the most basic strategic materials affecting human and national security. Import product shares are only a rough indicator of vulnerability because reliance on staples like grain makes a country more vulnerable than a diverse mix of luxury food items. The Danish company Maersk is the largest shipping company in the world, managing 76 ports and a transport fleet of more than 800 ships, almost 20 percent of the world's shipping capacity. It was disabled on June 27, 2017, by the NotPetya attack initially directed at Ukrainian power management and banking services in 2016, disrupting global supply chains (Greenberg 2019:151). National plans to manage ports and transport infrastructure are part of the supply picture; regional cooperation and sharing arrangements might also be necessary if the global food and energy supply becomes constricted by new factors like climate change. Dual-use military capabilities like engineering, transport, communications, cyber-security, and infrastructure management might be necessary for small country development in the face of Grey Zone threats and market disruptions.

Industry-related tactics. Tactics like hostile takeovers and share buyouts can be related to the expropriation and coercion of national industries. Advanced free-market economies are vulnerable because they are inherently competitive and adversarial. Market competition is one of the strengths of dynamic economies that create innovative solutions to technical problems. Small countries often invest public resources to stimulate innovation and local production. Must governments then thwart market forces that threaten to appropriate those publicly supported assets for national security? Or are claims of national security interest merely pretexts to support one group or another?

Masking is a tactic that conceals economic actors. It may be accomplished by operating through law firms with client confidentiality. Shell corporations with obscure ownership structures serve the same purpose. To the extent that the true economic actors are concealed, free and informed exchanges are replaced by manipulative exchanges that might verge on expropriation. New owners may want to eliminate competitors, strip assets, expropriate patents and technology, or sabotage national capabilities.

Pressure on company principals can be used to extract commercial intelligence, degrade company strategy, delay responses in time-sensitive transactions, or engineer hostile takeovers. Principals include the board, executives, and key shareholder representatives. Pressure could take the form of *kompromat*, harassing lawsuits, intimidation, and even intimidation or hostage-taking affecting family members. Using these tactics takes us out of free and fair exchange and into coercion and expropriation, but the motive might be criminal or purely commercial. Under these circumstances, the legal silos typical of security and defense intelligence, policing, and national security can leave small states woefully unprepared to respond until the damage has been done.

Criminal enterprise. Organized crime exists almost everywhere. Hostile interests can make alliances with organized crime, which can be difficult to detect. Illegal trade and smuggling of alcohol, tobacco, and gasoline reduces government revenue and can build habits of tax avoidance and networks of anti-government interests. Popular culture glorifying Robin Hood heroes can make these networks resilient and prone to cooperate with cross-border friends and relatives or, eventually, with foreign agents.

Drug trafficking has a more pernicious influence than other forms of illegal trade. The drug trade generates similar patterns of income concentration and distrust of authority but imposes human and social costs through addictions more destructive than alcohol and tobacco. Militarized or violent police responses reinforce alternative authority structures (Wolff 2015) and reduce public order and the government's ability to collect taxes (Flores-Macías 2018). European wars to force open China's market for opium were conceived in commercial terms as a solution to a balance of payments problem in a mercantilist age. They were perceived in China as attacks on society—part of a century of humiliation (Doshi 2021:28). This is so embedded in Chinese discourse that some have linked it to Chinese involvement in North America's opioid crisis (Battiloro 2018; Cooper 2021).

Human trafficking is probably less severe a threat than migrant smuggling. People are trafficked as commodities to be used as indentured labor in informal or illegal economies. This is criminal but seldom destabilizing. Smuggling migrants involve larger numbers without precise destinations. Large numbers of migrants in holding areas like Lampedusa (Italy), Lesbos (Greece), or Nauru (Australia) are costly and have a corrosive effect on public discourse. Waves of desperate migrants may prompt nativist reactions. They can be used in disinformation campaigns to unsettle domestic politics, as they were by the Russian Internet Research Agency during the Czech elections of 2016 (Jankowicz 2020:160).

Foreign support for most criminal enterprises can combine profitability with subversion. Counterfeiting and credit card fraud not only attack confidence in credit and currency but also provide funds to support anti-state activity. Gunrunning to secessionist and terrorist groups can be profitable. Loansharking and moneylending, supported by counterfeiting or fraud, can incur obligations, erode traditional finance, launder money, and provide funds for bribery and corruption of officials. Extortion and protection rackets can demonstrate government inadequacy. All these criminal enterprises help with the first task for anti-government groups—to demonstrate that the government is ineffective, fostering competition for control (Kilcullen 2013:116–167). Partnerships with organized crime are attractive for a foreign state seeking to erode governance. Even if they fail, they impose costs and may prompt an overreaction that undermines government credibility.

Organized criminal enterprise is not limited to gangs and non-state actors, whether supported by foreign governments or not. Transnational corporations violate national laws and regulations, evade taxes, and frequently pressure governments through legal and extra-legal means (Woodiwiss 2005; Antonopoulos and Papanicolaou 2018:81). When backed by a state or coerced by state agencies, major multinational corporations like US-based Microsoft and IBM, Russian-based Gazprom, French-owned Total, or Chinese Huawei can be powerful instruments to manipulate markets, collect information, and make money at the expense of other actors. Decisions about corporations—working with or against them—should involve security factors. Intelligence, police, and security leaders need education about and exposure to corporate culture and structure to enable understanding.

Markets, Grey Zones, and Modern War

Security and public services are the arms of a state, providing the muscle to respond to Grey Zone threats and modern wars when appropriate. Market mechanisms are the sinews connecting these muscles to the skeleton of national resources that sustain security and public services. The blood and oxygen needed by the muscles come from the myriad exchanges within Clausewitz's triangle of government, society, and the organizations of force. This biological analogy helps situate the power of markets, which are the most efficient way of matching resources to tasks. Markets are structured by states' authority and non-state actors' actions. Small states benefit significantly from free international trade, which allows them to live far beyond their resources. However, globalization can be mobilized to use trade for military advantage in service of an aggressive foreign policy. In discussing restraints upon imports, Adam Smith opined that defense was more important than prosperity (Smith 2016:365). Without defense, free exchange can turn to exploitation and expropriation. Understanding the risks and benefits of domestic and international markets is a professional responsibility for security leaders.

Sanctions and economic pressures are tools of coercion. Russia uses its neighbors' energy dependency to apply pressure, but economic pressures are inherently self-limiting over time. Because they are applied in a marketplace, states will find ways of reducing or offsetting dependency:

> …constantly exploiting the dependence of European customers for gas supplies, it encouraged them to diversify their sources of supply. When the Russians needed income as energy prices fell in 2014, it found that its market share had decreased. Turning financial and trade relationships into geopolitical instruments is rarely cost-free. Customers lose trust and market mechanisms fail.
>
> (Freedman 2019:181)

Freedman's observation fits easily with the concept of Grey Zones and modern war. The "positive peace" of free and fair exchange in markets that transcend boundaries can degenerate into a "negative peace" of incipient coercion and distrust as markets are weaponized to put pressure on adversaries.

What is true in energy markets is equally true for mining, finance, debt, and consumer electronics. When an economic tool is used coercively, actors aware of its use will seek alternatives to reduce the coercion. That awareness is crucial. Part of the strategy for using markets in Grey Zone conflicts and modern war is to take advantage of uncertainty, increase dependency over time, and obscure the extent of the dependency until it can be usefully exploited. Ideally, vulnerability and dependence can be exploited without the target becoming aware and countering. Full-spectrum engagement in the economy of a target country may involve investment, shareholder activism, loans to incur debt obligations, regulatory interference, bilateral or multilateral investment protection agreements, masking of involvement through legal obfuscation or shell companies, lawsuits, regulatory and political interference, disinformation, share-price manipulation, and more. This list has limited scope for engagement by police and security forces. The front line in Grey Zone market operations lies with business and the law. Given the West's adversarial legal system and lawyer-client privilege, the best lawyers are as likely—perhaps more likely—to be employed by hostile actors as by the state. Combating Grey Zone economic conflict, while preserving the strengths of open societies, requires engaging boardrooms, corporate lawyers, and legislators, and ensuring full transparency in coercive negotiations.

Key participants in efforts to counter Grey Zone economic conflict are non-governmental organizations and lawyers with government and private funding engaged in international and private sector consulting, research, and publication. The International Coalition against Illicit Economies is a Washington-based and US-funded organization to counter illicit cross-border economic threats. The ETAFIN project for mapping global risks began as an academic project at the University of Aberdeen and now appears to be supported by the Zurich Insurance

Group. The Conflict Information Consortium is a non-governmental organization based at the University of Colorado, which has established CRInfo, the Conflict Resolution Information Source, and Beyond Intractability. These efforts tend to reflect two cultures: a security-minded intelligence-and-policing approach to identifying and managing risks and a negotiation and positive-sum-game approach to resolving conflicts in pursuit of common interests. There is a risk that each culture will see the other as naïve and one-sided. Nevertheless, the two cultures reflect the "defensive" need to push back against coercion and violence and the "offensive" need to build common interests in pursuing positive peace through cultural and commercial ties across boundaries.

Conclusion

Markets are essential to prosperity and security. Unfortunately, they are also inherently competitive and always potentially adversarial. Sometimes it will not be possible to engage in fair and open exchange with market actors who have hostile, deceptive, or manipulative intent, or who are working for aggressive foreign interests. Intelligence and negotiation techniques designed to assess the limits of positive-sum market exchange are a starting point but need clear national strategic guidance and coordination across government and with allies. Governments of even small states have considerable power to manage markets through legislation, courts, restrictions, stockpiles, coordinated intelligence, and even bank raids like the one on the Hercegovacka Bank in 2001. Used judiciously with open markets, these powers can help defend against Grey Zone threats and move economic activity back toward the mutual gains of positive peace. There is a caution, however. The same tools can be used in the interests of politically connected individuals to enhance wealth and power at the expense of the public interest. Open markets and economic freedoms should be preserved as a source of prosperity, and arguments for constraints should be tested carefully.

References

Abrahamsen, R. and Williams, M. C. (2011), *Security Beyond the State: Private Security in International Politics*, Cambridge University Press, Cambridge.

Antonopoulos, G. A. and Papanicolaou, G. (2018), *Organized Crime: A Very Short Introduction,* Oxford University Press, Oxford.

Barojan, D. (2021), "Building Digital Resilience Ahead of Elections and Beyond", in S. Jayakumar, B. Ang, and N.D. Anwar (eds.), *Disinformation and Fake News*, Palgrave-Macmillan, Singapore, pp. 61–73.

Battiloro, C. (2018), "Fentanyl: How China's Pharmaceutical Loopholes Are Fuelling the United States Opioid Crisis", *Syracuse Journal of International Law and Commerce*, vol. 46, no. 2, pp. 343–376.

Black, J., Hashimzade, N. and Myles, G. (2017), *Oxford Dictionary of Economics,* Oxford University Press, Oxford.

Brauer, J. and Anderton, C. H. (2020), "Conflict and Peace Economics: Retrospective and Prospective Reflections on Concepts, Theories, and Data", *Defence and Peace Economics*, vol. 31, no. 4, pp. 1–23.

Brown, R. (2022), "Authoritarian Populism and its Sources", in *The Conservative Counter-Revolution in Britain and America 1980–2020*, Springer International Publishing, Cham, pp. 85–114.

Calof, J. (2017), "Reflections on the Canadian Government in Competitive Intelligence–Programs and Impacts", *Foresight*, vol. 19, no. 1, pp. 31–47.

Cooper, S. (2021), *Wilful Blindness: How a Network of Narcos, Tycoons and CCP Agents Infiltrated the West*, Optimum Publishing, Toronto.

Davies, P. (2019), *The Authorized History of British Defence Economic Intelligence: A Cold War in Whitehall, 1929–1990*, Routledge, London.

Denecé, E. (2016), "France: The Intelligence Services' Historical and Cultural Context", in de Graaff, B., Nyce, J. and Locke, C. (eds.), *Handbook of European Intelligence Cultures*, Rowman and Littlefield, Lanham, MD, pp. 135–146.

Doshi, R. (2021), *The Long Game: China's Grand Strategy to Displace the American Order*, Oxford University Press, Oxford.

Economist (2023a), "Britain's Semiconductor Strategy Shows the Bind the Country is in", *Economist* 25 May. https://www.economist.com/britain/2023/05/25/britains-semiconductor-strategy-shows-the-bind-the-country-is-in

Economist (2023b), "Homeland Economics Will Make the World Poorer: Industrial Policy and Protectionism Could Endanger Trade, without Making Western Economies Safer", Special Report: Homeland Economics, *Economist* 2 October. https://www.economist.com/special-report/2023/10/02/homeland-economics-will-make-the-world-poorer

Flores-Macías, G. (2018), "The Consequences of Militarizing Anti-drug Efforts for State Capacity in Latin America: Evidence from Mexico", *Comparative Politics*, vol. 51, no. 1, pp. 1–20.

Freedman, L. (2019), *Ukraine and the Art of Strategy*, Oxford University Press, Oxford.

Gerstel, D. (2016), "Motivations of the Kremlin: The Crimean Annexation as a Diversionary Conflict", *Swarthmore International Relations Journal*, vol. 1, no. 1, pp. 31–36.

Greenberg, A. (2019), *Sandworm: A New Era of Cyberwar and the Hunt for the Kremlin's Most Dangerous Hackers*, Anchor Books, New York.

Hairong, Y. and Sautman, B. (2010), "Chinese Farms in Zambia: From Socialist to "Agro-Imperialist" Engagement?", *African and Asian Studies*, vol. 9, no. 3, pp. 307–333.

Holmes, K. (2012), *Foreign Ownership Control and Influence in the 21st Century: Analysis of Risk, Technical and Organizational Requirements*, 30 April, Industrial Security Sector, Public Works and Government Services.

Hristov, J. (2014), *Paramilitarism and Neoliberalism: Violent Systems of Capital Accumulation in Colombia and Beyond*, Pluto Press, London.

Jankowicz, N. (2020), *How to Lose the Information War: Russia, Fake News, and the Future of Conflict*, I. B. Taurus, London.

Johnson, L. K. (2010), *The Oxford Handbook of National Security Intelligence*, Oxford University Press, Oxford.

Jones, A. (2008), "Industrial Espionage in a Hi-tech World", *Computer Fraud & Security*, vol. 2008, no. 1, pp. 7–13.

Kilcullen, D. (2013), *Out of the Mountains: the Coming Age of the Urban Guerilla*, Oxford University Press, Oxford.

Losurdo, D. (2014), *Liberalism: A Counter-History*, Verso, London.

Milanovic, B. (2019), *Capitalism, Alone: The Future of the System that Rules the World*, Belknap Press, Harvard University, Cambridge, MA.

Miller C. (2022), *Chip War: The Fight for the World's Most Critical Technology*, Simon & Schuster, New York.

Mirowski, P. (2013), *Never Let a Serious Crisis Go to Waste: How Neoliberalism Survived the Financial Meltdown*, Verso, London.

Mozur, P. and Hopkins, V. (2023), "Ukraine's War of Drones Runs into an Obstacle: China", *New York Times*, 30 September. https://www.nytimes.com/2023/09/30/technology/ukraine-russia-war-drones-china.html

OCHA Relief Web (2001), *Bosnia-Herzegovina: Violence in Mostar*, Report by UK Foreign and Commonwealth Office, 6 April. https://reliefweb.int/report/bosnia-and-herzegovina/bosnia-herzegovina-violence-mostar.

Palan, R., Murphy, R. and Chavagneux, C. (2013), *Tax Havens: How Globalization Really Works,* Cornell University Press, Ithaca, NY.

Perkins, J. (2004), *Confessions of an Economic Hitman*, Dutton, Penguin Group, New York.

Perkins, J. (2007), *The Secret History of the American Empire: Economic Hit Men, Jackals, and the Truth about Global Corruption*, Penguin, New York.

Piketty, T. (2014), *Capital in the Twenty-first Century*, Harvard University Press, Cambridge, MA.

Potter, E. H. ed. (1998), *Economic Intelligence and National Security*, Carleton University Press, Ottawa, Canada.

Reid, M. (2015), "A Comparative Approach to Economic Espionage: Is Any Nation Effectively Dealing with this Global Threat?", *University of Miami Law Review*, vol. 70, no. 3, pp. 757–825.

Rickards, J. (2012), *Currency Wars: the Making of the Next Global Crisis*, Penguin Publishing Group, New York.

Sen, G. (1984), *The Military Origins of Industrialisation and International Trade Rivalry*, Frances Pinter, London.

Silwal, S. B., Anderton, C. H., Brauer, J., Coyne, C. J. and Dunne, J. P. (2021), *The Economics of Conflict and Peace: History and Applications*, Cambridge University Press, Cambridge.

Smith, A. (2016), *The Wealth of Nations,* Titan Books, London.

Stewart, C. (2023), "Analysis: 21st Century Tech Will Give Ukraine a Military-Industrial Edge against Russia's Soviet-Era Arms", *Kyiv Post*, 18 September. https://www.kyivpost.com/analysis/21744.

Stiglitz, J. (2002), *Globalization and Its Discontents*, W.W. Norton & Company, New York and London.

Turchin, P. (2023), *End Times: Elites, Counter-Elites, and the Path of Political Disintegration*, Penguin Press, London.

UNCTAD (2023), *Investment Policy Hub, Huawei Technologies Co., Ltd. v. Kingdom of Sweden, ICSID Case No. ARB/22/2.* https://investmentpolicy.unctad.org/investment-dispute-settlement/cases/1208/huawei-v-sweden

Van Apeldoorn, B. and De Graaff, N. (2016), *American Grand Strategy and Corporate Elite Networks: The Open Door Since the End of the Cold War*, Routledge, London.

Wolff, M. J. (2015), "Building Criminal Authority: A Comparative Analysis of Drug Gangs in Rio de Janeiro and Recife", *Latin American Politics and Society*, vol. 57, no. 2, pp. 21–40.

Woodiwiss, M. (2005), *Gangster Capitalism: The United States and the Global Rise of Organised Crime*, Constable, London.

Zuboff, S. (2019), *The Age of Surveillance Capitalism: The Fight for a Human Future at the New Frontier of Power*, Profile Books, London.

6 Violent Non-state Actors in the Modern Conflict Environment

Tarik Solmaz

Introduction

In May 2024, Israel and Sweden jointly accused Iran of using criminal gangs as terrorist proxies in Stockholm, including an attempted attack on the Israeli Embassy (FDD 2024). Was this an act of war or criminal violence? Violent non-state actors (VNSAs), or non-state armed actors, are a feature of Grey Zone conflict and modern wars. A VNSA is any armed group not part of a state's formal security forces. Such actors engage in violent activities to pursue their objectives (Hofmann and Schneckener 2011:603; Berti 2016:1; Ezrow 2017:37–38). This chapter links the preceding description of markets (Last, Chapter 5) with Part IV about preserving society, social cohesion, and human security. No state is isolated from markets, which wield substantial influence over security. Market-related Grey Zone challenges include loss of access to or control of vital assets, attacks on banking and currency systems, market manipulation, economic espionage, collusion with criminal disorder, intimidation and dispossession for foreign benefit, and disruption of political stability. These challenges intersect directly with VNSAs, often the enforcers to support economic depredations.

Terrorists, insurgents, paramilitaries, militias, warlords, gangs, drug cartels, private military companies (PMCs), and mercenaries are VNSAs. However, these groups differ in their use of force, governance roles, and interactions with states. We group VNSAs into three categories according to their principal motivations: *Political*, *criminal*, and *commercial.* Understanding principal motivations may help manage or reduce VNSA activities by seducing away members and redirecting the organizations to support stable peace. These motivations inevitably overlap; criminal and commercial VNSAs both pursue profit. Ideologues also need funds. Secondary parameters include VNSA methods, structure, and legitimacy.

The rest of this chapter considers the growth of VNSAs in the Grey Zone, the intersection and mutual reinforcement of political, criminal, and commercial motivations, and their implications for Grey Zone threats and modern wars.

VNSAs in the Grey Zone

VNSAs are no new phenomenon (Williams 2008:5; Lemke 2008:775; Strachan 2019:172). When interstate wars fell and intrastate wars rose during the Cold War, VNSAs were important actors (Council on Foreign Relations 2023). Their proliferation often accompanies the weakening of the state (see Kazmierczak et al., Chapter 16). Weak states and state failures to monopolize violence open space for VNSAs (Fund for Peace 2023:49). Protracted social conflicts, porous borders, corruption, poverty, weak governance, and inability to provide public services are characteristics of failed states (Helman and Ratner 1992:3; Rotberg 2003:5–9). The Fragile

DOI: 10.4324/9781003428701-9

Table 6.1 Categories and Indicators of Fragile States Index

Category	*Indicators*
Cohesion indicators	C1: Security apparatus C2: Factionalized elites C3: Group grievance
Economic indicators	E1: Economic decline E2: Uneven development E3: Human flight
Political indicators	P1: State legitimacy P2: Public services P3: Human rights
Social and cross-cutting indicators	S1: Demographic pressures S2: Refugees and Internally Displaced Persons (IDPs) X1: External intervention

Source: The Fund for Peace Website, The Fragile State Index Indicators, https://fragilestatesindex.org/indicators/

State Index, released annually by the *Fund for Peace*, uses 12 key indicators (Table 6.1) to assess and measure the risk of state failure. These categories fall into four broad categories: Cohesion, Economic, Political, and Social (The Fund for Peace, n.d.).

Modern conflicts, as outlined in Chapter 1, defy traditional distinctions between war and peace. This undermines the norms and rules that underpin stability. Grey Zone activities are not confined to states; they can also be wielded to advance the interests of VNSAs. The existence of VNSAs pushes stable peace toward negative peace and war, depending on the persistence and intensity of their activity. Levels of physical violence in a Grey Zone conflict may be indistinguishable from war. The Positive Peace Framework includes a high-functioning government, secure business environment, equitable distribution of resources, acknowledgment of the rights of others, positive relations with neighbors, free dissemination of information, high levels of human capital, and low levels of corruption (Vision of Humanity 2024). The influence of VNSAs jeopardizes each of these components.

Criminal gangs gain funds and power through the pernicious impact of drugs. Addiction erodes social cohesion and human security, with implications for national security. China's support for Fentanyl exports to the United States has destroyed communities and destabilized conventional politics (Rusin 2019:5). Political non-state actors engage in strategies of "competitive control," weakening cohesion, increasing the risk of failed states, and eroding international security. Armed groups that dominate a population can provide more consistent and predictable governance than a distant government (Kilcullen 2013:126). Legitimate businesses that evade or manipulate laws with payoffs to police, regulators, or politicians undermine the integrity of institutions (Briscoe and Goff 2016:27). Commercial interests generate violence that undermines human security. Multinational corporations hire paramilitary forces to intimidate union labor and force peasants off the land, exacerbating inequality. Dispossessed people are available and motivated to join narco-gangs and guerrilla groups (Hristov 2014).

The term "violent non-state actor" encompasses many groups with distinct characteristics. A typology of VNSAs helps security professionals to understand and manage this part of the security landscape.

Confronting the Classification Problem: Intersecting Motivations

Political, criminal, and commercial motives overlap. Criminality is defined by laws, which are created through politics. Legitimate business interests, finding their profitability constrained, will use money and social networks to change laws to increase profitability. In the illegal economy, violence and intimidation are part of the business model for controlling market share and supply, but legal and illegal enterprises are linked. The same gangs hired by multinationals to intimidate workers may earn revenue from extortion, kidnapping, and assassination (Hristov 2014:4). Terrorist groups like the Red Brigades robbed banks to finance their operations (Orsini and Nodes 2011:18). The Taliban earned revenue from the drug trade while suppressing drug use at home (O'Donnell 2021).

Criminally motivated VNSAs seek to establish alternative political and social orders, filling a void to establish parallel government functions. They control market entry and challenge the government's monopoly on enforcement (Ezrow 2017:144). Cartels in Mexico and Guatemala wield control that defines "good" and "bad" members of society, excluding or expelling those who resist their authority (Kloppe-Santamaría 2021).

PMCs like US-based Blackwater or Russia-based Wagner Group are profit-motivated but support state goals. The Wagner Group follows political direction from the Kremlin even when contracted to other states (UK House of Commons 2023:7) and is facilitated by the Russian military intelligence (GRU) (Østensen and Bukkvoll 2018:25). It has operated in support of Russian interests in Ukraine, Syria, Mali, Libya, and across Africa (Blazakis et al. 2023:6; UK House of Commons 2023:3; Bowen 2023) and should be seen as a Russian surrogate rather than an independent commercial entity.

When do VNSAs' motives shift? VNSAs tend to be driven by profitability and access to future opportunities. They sometimes expand operations to outflank their opposition and preserve their freedom of movement. If VNSAs have principal motives, they can be discerned in their actions over time but these are likely to change. Changes in operational circumstances will generate changing goals, methods, and strategies. An insurgent organization motivated by political grievances might use criminal activities like smuggling to support the insurgency; smuggling remains a way of life after the insurgency fades. The Islamist group al-Shabaab in Somalia is part of an ecosystem of piracy and acacia wood charcoal smuggling (Walker 2014:41–48; Bolognesi et al. 2015:40–49). The primary motive for al-Shabaab is probably ideological, but former fishing communities that have turned to piracy and smuggling support the cause out of survival interest.

By aligning with causes, criminally motivated VNSAs may gain protection and support from political organizations or states, gaining sanctuary, resources, and a shield against law enforcement operations. In its campaign in Ukraine after 2014, the Russian Federation employed the Night Wolves, a nationalist motorcycle gang, as a proxy force. The Night Wolves established roadblocks, coordinated demonstrations, gathered intelligence, disseminated propaganda, intimidated local authorities, and exerted influence over regional events (Galeotti 2016:297; Lauder 2018:9; Bouwmeester 2021:500).

There is no reason to expect that VNSAs should stay within the boundaries of labels we might attach to them. The classification is useful to develop tactics to constrain VNSAs. If they exist because they are profitable, their profitability might be eroded. If they are politically motivated, support might be undermined. If they are criminal, law enforcement might be more effective. All tactics, however, affect the motivation and evolution of VNSAs with which states must contend.

Comparing the Convergence of VNSAs

The essence of political VNSAs lies in seeking power, establishing or challenging political authority, and pursuing political agendas (Berti 2013:11–12). Criminal VNSAs seek financial gain through illegal activities (Williams 2008:15–16; Ezrow 2017:144). Commercial VNSAs

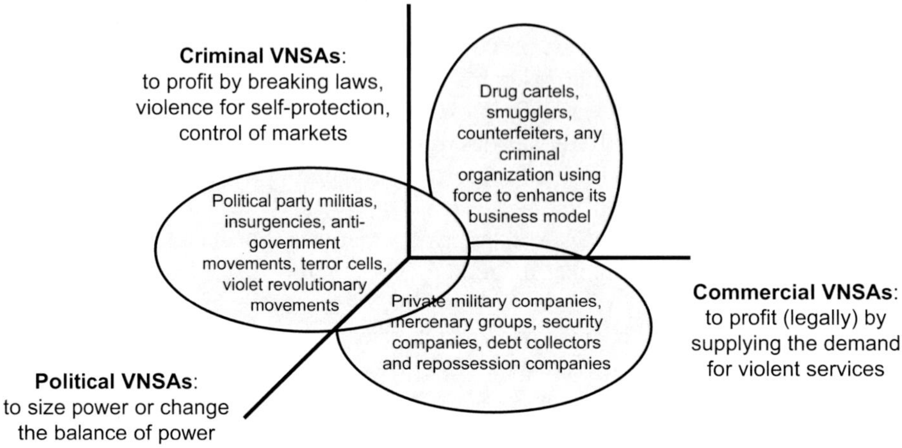

Figure 6.1 Converging Motivations for VNSAs.

Source: Author's elaboration.

seek economic profit through the provision of military and violent or coercive security services (Ezrow 2017:148; McFate 2019:7). These motives are shown as three axes in Figure 6.1. While criminal and commercial VNSAs seek profit, they have different motivations concerning the law and political power. Debt collectors and security firms serving commercial clients operate mostly within the law to obtain contracts. Governments may be the most lucrative clients for PMCs. Criminal enterprises like counterfeiters and smugglers differ from their legal counterparts in not being able to rely on the policing and security afforded by the state. Their violent enforcement activity is a form of private policing outside the law, secondary to their main business. For other criminal VNSAs, violent enforcement is central to their business model. Gang protection rackets fall in this category: pay or suffer. However, protection rackets may look like government enforcement if party militias use coercive measures to get out the vote or enforce policy. Politics often brings together political entrepreneurs and violent specialists to do business (Tilly 2003:Chapter 3).

As Figure 6.1 illustrates, political, criminal, and commercial VNSAs converge in practice. Scope of operations and legitimacy are further criteria that help distinguish between their operations. The scope of operations can be described as a spectrum of violence and geographical reach. Legitimacy can be conferred by legal standing, endorsement, tacit acknowledgment, or covert partnerships.

Political VNSAs range from global to regional and local and engage in a spectrum of violence, from large-scale murder to intimidation and influence operations, in efforts to seize power or influence behavior. Political theories offer prescriptions for violent action to seize power, ranging from anarchist "propaganda of the deed" to the progression from peasant insurgency to conventional war. Some political VNSAs are incompatible with existing governments, while others are prepared to make overt or covert alliances of convenience. Political VNSAs seeking the overthrow of governments forego legitimacy conferred by state recognition for the legitimacy conferred by their constituents or believers. Combining the stick of coercion with the carrot of services, political VNSAs like Hezbollah and Hamas gain widespread legitimacy and deal directly with state governments (Berti 2013). Other political VNSAs begin life as party militias. Privileged access to politicians in power is evidence of legitimacy. The acceptable extent of government-sponsored violence limits their scope of action. The Basij ("mobilization") is a paramilitary volunteer militia associated with Iran's theocracy and a responsive tool of government repression (Golkar 2015). It is less formally linked to power than the Iranian Revolutionary Guard Corps (IRGC), which pursues Iran's revolution with Hamas, Hezbollah, and now criminal gangs in Sweden (FDD 2024).

Criminal VNSAs profit from breaking the law, but for some, the violence is incidental, and for others, it is central. Money laundering, smuggling, fraud, and counterfeit production may need violent support to protect markets and enforce contracts. Contract killing, arson, piracy, and protection rackets are inherently violent. The legitimacy of criminal VNSAs is not in direct proportion to their violence. Pirates and contract killers may operate with tacit approval, while insurance fraud may be ruthlessly policed.

Commercial VNSAs like G4S and the Russian Wagner Group operate globally, providing data management, security, training, logistical support, and combat services (Abrahamsen and Williams 2011; McFate 2019:7). They generally operate within legal frameworks or exploit poorly regulated areas, presenting themselves as legitimate businesses providing security services. Their legitimacy and share price decline when they engage in illegal activities. They may stray into politics when providing close personal protection services or neighborhood security with political overtones.

How should small states understand the converging and overlapping motives of VNSAs, their methods, organization, sources of legitimacy, and sponsorship patterns? How should they confront or use VNSAs when planning policing and security operations in a complex security landscape?

Confronting VNSAs in the Grey Zone

VNSAs have a different impact on weak states, states facing external pressure, and effective states with the potential to use them. In all cases, VNSAs are most likely to be self-serving or allied to more powerful interests. They are, therefore, logical intelligence targets for most countries. What states should know and how they should use that knowledge is related to the scope, organization, legitimacy, and sponsorship of the three types of VNSAs identified.

Confronting the Scope of VNSAs—Violence and Reach

The scope of VNSA operations can be measured according to capacity for violence and geographic reach. Commercial VNSAs, like the global business of private security companies, have extensive capacity to manage low levels of violence but do not inflict much violence. They are probably net contributors to human and national security, even given their growing role in data management and surveillance (Abrahamsen and Williams 2011). To work with them effectively, states need a sound legislative framework and adequate expertise to provide supervision. The same is true for PMCs with a greater capacity for violence.

Political VNSAs with internationalist ideologies have a global reach and capacity to inspire lethal violence, often widespread. Communism and anarchism are in decline. Sunni Jihadism is global, while Shiite Jihadism is strongest in Iran's hinterland (Hezbollah, Hamas, and Houthis). Partnering with criminal gangs creates new challenges demanding creative policing (FDD 2024). Political VNSAs like party militias, ethnic or religious separatist groups, and small terrorist groups tend to have local scope and limited capacity for violence.

Criminal VNSAs may kill more people than civil or interstate wars in protracted "crime wars" (Muggah and Sullivan 2018). The Brazilian Forum for Public Security claims that more people died from criminally motivated violence in Brazil than in the Syrian Civil War in 2015

(Worley 2016). Violent policing is unlikely to reduce the death toll. The violence and reach of VNSAs might be best addressed by strengthening the social cohesion of the state, by building trust in government and by providing effective policing and services (see Part IV).

Confronting the Organization of VNSAs—Stakeholders and Shadows

Political and criminal VNSAs generally make decisions in the shadows. They need covert and compartmentalized structures for security. Their stakeholders are internal to the organization, plus a few trusted outsiders with privileged communications. Their supporters may constitute a wider circle of admirers or beneficiaries, and this secondary layer of stakeholders and knowledgeable insider-outsiders can be a valuable source of access and intelligence for state security.

Commercial and criminal VNSAs tend to be constrained by costs and motivated by benefits. Commercial organizations will be relatively open to their stakeholders—often shareholders or government sponsors—but both criminal and commercial organizations will guard their trade and operational secrets jealously.

Political VNSAs calculate costs and benefits in ideational rather than material terms. Strategies of atrocity work because they trigger a violent reaction, and liberal states are held to a different standard than terrorist groups, creating a form of no-win asymmetry (Wilkinson 1977). The closer a group comes to exercising political power, the more important legitimacy becomes.

Confronting the Legitimacy of VNSAs—Reputational Damage

Legitimacy allows VNSAs to achieve their objectives with minimum resources. For the criminal VNSA, Robin Hood legends about redistributed wealth are helpful. The legitimation of greed (generating wealth) helps criminal and commercial VNSAs and nonviolent corporations that stray close to the boundary of illegal operations. Ideals of corporate social responsibility help constrain the behavior of regular corporations, including those trafficking in security and military action.

Reputational damage concerns all VNSAs, but different reputational attacks damage them. "Honor among thieves" is important because contracts within criminal VNSAs demand reliability. Criminal organizations' severe internal violence and honor codes keep transaction costs low. Attacking the personal reputations of criminal actors can diminish their power. Reliability and effectiveness are central to both criminal and political VNSA business models. A reputation for predictability is important for political VNSAs controlling territory or people; subject populations will be safe if they support and conform and will suffer if they do not. The legitimacy of a political VNSA in control depends on its ability to deliver order. A political VNSA seeking control, however, must be seen doing something for the cause. Is the political party providing any benefit to the people its militia rounds up for donations, protests, and votes? These benefits can be social and psychological more than material. Spontaneous celebrations by repressed populations when their fighters strike back at perceived tormentors are evidence of the cost-benefit calculus at work for extremist political VNSAs. VNSA reputations are further enhanced by material benefits like pensions for suicide bombers' families paid by foreign contributors. That these payments are denied in defiance of evidence suggests some reputational damage within the community of states and international civil society (Human Rights Watch 2002:104).

Confronting the Sponsorship of VNSAs—Deniable Proxies

For all their variety, VNSAs' deniability makes them ideal allies and dangerous enemies in the waging of Grey Zone conflicts and modern war. During the Cold War, delegation of war to third-party forces became the rule rather than the exception (Mumford 2013:12). States preferred covert and indirect attacks faced with the risk of escalation to nuclear war. US military supremacy has created similar incentives for states or non-state actors that would oppose the dominant order. The first quarter of the 21st century has also been dominated by proxy warfare, with a wider range of potential allies and new configurations of support and deniability. Proxy warfare was previously defined as a state actor's reliance on a non-state armed group or another state (Deutsch 1964:102; Duner 1981:353–354; Tierney 2021:52). The use of proxies has expanded beyond states; non-state actors are also using proxies in their attacks on states and rival VNSAs.

Since the Cold War, Western support for regime change and peaceful transitions has relied on nonviolent civil society organizations, human rights groups, and investment and development initiatives. Russia and Iran have pushed back with ecosystems of proxy alliances, including criminal, political, and commercial VNSAs. Their deniability has been a major asset in waging war below the threshold in the Grey Zone.

Iran's ecosystem includes Hezbollah in Lebanon, Hamas operating out of the Gaza Strip, Shiite Houthi forces in Yemen, and expatriate allies of the regime in Western countries. Tying these groups together, the Iranian Revolutionary Guard Corps (IRGC), or *Pasdaran*—a branch of the Iranian Armed Forces established after the 1979 Revolution—provides weapons, doctrine, training, intelligence, and strategic communication through Quds Force (QF)—the IRGC's umbrella for foreign fighters (Keshavarz 2023:68–71; Azizi 2020). Hezbollah is actively involved in Lebanese national politics, building a Shia Muslim constituency through social services and education (Berti 2013). It commands advanced military capabilities with regional reach (Hoffman 2007:35–43) and is developing offensive cyber capabilities (Clarke 2017). The Swedish-Israeli accusation of May 2024 indicates a new tactic in which QF weapons and training are mobilized opportunistically through links to organized crime gangs.

Russia's ecosystem includes a network of organized criminal organizations across the former Soviet Union with political connections in Moscow and the capitals of the successor republics (Galeotti 2018). Russian offensive cyber capability has been developed by contracting private commercial and criminal groups and using intelligence service (GRU) pressure to compel cooperation with the state (Carlin and Garrett 2018:398). The increase in Russian sabotage in Europe is attributed to the GRU, and gang members with covert support (Economist 2024). At war in Ukraine, faced with dissent in Belarus, and further afield in Africa, Russia drew on the Wagner Group. A year after the Prigozhin coup, Bloomberg News reported that PMC Redut, a GRU-controlled firm, would likely replace Wagner's functions.

Political, commercial, and criminal VNSAs are evident in the Russian and Iranian ecosystems for Grey Zone warfare. Whether VNSA assets are deployed as ongoing irritants or as preludes to conventional military attacks, VNSAs erode their targets' capacity to develop positive peace characteristics. In the case of weak or fragile states like Sudan, Central African Republic, Libya, Mozambique, Mali, and Burkina Faso—in all of which the Wagner Group was active—VNSA activity can erode cohesion, hasten human flight and economic decline, and open the door to further external intervention (Table 6.1).

Conclusion

Political, criminal, and commercial VNSAs demonstrate great variety and adaptability. They operate on their own account and in ecosystems established by states or other non-state groups to pursue deniable activities at home or abroad for profit and power. They operate in the legitimate and illegal markets for violence. Organized crime can erode state authority and legitimacy. Privatizing public security services can reduce the capacity of a state to govern. Political violence by party militias, ethnic or religious zealots, and criminal gangs can reduce public political life to disorder. As these motivations converge and are manipulated by hostile external actors, positive peace and good governance disappear into the Grey Zone, which may be a prelude to conventional war or part of a continuing war nexus.

References

Abrahamsen, R. and Williams, M. C. (2011), *Security Beyond the State: Private Security in International Politics*. Cambridge University Press, Cambridge.

Azizi, A. (2020), *The Shadow Commander: Soleimani, the US, and Iran's Global Ambitions*. Bloomsbury, London.

Berti, B. (2013), *Armed Political Organizations: From Conflict to Integration* John Hopkins University Press, Baltimore, MD.

Berti, B. (2016), "What's in a Name? Re-conceptualizing Non-state Armed Groups in the Middle East", *Palgrave Communication*, vol. 2, no. 16089, pp. 1–8.

Blazakis, J., Clarke, C. P., Fink, N. C. and Steinberg, S. (2023), *Wagner Group: The Evolution of a Private Army*. The Soufan Center. https://thesoufancenter.org/research/wagner-group-the-evolution-of-a-private-army/

Bolognesi, M., Vrieling, A., Rembold, F. and Gadain, H. (2015), "Rapid Mapping and Impact Estimation of Illegal Charcoal Production in Southern Somalia Based on WorldView-1 Imagery", *Energy for Sustainable Development*, vol. 25, pp. 40–49. https://doi.org/10.1016/j.esd.2014.12.008.

Bouwmeester, A. J. H. (2021), "The Art of Deception Revisited (Part 2): The Unexpected Annexation of Crimea in 2014", *Militaire Spectator*, vol. 190, no. 10, pp. 494–507.

Bowen, A. S. (2023), Russia's Wagner Private Military Company (PMC), *Congressional Research Service*. https://www.congress.gov/crs-product/IF12344

Briscoe, I. and Goff, D. (2016), *Protecting Politics: Deterring the Influence of Organized Crime on Elections*. International IDEA, Stockholm, Sweden and Clingendael Institute, The Hague, Netherlands. https://www.idea.int/sites/default/files/publications/protecting-politics-deterring-the-influence-of-organized-crime-on-elections.pdf

Carlin, J. P. and Garrett, M. G. (2018), *Dawn of the Code War: America's Battle against Russia, China, and the Rising Global Cyber Threat*. PublicAffairs, New York.

Clarke, C. P. (2017), How Hezbollah Came to Dominate Information Warfare. *The Jerusalem Post*, 17 September. https://www.jpost.com/Opinion/How-Hezbollah-came-to-dominate-information-warfare-505354

Council on Foreign Relations (2021), Mexico's Long War: Drugs, Crime, and the Cartels. https://www.cfr.org/backgrounder/mexicos-long-war-drugs-crime-and-cartels

Council on Foreign Relations (2023), Understanding Intrastate Conflict. https://education.cfr.org/learn/reading/understanding-intrastate-conflict

Deutsch, K. W. (1964), "External Involvement in Internal Wars", in H. Eckstein (ed.), *Internal War: Problems and Approaches*, Free Press of Glencoe, New York, pp. 100–110.

Duner, B. (1981), "Proxy Intervention in Civil Wars", *Journal of Peace Research*, vol. 18, no. 4, pp. 353–361.

Economist (2024), "Europe: Active Measures—Russia Is Ramping Up Sabotage in Europe", *Economist*, 12 May. https://www.economist.com/europe/2024/05/12/russia-is-ramping-up-sabotage-across-europe

Ezrow, N. (2017), *Global Politics and Violent Non-state Actors*. Sage Publications Ltd, London.

FDD (2024), *Israel, Sweden: Iran Using Criminal Gangs for European Terror,* Foundation for the Defence of Democracy (FDD), 30 May. https://www.fdd.org/analysis/2024/05/30/israel-sweden-iran-using-criminal-gangs-for-european-terror/

Galeotti, M. (2016), "Hybrid, Ambiguous, and Non-linear? How New Is Russia's 'New Way of War'?", *Small Wars & Insurgencies*, vol. 27, no. 2, pp. 282–301.

Galeotti, M. (2018), *The Vory: Russia's Super Mafia*. Yale University Press, New Haven, CT.

Golkar, S. (2015), *Captive Society: The Basij Militia and Social Control in Iran*, Columbia University Press, New York.

Helman, G. B. and Ratner, S. R. (1992), "Saving Failed States", *Foreign Policy*, vol. 89, pp. 3–20. https://doi.org/10.2307/1149070

Hoffman, F. G. (2007), *Conflict in the 21st Century: The Rise of Hybrid Wars*. Potomac Institute for Policy Studies, Arlington, VA. https://potomacinstitute.us/reports/19-reports/1163-conflict-in-the-21st-century-the-rise-of-hybrid-wars

Hofmann, C. and Schneckener, U. (2011), "Engaging Non-state Armed Actors in State- and Peace-building: Options and Strategies", *International Review of the Red Cross*, vol. 93, no. 883, pp. 603–621.

Hristov, J. (2014), *Paramilitarism & Neoliberalism: Violent Systems of Capital Accumulation in Colombia and Beyond*, Pluto Press, London.

Human Rights Watch (2002), Erased in a Moment: Suicide Bombing Attacks Against Israeli Civilians. https://www.hrw.org/reports/2002/isrl-pa/

Keshavarz, A. (2023), *The Iranian Revolutionary Guard Corps: Defining Iran's Military Doctrine*, Bloomsbury Academic, London.

Kilcullen, D. J. (2013), *Out of the Mountains: The Coming Age of the Urban Guerilla*, Oxford University Press, Oxford.

Kloppe-Santamaría, G. (2021), *The Politics of Criminal Violence in Mexico and Northern Central America*. Wilson Center, 9 December. https://www.wilsoncenter.org/article/politics-criminal-violence-mexico-and-northern-central-america

Lauder, M. A. (2018), "'Wolves of the Russian Spring': An Examination of the Night Wolves as a Proxy for the Russian Government", *Canadian Military Journal*, vol. 18, no. 3, pp. 5–16.

Lemke, D. (2008), "Power Politics and Wars without States", *American Journal of Political Science*, vol. 52, no. 4, pp. 774–786.

McFate, S. (2019), *Mercenaries and War: Understanding Private Armies Today*. Washington, DC: National Defense University Press, National Defense University Press, Washington, DC.

Mumford, A. (2013), *Proxy warfare: War and Conflict in the Modern World*. Polity Press, Cambridge.

Muggah R., and Sullivan, J. P. (2018), "The Coming Crime Wars", *Foreign Policy*, 21 September. https://foreignpolicy.com/2018/09/21/the-coming-crime-wars/

National Commission on Terrorist Attacks Upon the United States (2004), *The 9/11 Commission Report. Executive summary*. National Commission on Terrorist Attacks upon the United States, Washington, DC.

O'Donnell, L. (2021), "The Taliban Are Breaking Bad", *Foreign Policy,* 19 July. https://foreignpolicy.com/2021/07/19/taliban-expanding-drug-trade-meth-heroin/

Orsini, A. and Nodes, S.J. (2011), *Anatomy of the Red Brigades: The Religious Mindset of Modern Terrorists*, Cornell University Press, Ithaca, NY.

Østensen, Å. G. and Bukkvoll, T. (2018), Russian Use of Private Military and Security Companies - The Implications for European and Norwegian Security. Norwegian Defence Research Establishment (FFI) Report 18/01300. https://www.cmi.no/publications/6637-russian-use-of-private-military-and-security

Rotberg, R. I. (2003), "Failed States, Collapsed States, Weak States: Causes and Indicators", in R. I. Rotberg (ed.), *State Failure and State Weakness in a Time of Terror*, pp. 1–26, Brookings Institution Press, Washington, DC.

Rusin, N. L. (2019), *Illicit Chinese Fentanyl: America's Next National Security Crisis*, US Naval War College, Newport, RI.

Strachan, H. (2019), "Strategy in Theory; Strategy in Practice", *Journal of Strategic Studies*, vol. 42, no. 2, pp. 171–190.

The Fund for Peace (2023), Fragile States Index Annual Report 2023. https://fragilestatesindex.org/wp-content/uploads/2023/06/FSI-2023-Report_final.pdf

The Fund for Peace (n.d.), Indicators. https://fragilestatesindex.org/indicators/

Geneva Centre for the Democratic Control of Armed Forces (2015), *Armed Non-State Actors: Current Trends & Future*. DCAF Horizon 2015 Working Paper no. 5. DCAF and Geneva Call, Geneva. https://www.dcaf.ch/sites/default/files/publications/documents/ANSA_Final.pdf

Tierney, D. (2021), "The Future of Sino-US. Proxy War", *Texas National Security Review*, vol. 4, no. 2, pp. 49–73.

Tilly, C. (2003), *The Politics of Collective Violence*, Cambridge University Press, Cambridge.

Vision of Humanity (2024), *The Eight Pillars of Positive Peace*, Institute for Economics & Peace. https://www.visionofhumanity.org/the-eight-pillars-of-positive-peace/

Walker, C. (2014), "Somalia's Security Is Tied to Its Natural Resources: Al-Shabaab's Hold on Conflict Charcoal Raises Alarms: East Africa - Issue in Focus", *Africa Conflict Monthly Monitor*, vol. 1, pp. 41–48. https://journals.co.za/doi/abs/10.10520/EJC147672

Wilkinson, P. (1977), *Terrorism and the Liberal State*, Macmillan Press, London.

Williams, P. (2008), *Violent Non-state Actors and National and International Security*. International Relations and Security Network. https://www.files.ethz.ch/isn/93880/VNSAs.pdf

Worley, W. (2016), "Brazil Saw More Violent Deaths than Civil-War Torn Syria in 2015, Report Says", *Independent,* 28 October. https://www.independent.co.uk/news/world/americas/brazil-deaths-violent-crime-syria-police-brutality-report-brazilian-forum-for-public-security-a7386296.html

Design Problem 1

Problem Finding with Divergent Thinking—Horizon Scanning and Weak Signals

Archipelago of Design (AoD)

Oliver Jones, Kassie Miedema, and Philippe Beaulieu-Brossard

Introduction

Section I of *Modern War and Grey Zones: Design for Small States* describes the contemporary conflict environment: the war nexus, climate change, reliance on space, the vulnerability of cyber systems, pervasive market influences, and the role of violent non-state actors allied to state interests. Divergent thinking to elaborate problems, horizon scanning, and early identification of weak signals are valuable strategic planning and risk management design tools.

The first step in any design exercise is to define the problems or issues that need to be addressed. Avoid taking the initial problem statements for granted in order to expand the range of possibilities by divergent exploration of the problem. This means expanding the scope of design inquiry, gathering information, and envisioning or brainstorming many possibilities. Readers should select a topic that will enable them to develop an evolving understanding of a complex set of interrelated Grey Zone issues while doing the four design exercises. The design tools can be applied iteratively through divergent and convergent approaches to problem definition and solution-finding around the chosen topic. To aid this process, it is helpful to formulate a challenging question without an objective, ready-to-hand answer to orient the design inquiry. Three examples of challenging design questions are:

- How to reduce vulnerability to state-sponsored transnational organized crime,
- How to build social resilience against disinformation, and
- How to manage dependence on technology controlled abroad.

Each Design Problem or Design Exercise illustrates a process readers can adapt to their challenging question. This exercise introduces *horizon scanning* and *weak signals*, encouraging approaches that gather data and information from atypical sources to scope emerging issues.

Method

Horizon scanning is an intelligence-gathering method that systematically identifies emerging trends, potential threats, and opportunities in the security environment. Scanning for weak signals is the first step in a horizon-scanning exercise (AoD 2022:33; 2024:22–30). What is a weak signal?

Weak signals are the earliest signs of change when the overall pattern or trend they point to is not yet evident. In other words, weak signals hint at possibilities that may unfold in the future. Weak signals include data points, behaviors, scientific findings, news items, or cultural phenomena, which can be from social, technological, economic, environmental, political, or

DOI: 10.4324/9781003428701-10

cultural fields. These signals are a source of information about emergent possibilities. They can help strategists and planners anticipate and adjust to those changes if they can be translated into legible indicators of an emerging challenge or problem with the potential to disrupt ongoing plans or policies.

This exercise encourages paying attention to the fringes, to voices and trends that don't fit in with established fields of view. There's no easy way of telling where a weak signal may come from. Those looking for weak signals must be curious and open to unexpected inputs like blog posts, newspapers, and information feeds on the extreme ends of political spectrums. As a rule of thumb, weak signals should not get more than about 500 results on a search engine.

Exercise—Weak Signals

Scan available internet sources for weak signals following the Russian invasion of Ukraine in 2022 by narrowing the selected timeframe on the search engine. This constrains the search for events that followed Russia's invasion. Search parameters can be changed in the "tools" setting on Google. Think of the problem from multiple perspectives. How do bystanders and other parties affected by the conflict perceive the conflict after the invasion? Avoid limiting the search to obvious actors to counter your biases and the biases of your common sources. What are some players not directly linked to the Russia-Ukraine conflict? How is the ongoing conflict viewed in the Global South, Brazil, India, China, or the Organization of the Islamic Conference? On what assumptions did these perspectives rest? It may be helpful to sort these signals by frameworks like STEEPV (social, technological, environmental, economic, values) or PMESII (political, military, economic, social, information, infrastructure). This will help to identify blind spots in the search, which may be fruitful to explore further. Considering multiple perspectives will help you to appreciate the complexities of competing and conflicting viewpoints.

Seek out diverse sources and types of information. Consider news outlets, social media, radio stations, podcasts, streaming playlists, popular culture, technical magazines, blogs, art and literature, activist statements, economists, business analysts, and farmers. Narrow down the findings to three to five signals by selecting examples that don't fit with current assumptions. What might these signals hint at? Do they suggest a change in the environment? What normative expectations do these signals complicate or change?

Share the signals gathered with a small group and try to cluster any similar signals to identify possible trends over the period of the horizon scan. Are there any corresponding relationships between these weak signals? Do any of these relationships between signals suggest broader trends?

Group Exercise—Applying Weak Signals and Horizon Scanning

Analyzing weak signals may provide early warning of potential conflict and escalation of existing tensions. Horizon scanning may look for signals such as changes in popular culture, political rhetoric, military build-ups, and alliance changes, which often precede major geopolitical events. Divergent thinking and challenging assumptions can reduce the likelihood of errors and omissions, like the failure to predict the 1989 fall of Communism and the 2008 financial crash.

Task. After Russia invaded Ukraine in 2022, the signals indicating changes in arenas like transnational crime, the information domain, and technology sharing might be stronger than before 2022. Are these signals consistent with prior assumptions? Are there any trends in these areas that complicate current assumptions? What can we tell about using weak signals and horizon scanning by looking at open sources related to the Ukraine-Russia conflict after 2022?

Consider

- Transnational crime: Are there indications of increased cooperation, realignments, or rifts between criminal networks and state actors in Eastern Europe? What might these signals suggest about the evolution of organized crime in conflict zones?
- Information domain: How has the tone or frequency of information campaigns shifted in different regions post-2022? What narratives are prevalent, and which actors are amplifying them?
- Technology sharing: Are there signs of emerging technological alliances or restricted access to certain technologies in response to the conflict? How does this impact actors who may not be party to the conflict, like neutral or Global South actors?
- Economic interdependencies: Have new economic dependencies emerged due to sanctions, resource shortages, or supply chain disruptions? What trends are visible in markets and trade routes, especially in non-Western economies?
- Popular culture: Are there shifts in how popular culture—music, television, and social media—reflects attitudes toward the conflict or its actors? How might these shifts indicate broader social sentiments?

Implications. What are some of the implications of this exercise? Are there limitations like language barriers that might be influencing them? Do certain weak signals gain strength over time? How might the following functions be linked to horizon scanning for Grey Zone threats?

- **Early recognition:** Identify early indicators of destabilizing factors, from cyber risks to economic shifts.
- **Preemptive campaigns:** Develop proactive social media strategies based on potential emerging narratives.
- **Strategic communications:** Refine messaging in anticipation of evolving narratives in conflict zones.
- **Cyber defense readiness:** Adjust cybersecurity strategies in response to newly observed vulnerabilities.
- **Humanitarian preparedness:** Forecast and prepare for humanitarian challenges linked to the observed weak signals.

See the Archipelago of Design website for additional resources, https://aodnetwork.ca, as well as the *Horizon Scanning Report* published by the NATO Cooperative Cyber Defence Centre of Excellence for examples of this approach to strategy. For those developing Design programs in professional military education, consider the podcast *The Bold and Messy Roller-coaster Ride of Design Education for Military Officers* (Dupont 2024).

References

Archipelago of Design (2022), *Collaborative Innovative Thinking by Design for the Canadian Armed Forces*. Archipelago of Design. https://aodnetwork.ca/collaborative-innovative-thinking-by-design-for-the-canadian-armed-forces/

Archipelago of Design, Kerr, H. & Miedema, K. (2024), *Safeguarding the Future: A Foresight Handbook for Defence and Security Professionals*. Archipelago of Design. https://aodnetwork.ca/safeguarding-the-future/

Dupont, D. (Host) (2024), *Tension Podcast Ep. 7: The Bold and Messy Roller-Coaster Ride of Design Education for Military Officers* [Audio Podcast Episode]. In *Tension Podcast*. Archipelago of Design. https://aodnetwork.ca/tension-podcast-ep-7/

Further Reading

Archipelago of Design (2023), *Breakaway: Reframing to Prevail*. Archipelago of Design, https://aodnetwork.ca/breakaway-reframing-to-prevail/

Ertan A., Floyd K., Pernik P., and Stevens T. (eds.)/ NATO CCDCOE (2021), *Cyber Threats and NATO 2030: Horizon Scanning and Analysis*. NATO Cooperative Cyber Defence Centre of Excellence (CCDCOE) https://ccdcoe.org/library/publications/cyber-threats-and-nato-2030-horizon-scanning-and-analysis/.

Section II

Policy, Strategy, Operations, and Tactics

7 From Policy to Strategy and Tactics for the Grey Zone and Modern Wars

Jeffrey V. Gardner, Marzena Żakowska, and David Last

Introduction

The primary responsibility of any nation is protecting the state's sovereignty and people. Most states face constraints on their autonomy and must make difficult choices to protect vital interests in the face of demands by larger neighbors or major powers. These choices are increasingly complex as powerful states use Grey Zone operations to manipulate and undermine allies and adversaries. How do small countries establish policy, plan strategy, and develop the operational and tactical capabilities to pursue their strategy and achieve their policy ends? In the world of Grey Zone threats and modern wars that attack societies below the threshold of violence, effective intergovernmental cross-sectoral coordination is essential for implementing national security.

Policy Precedes Strategy

Policy describes what should be achieved, and strategy articulates how it should be achieved against opposition or adversity. National security policies are typically articulated in government statements or White Papers reflecting government preferences and priorities. These preferences and priorities emerge from the governing processes, in which legislative and executive roles vary according to the constitution and functioning of a country's government. The security policies of democracies will probably see less divergence between national and human security, while those of autocracies may see less divergence between national and regime security.

National security is broader than military security, typically including anti-terrorism and anti-crime policies, cybersecurity, environmental and health security, and responses to civil emergencies (Bailes 2015). It serves as an overarching concept most concerned with preserving the institutions of the state and autonomous decision-making but must also address the preservation of populations and human security, without which states and regimes fail (Reveron and Mahoney-Norris 2021:3, 34). The balance of these ends is a matter of national policy.

National security and defense policies are typically articulated in policy documents promulgated by central governments. Switzerland's Centre for Security Studies at ETH Zurich has compiled over 200 publicly accessible defense policy documents, White Papers, and reviews for more than 60 countries (Centre for Security Studies 2015). These policy documents assess domestic and international security challenges and articulate aims combining existing and anticipated capabilities.

Military and police staff colleges in many countries and many business schools teach that there is a hierarchy of policy, strategy, operations, and tactics. This hierarchy privileges top-down decision-making: leaders establish policies, devise strategies to achieve them, assign resources

DOI: 10.4324/9781003428701-12

to the conduct of operations, and employ tactics to overcome adversity or adversaries. At each level, organizational processes are established: strategic design, operational planning, and tactical estimates. The values and beliefs of security organizations (i.e., their organizational culture) generally reinforce this hierarchy. However, many people and organizations responding to Grey Zone threats have different values and beliefs. Legislative bodies, public administrations, emergency responders, civil society organizations, and the private sector are all part of the necessary whole-of-society response to threats below the threshold of war. The political-military hierarchy of policy, strategy, operations, and tactics must mesh with the decision-making web in the economic and social structures attacked by Grey Zone threats. The industrial-era management structures taught at staff colleges and business schools and practiced by major powers may be ill-suited to achieve this.

National security policy goes well beyond defense policy. It encompasses the governance and management of policing, security intelligence, health, education, welfare, and intercommunal relations for social cohesion. These connections can be easily overlooked in large states with overwhelming defense industrial establishments and thousands of defense planners. The firm separation of policy from strategy articulated in US doctrine and military management literature (Neill et al. 2017) is less obvious in smaller countries. National policies precede strategies, but there is a fluid relationship between them.

National Security Policy Choices

States live in neighborhoods of larger and smaller states. Geographical proximity and historical, cultural, economic, and political ties influence a region's security dynamics. The idea that a community of states with common values and objectives can largely set aside military conflict is well-established (Deutsch et al. 1957). Regional security communities emphasize the development of trust and confidence between neighbors. Regional security complexes emphasize the interconnected security dynamics of a region (Buzan and Waever 2003:18). Grey Zone conflicts and modern wars create new ways of pursuing conflicts and place new demands on national security policies to address challenges like survival migration, climate change, social fragmentation, information wars, and welfare in a global economy. Military threats overlay these new threats and transcend the boundaries of regional security complexes. National security policy is a survival strategy tailored to changing neighborhoods.

States may be small in terms of population, territory, or GDP. They may also be small in terms of others' perceptions or institutional capacity or in relation to other actors for some purposes, such as trade, investment, or military capacity (Long 2022:7). These variations affect national policies and the sources of power on which policies rest. Intrinsic power is derived from the characteristics of the state. Kuwait's oil wealth, or Panama's canal, are intrinsic power sources and can be used to threaten or promise. Derivative power stems from relationships with great powers. Canada's proximity and importance to the United States generate derivative power, which may be exercised through lobbying, framing, and manipulation of patronage. Collective power derives from joining forces with other small states, building institutions, norms, and coalitions. The creation of the International Criminal Court (ICC) represented a victory for collective power (Long 2022:60).

Small states have four main alignment options, either within alliances or outside them. *Bandwagoning* involves combining forces in alliances or coalitions of convenience to pursue common goals; this is a "with us or against us" strategy (Vaicekauskaitė 2017). *Hedging*, in contrast, involves insurance-seeking behavior, pursuing opposing measures with different actors (Kuik 2021). *Balance of power* entails contributing to one party to avoid being overpowered by another. Finally, *isolation* involves attempting to avoid engagement in conflicts. Faced with the external shocks of 9/11, Russia's invasion of Georgia in 2008, the Da'esh Caliphate, and Russia's attack on Ukraine in 2014, small Nordic states (Denmark, Sweden, Norway, and Finland) adopted each of these strategies at different times (Edström, Gyllensporre and Westberg 2019:183). After Russia's 2022 invasion of Ukraine, Finland and Sweden moved from hedging to bandwagoning, seeking NATO membership. Austria, Ireland, and Switzerland, however, maintain neutrality for reasons that can be described in a Realist framework (Simpson 2018).

Strategy in the Grey Zone

Strategy comes from the Greek word *strategos*, the art of leading an army. Like its cousin, intelligence, it has accrued many adjectives in migrating from military to civilian spheres; there is a strategy to everything (Heuser 2010). The highest level for states is Grand Strategy—using all instruments of power in peace and war to achieve national objectives (Hamilton 2017:6). Physical and political geography frame debate about policy and Grand Strategy, including opportunities, challenges, and dangers (Edström, Gyllensporre and Westberg 2019:23), but culture, ideologies, and elite interests also feature. Grand strategies can be thought of as the survival strategies of states, paring questions down to the essential elements of enmity, alignment, peace, and war (Kassab 2018:9). Most states must approach these questions with an eye on their relations with neighbors and major powers. Central to Grey Zone conflict is defining the onset of war. If the vital interests of a state can be attacked below the threshold of war, how should a national strategy respond?

A strategy is a type of problem-solving. It is a method to address challenges or threats. The kernel of strategy is a diagnosis of the problem, a guiding policy to overcome it, and a set of coherent actions to achieve the objective (Rumelt 2011). Theories of strategy taught in the US Army War College emphasize using power to advance national interests (Yarger 2006:1). This is not a dominant theme in the national security policies of the 60 countries and 204 policy documents compiled by ETH Zurich's Centre for Security Studies. Rather, we find consistent references to domestic and international contributions of military forces, cooperation with society, assistance to civilian authorities, and disaster assistance. Many national policy documents refer to international deployments, usually in the context of Alliance, UN, or regional operations. Other common references include national cohesion and unity, the rule of law, and the protection of institutions. All of these are targets of Grey Zone threats.

The Zurich Centre's policy documents date from 1998 to 2015, postdating the Cold War. They frequently comment on the amorphous, unstable, multipolar international environment, new threats, and environmental, social, and economic uncertainties. These are not construed as exclusively military problems, even in defense department documents. Strategy in the Grey Zone is a whole-of-society problem.

Hierarchy or Web of Strategies?

If a nation's grand strategy is the overarching concept for national security across all means and domains, subordinate national-level strategies naturally follow. While hierarchy suggests one strategy is subordinate to another, a web suggests mutual dependence. Common human and national security strategies include health, education, welfare, housing, immigration, and more specialized strategies for anti-terrorism, anti-crime, cybersecurity, environmental and health security, and responses to civil emergencies. Health, housing, and immigration strategies are interdependent. Treating public welfare as a component of national security refers to the social contract that legitimizes the political and economic order. Religions, ideologies, and accepted narratives can sustain political and economic orders that leave many poorly served so that national security can diverge from human security.

The thread uniting the hierarchy of strategies is national interest. Like grand strategy and security, "national interest" is a contested concept. If states define their national interests in terms of survival as independent political units, then their grand strategies promote survival against threats not limited to military attacks (Kassab 2018:59). The priority of subordinate strategies—health, education, welfare, domestic security, or external defense, for example—is determined by the factors most threatening state survival. These will vary. Common interests for small states have been identified as the protection of citizens, political independence, territorial integrity, and economic prosperity (Edström, Gyllensporre and Westberg 2019:21). The possibility of a military attack above the threshold of war makes military defense the top priority. Bandwagoning with a major power generally lowers the priority of military responses.

Grey Zone threats and modern wars waged below the threshold of direct attack undermine assumptions about the military defense of interests. During the Cold War, American political scientist Donald Neuchterlein described a matrix of survival, vital, major, and peripheral interests in four categories: homeland, economic well-being, favorable world order, and promotion of values (Sklenka 2007:4). Survival interests like the willingness of groups to live together in a state can be eroded without military attack and cannot be defended with military force. Small states cannot pursue a favorable world order with military means; they must deal with climate change impacts, survival migration, and economic effects of globalization, which threaten their stability and well-being. How are they to prioritize strategies and resources to address these threats?

Strategic Threat Assessments

Guidelines for strategic threat assessments developed for defense policy and military strategy focus, not unreasonably, on military commanders' assessments of military threats. The process is reasonable, but the focus is no longer adequate. Rumelt's *kernel of strategy approach* begins with a diagnosis. In the era of Grey Zone threats and new wars, strategic threat assessments must address political, economic, social, technological, legal, and environmental factors (PESTLE). NATO doctrine reflects this in Joint Publication 01 (2017:1–5). Beyond NATO, the Organization of American States' *Guidelines on Developing National Defense Policy and Doctrine Papers* (OAS 2002) suggest outlining the security landscape, including domestic and international aspects and traditional and non-traditional security threats and factors.

Multiple ministries and government agencies contribute to assessing internal and external factors to diagnose the overarching strategic environment. In today's context, assessing vulnerability and resilience for any state must encompass a wide spectrum of non-military threats and risks. These range from intentional and man-made, such as terrorism, crime, and cyber-attacks, to purely natural events like disasters and pandemics. Additionally, economic, financial, and social vulnerabilities are integral aspects that shape a state's security landscape (Bailes 2011).

The European Union has crafted an effective approach to National Risk Assessment (NRA) summarized in Figure 7.1. It follows international standards and is focused on disasters but can be applied to the full spectrum of threats (Poljanšek, Valles and Ferrer 2019). Different ministries and organizations conduct different assessments, and these must be reconciled.

Ends, Ways, and Means or Objectives, Concepts, and Resources

Ends refer to the persistent objectives that a state seeks to achieve. Ways are the different approaches or options to achieve those objectives. Means are the capabilities and resources that will be used

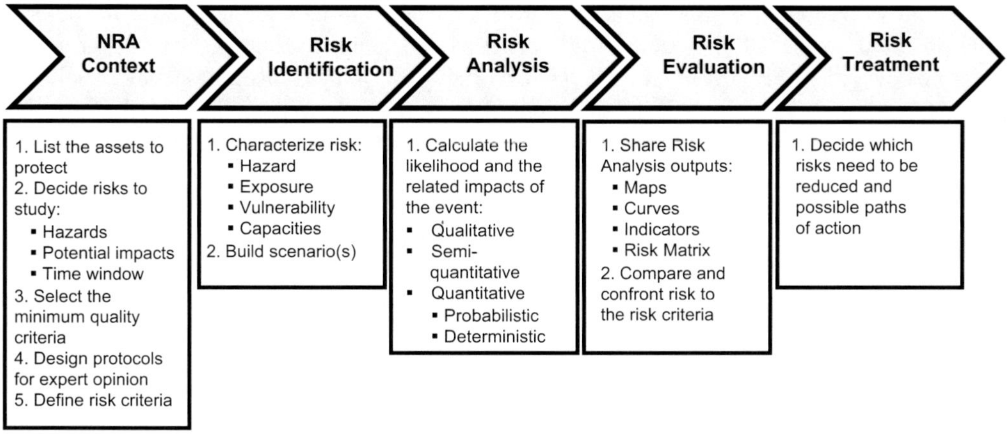

Figure 7.1 National Risk Assessment (NRA).

Source: Poljanšek, Valles and Ferrer (2019).

to achieve them. The terms correspond to objectives, concepts, and resources. A national strategy defines what is to be achieved (ends or objectives), how it is to be achieved (ways or concepts), and what will be used to execute the strategy (means or resources). Generations of staff officers have exercised their creativity in exploring and diagramming the relationships between objectives, concepts, and resources. All conclude that these three must be coherent for a strategy to work. Objectives without a realistic way to achieve them or the resources to implement the strategy will fail. Yet, the most common problem for most countries is inadequate resources to execute a strategy.

Perennially inadequate resources indicate that objectives and concepts need to be reconsidered. Military planners who cannot reconcile themselves to the available forces should focus on other instruments that contribute to strategic objectives; for most countries, traditional military power may be marginal or irrelevant to defending vital national interests or the state's survival. The Grey Zone and modern wars have accentuated this without making military power irrelevant. The instruments of power are not only resources or means; they also provide concepts or ways to achieve objectives. They might indicate ends or objectives themselves. Lawfare may be a means of pursuing objectives, but undermining the rule of law or confidence in the legal system might be a strategic objective for a hostile power.

With this in mind, we can compare two ways of thinking about the resources available to strategists in the Grey Zone. PESTLE and DIMEFIL are summarized in Table 7.1. These are not equivalent. On the left, DIMEFIL describes instruments of national power and influence described in US sources. On the right, PESTLE lists factors in conducting national security assessments for states facing Grey Zone threats. These factors can be approached with offensive or defensive operations in mind. Combining the DIMEFIL instruments and PESTLE vulnerability factors, planners can derive strategies that are part of a hierarchy of responses to achieve national security, human security, and the management of the international order. Defining the threshold for war—a fundamental change in relations with another state—is always a consideration.

Assessing a Strategy

Strategies are creative acts of hypothetical analysis: if we do this, then that will happen (Hoffman 2020). The range and complexity of factors listed in Table 7.1 and their continuous and complex interaction demand multiple strategies and contingency plans.

Feasibility, acceptability, and suitability are tests to determine whether a strategy belongs in the potential arsenal of approaches to problems. The feasibility test asks, *can the action or strategy as conceived be carried out with the resources available*? If resources are not available, the strategy is moot. The acceptability question asks, *are the costs of the action or strategy justified by the importance of the effect sought*? Finally, the suitability test asks, *will this action or strategy achieve the desired objective*? (Yarger 2006).

Any strategy may be inadequately resourced, ill-conceived, or ineffective in achieving the intended objectives. Adversary strategies or adversity will play a role in this. Red teaming and alternative analysis are tools to test a strategy. Mimetic, or traditional, red teaming involves mimicking a possible adversary to test a strategy or organizational vulnerability. Diegetic red teaming, or alternative analysis, involves setting up separate cells to rethink a strategy and test assumptions (Lauder, Eles and Banko 2011:10). With the rise of Grey Zone threats, red teaming across the broad spectrum of factors in Table 7.1 is an essential step to assess strategy.

Operationalizing Strategy: Actions at the Operational and Tactical Levels

A national strategy is more a living process than a document. Approval is the starting point for further analysis, planning, implementation, and adjustment to evolving circumstances. Ministers and government officials deploy resources to implement national goals. Assessment and adaptation are important at each level. The image of an omniscient source of national strategy and a cascading response by subordinate levels of government is inappropriate for small states responding to Grey Zone threats and modern war; adaptation goes up, down, and laterally as different agencies consider the implications of new factors and situations.

Ireland's National Implementation Plan for Sustainable Development Goals is one example of an evolving national strategy. The 2022 document lists 5 strategic objectives, 51 actions, and 119 individual measures related to Sustainable Development Goals agreed upon internationally at the United Nations. Every government department was involved, with statements from the Prime Minister (Taoiseach), his deputy (Tánaiste), and ministers for environment, education, agriculture, overseas development, heritage, procurement, health, and finance (Martin, Varadkar and Ryan 2022). The keystone document, *Ireland's Second National Implementation Plan for Sustainable* Development Goals, 2022–*2024*, mentions contributions to international peace through women, peace, and security as an example of its international approach to sustainable development goals (Government of Ireland 2022:24). The vertical integration of policies from global to European down to local authorities is matched by horizontal integration of ministries and fields of government authority. National strategies for sustainable development can be seen as efforts to achieve positive peace, with each component increasing resilience against Grey Zone threats. Most action items fall in the interconnected categories of economic, social, and environmental factors listed in Table 7.1. An economic recovery plan, housing strategy, social inclusion, and well-being framework bolster national cohesion and human security.

National strategies for election security, cyber defense, border security, and legal integrity are among the many demands of Grey Zone threats. There is growing awareness that international cooperation and mutual support for sustainable development are also required for national cybersecurity strategies, extending beyond states to the private sector and non-state actors (Oruj 2023). Small states facing domestic labor pressures, ethnic affinities that transcend borders, and regional economic and political dynamics need regional strategies to manage border security and develop protocols on border access (Omoyeni 2023:40). Adversarial action can disrupt these strategies and demand adaptation. Russia's involvement in West Africa and withdrawal from ECOWAS by Niger, Mali, and Burkina Faso require not only a revision of border access

Table 7.1 Instruments of Influence and Factors in Security

DIMEFIL *Instruments of national power and influence* *(How can a state attack or be attacked)*	*PESTLE* *Factors in Grey Zone Threats* *(State assessment for defense and security)*
Diplomatic – Representation – Negotiation – Implementation Informational – Perceive – Inform – Manipulate Military – Force – Threat of force – Force enabling Economic – Assistance – Trade – Finance Financial – Formal and informal financial systems – Restrict or enable payments – Target individuals or corporations with sanctions, embargoes, and restrictions Intelligence – Clarify of create ambiguity --Leak intelligence – Shape options for allies Law-enforcement – Arrest, detention, and lawsuits – International arrest warrants – Rule-of-law and policing assistance	PoVlitical – Governance structures, succession – Elite networks, concentration – Three faces of power (Lukes 1975) – Threat or use of force Economic – Production, consumption, and supply – Employment – Income distribution, equality – Quality of life Social – Affiliation, disposition – Cohesion, identity, and polarization – Ethnic/linguistic/religious tensions – Migration/refugee crises – Public health – Mass media, social media, and information – Demographic shifts Technological – Ownership and intellectual property – Cybersecurity, digital threats – Surveillance, intelligence tools – Impact of AI, quantum computing – Secure encryption – Communication networks, infrastructure Legal – International treaties and agreements – National security/human rights laws – Data protection and privacy – Law enforcement powers – Emergency response laws – Immigration and border security laws – Martial law and military law – Legislative oversight Environmental – Climate change, energy security – Pollution, marine pollution – deforestation, biodiversity, and land degradation – Natural disasters – Illegal wildlife trade

Source: Table compiled by Authors from Air University Library (2024), Arkman (2020) and source documents for Tabular Summary of Grey Zone Threats (Appendix).

protocols in remaining ECOWAS states but also a wider re-examination of defense and development strategies, including domestic security arrangements (Heldt 2024). Electoral and legal integrity strategies fall within the OECD strategy for public integrity. Whole-of-society strategies extend beyond the public sector to encompass the private sector and civil society organizations (OECD 2020). Risks to public integrity not only can emerge from within the government and from political actors but can also be triggered by adversarial actions. False accusations, erroneous whistleblowing, deepfakes, and disinformation erode public trust.

The political divisions of federal states may confound tactical actions to implement national strategies. Even unitary states face the silos of local jurisdictions which demand acquiescence if not active cooperation. Much of the information and resources necessary for any national strategy may be controlled at local levels, sensitive to local incentives, dependent on inadequate local resources, and subject to local leadership and political influence. National assets like military forces, border services, and security intelligence intersect with the web of local assets like police and fire services, health, education, and welfare services, privy to important local knowledge. Access to information and privacy laws can make sharing information difficult.

Interagency Collaboration

Few plans can be implemented by a single government ministry. National security requires communication, coordination, and cooperation across the public, private, and civil society sectors. The complex security environment involves multiple evolving interactions, demanding enduring, scalable, and transformative partnerships between diverse actors. The boundaries between sectors have become blurred as governments pull back from some business and civil society governance responsibilities (Kwibisa and Majzoub 2017:1).

Unfortunately, most countries do not have cohesive intergovernmental and cross-sector collaboration. Explanations include legal mandates, policy conflicts, procedural differences, security sensitivities, physical/virtual impediments, risk aversion, inadequate incentives, organizational culture, and personality issues (US Army Asymmetric Warfare Group 2009). These challenges can be grouped as technical, organizational, and political factors (Dawes 1996:384).

Barriers to effective interagency coordination are not insurmountable. Intelligence fusion centers enable collaboration on specific threats such as money laundering, counterterrorism, or border security. They began by combining and analyzing threat information but have expanded into coordinating national and international security and intelligence cooperation and implementing national action plans (van der Veer, Bos and van der Heide 2019:13).

Finland has developed a model for effective intergovernmental and cross-sector collaboration below the national level, pushing security coordination down to the lowest level. Each of Finland's 18 regions has a security cluster in which police, rescue services, border guards, defense forces, private businesses, and civil society actors like the Red Cross and church officials meet regularly to coordinate actions. Whether in peace or crisis, the rescue service calls participants to regular meetings as the key facilitator. Joint plans are discussed for exercises or during crises, and the competent authority leads cross-sector execution. The security cluster in South Karelia in Finland's southeast bordering Russia includes customs, police, defense forces, church, Red Cross, social and health institutions, municipal authorities, and rescue services. For coordination to work, meetings are held regularly, key players are educated on specialist courses, and trust is built in regular exercises. Citizens are a vital part of the security concept (Valtonen 2017).

Conclusion

Small states have some advantages over their larger neighbors in pursuing whole-of-society security in the Grey Zone below the threshold of war. They grapple with factors as many and varied as those confronting large states, but their key players are more likely to know each other. Coalitions, alliances, and bilateral arrangements can expand their potential responses. Many of their engagements are likely to be optional. They can draw on intrinsic, derivative, and collective power and choose between bandwagoning alliances, hedging, balancing, and isolation strategies. Grand strategies are the survival plans of small states; successful strategies engage the whole of society. From the grand strategy, a web of related strategies addresses security aspects for which there is no military defense. Defense policy often entails deployment at home and signaling with small discretionary expeditions abroad. Holistic National Risk Assessment (NRA) and consideration of threats and opportunities across the PESTLE and DIMEFIL spectra fuel a continuous creative process of generating and assessing strategies. Making a strategy work in the Grey Zone requires collaboration and sharing across organizational lines and between government, the private sector, and civil society organizations.

References

Air University Library (2024), DIMEFIL: *Instruments of Power*, Fairchild Military Library Guides. https://fairchild-mil.libguides.com/dimefil

Arkman Kubilay M. (2020), "Defence Management and PESTLE Analysis", *Ante Portas - Security Studies*, vol. 14, no. 1, pp. 93–102.

Bailes, A. J. K. (2011), *New Security Challenges of Small States*, European Consortium for Political Research. https://ecpr.eu/Events/Event/PanelDetails/1589

Bailes, A. J. K. (2015), *Parliaments and National Strategy Documents: A Comparative Case-Study from the Nordic Region*, Policy Paper no. 36, Geneva Centre for the Democratic Control of Armed Forces (DCAF). https://www.dcaf.ch/parliaments-and-national-strategy-documents-comparative-case-study-nordic-region

Buzan, B. and Waever, O. (2003), *Regions and Powers: The Structure of International Security*, Cambridge University Press, Cambridge.

Center for Security Studies at ETH Zurich (2015), *Defense White Papers and National Security Strategies.* https://css.ethz.ch/en/services/digital-library/series.html/154839

Dawes, S. S. (1996), "Interagency Information-sharing: Expected Benefits, Manageable Risks", *Journal of Policy Analysis and Management*, vol. 15, no. 3, pp. 377–394.

Deutsch, K. W., Burrell, S. A., Kann, R. A., Lee, M. Jr, Lichterman, M., Lindgren, R. E., Loewenheim, F. L. and van Wagenen, R. W. (1957), *Political Community and the North Atlantic Area: International Organization in the Light of Historical Experience*, Princeton University Press, Princeton.

Edström, H., Gyllensporre, D. and Westberg, J. (2019), *Military Strategy of Small States: Responding to External Shocks of the 21st Century*, 1st ed., Routledge, New York.

Government of Ireland (2022), *Ireland's Second National Implementation Plan for Sustainable Development Goals, 2022–2024.* https://www.gov.ie/en/department-of-the-environment-climate-and-communications/publications/national-implementation-plan-for-the-sustainable-development-goals-2022-2024/

Hamilton, R. E. (2017), *National Security Policy and Strategy*, US Army War College, Department of National Security and Strategy, Carlisle, PA.

Heldt, A. (2024), *Russian Advance in Africa? Interview on the Planned ECOWAS Exit of Niger, Mali and Burkina Faso*. Friedrich Naumann Foundation for Freedom. https://www.freiheit.org/sub-saharan-africa/russian-advance-africa

Heuser, B. (2010), *The Evolution of Strategy: Thinking War from Antiquity to the Present*, Cambridge Univer sity Press, Cambridge.

Hoffman, F. G. (2020), "The Missing Element in Crafting National Strategy: A Theory of Success", *Joint Force Quarterly*, vol. 97, no. 2, pp. 55–65.

Kassab, H. S. (2018), *Grand Strategies of Weak States and Great Powers*, Palgrave Macmillan, Cham.

Kuik, C. C. (2021), "Getting Hedging Right: A Small-state Perspective", *China International Strategy Review*, vol. 3, no. 2, pp. 300–315.

Kwibisa, N. and Majzoub, S. (2017), *Challenges Faced in Inter-organizational Collaboration Process: A Case Study of Region Skåne*, Malmö University, Malmö, Sweden.

Lauder, M. A., Eles, P. and Banko, K. (2011), *The Glaucus Factor: Red Teaming as a Means to Nurture Foresight*, *Canadian Army Journal,* vol. 14, no. 1, pp. 45–59.

Long, T. (2022), *A Small State's Guide to Influence in World Politics*, Oxford University Press, Oxford.

Lukes, S. (1975), *Power: A Radical View*, Humanities Press, New Jersey.

Martin, T. M., Varadkar, T. L. and Ryan, E. (2022), *Cross-government Launch of New National Implementation Plan for the Sustainable Development Goals 2022–2024.* https://enterprise.gov.ie/en/news-and-events/department-news/2022/october/cross-government-launch-of-new-national-implementation-plan-for-the-sustainable-development-goals-2022-2024.html

NATO Standardization Office (2017), Allied Joint Doctrine, Allied Joint Publication 01 (AJP-01).

Neill, M., Taliaferro, A. C., Tillman, M. E., Morgan, G. D. and Hinkle, W. P. (2017), *Defense Policy and Strategy*, in M. Neill, A. C. Taliaferro, M. E. Tillman, G. D. Morgan, and W. P. Hinkle (eds.), *Defense Governance and Management: Improving the Defense Management Capabilities of Foreign Defense Institutions,* Institute for Defense Analyses, Paper NS P-5350 (March 2017), pp. 1–22. https://www.jstor.org/stable/resrep22899.1

OAS (2002), *Adoption of the Guidelines on Developing National Defense Policy and Doctrine Papers*. Permanent Council of the Organization of American States, Committee on Hemispheric Security. https://2009-2017.state.gov/p/wha/rls/70119.htm

OECD (2020), *Public Integrity Handbook*, OECD Publishing. https://www.oecd.org/en/publications/oecd-public-integrity-handbook_ac8ed8e8-en.html

Omoyeni, T. J. (2023), "Economic Community of West African States (ECOWAS) Protocol on Free Movement and National Security: A Case Study of Nigeria", *International Journal of Development and Economic Sustainability*, vol. 11, no. 3, pp. 39–54.

Oruj, Z. (2023), "Cyber Security: Contemporary Cyber Threats and National Strategies", *Distance Education in Ukraine: Innovative, Normative-Legal, Pedagogical Aspects*, vol.1, no.2, pp. 100-116.

Poljanšek, K., Valles, A. C. and Ferrer, M. M. (2019), *Recommendations for National Risk Assessment for Disaster Risk Management in EU: Approaches for Identifying, Analysing and Evaluating Risks*. The European Commission's Joint Research Centre (JRC).

Reveron, D. S. and Mahoney-Norris, K. A. (2021), *Human and National Security:* Understanding Transnational Challenges, Routledge, New York.

Rumelt, R. P. (2011), *Good Strategy, Bad Strategy: The Difference and Why It Matters*, Crown Publishing, New York.

Simpson, A. W. (2018), "Realism, Small States and Neutrality", in D. Orsi, J. R. Avgustin, and M. Nurnus (eds.), *Realism in Practice: An Appraisal*, E-International Relations, pp. 119–130. https://www.e-ir.info/wp-content/uploads/2018/01/Realism-in-Practice-E-IR.pdf

Sklenka, S. D. (2007), *Strategy, National Interests, and Means to an End*, US Army War College, Strategic Studies Institute, Carlisle, PA.

US Army Asymmetric Warfare Group (2009), *Interagency Teaming to Counter Irregular Threats*, Fort Meade. https://publicintelligence.net/ufouo-u-s-army-interagency-teaming-to-counter-irregular-threats-handbook/

Vaicekauskaitė, Ž. M. (2017), "Security Strategies of Small States in a Changing World", *Journal on Baltic Security*, vol. 3, no. 2, pp. 7–15.

Valtonen, V. (2017), *The Finnish Concept of Comprehensive Security*, The Security Committee of Finland, 14 October. https://www.defmin.fi/files/3827/Valtonen_2017_06_14_FI_Concept_for_Comprehensive_Security_Valtonen.pdf

Veer van der, R., Bos, W. and van der Heide, L. (2019), *Fusion Centres in Six European Countries: Emergence, Roles and Challenges*, Report (February 2019), International Centre for Counter-Terrorism – The Hague (ICCT). https://icct.nl/publication/fusion-centres-six-european-countries-emergence-roles-and-challenges

Yarger, H. R. (2006), "Toward a Theory of Strategy: Art Lykke and the Army War College Strategy Model", in J.B. Bartholomees, Jr. (ed.), *US Army War College Guide to National Security Policy and Strategy*, US Army War College, Carlisle, PA, pp. 107–113.

8 Crisis Management in the Grey Zone

Ilkka Tikanmäki and Harri Ruoslahti[1]

Introduction

War is armed political violence implying kinetic attack and national effort. Hybrid warfare combines military and non-military means with regular and irregular conflict. It can remain below the threshold of recognized war within the Grey Zone of conflict. A crisis is a time of intense difficulty or danger, but the word originates from the Greek word *krisis* which means decision. Since the 17th century, it has been linked to the idea of a decisive point. Crisis management in the Grey Zone is characterized by managing the risk of escalation to war while minimizing the damage to human and national security from adversary action below the threshold of national mobilization for war. The decisive point might be when Grey Zone threats accumulate to cause state collapse or enable an invasion, but crisis management requires intervention well before this decisive point. Management is continuous in the war nexus described in Chapter 1. A second threat category is that crisis management of natural adversity might distract authorities, opening vulnerabilities to hostile actors.

This chapter considers the types and levels of crises and strategies for detecting, deterring, and responding to crises when hostile actors seek to exacerbate them. A hostile actor may use natural adversity to weaken a target or launch a conventional attack or generate crises to achieve the same aim. Small states face distinct challenges due to limited resources and geopolitical vulnerability. Some states have concentrated security infrastructure with little redundancy. They may not be able to count on allies or neighbors for support. A strategy to ensure effective crisis management includes all-hazards intelligence and surveillance (Chapter 9), cyber-resilience (Chapter 4), public awareness and education, robust legal and regulatory frameworks (Chapter 14), interagency coordination, crisis communication strategies, and community engagement. International support is critical. Small states in Europe can count on the EU and NATO. Other regions may need to develop cooperation mechanisms through regional organizations. The evolving nature of hybrid threats demands continuous adaptation, regular training, and awareness of new technology.

Crises and Crisis Management

A crisis is an "uncertain situation possessing latent risks and opportunities that must be resolved within a given timeframe" (Canyon 2020:6). Crises can create threats to public safety, financial loss, and loss of reputation (Coombs 2007). Effective management of natural or man-made catastrophic events is increasingly important (Moon et al. 2020), and humanitarian crises require a timely response to reduce human suffering and maintain the legitimacy of government. The central themes of crisis management for small states facing hybrid threats are effective

DOI: 10.4324/9781003428701-13

coordination, information sharing, and interagency cooperation within states and a regional international community.

Crises may arise from *adverse circumstances* such as fires, floods, famines, or storms. They may arise from *intentional actions* by adversaries, such as sabotage, subversion, or cyberattack. A third category between adversity and adversarial action arises when a hostile actor seeks to *take advantage of a crisis* of adversity, undermining the response or magnifying the effects in its interests. A challenge with identifying opportunistic adversarial contributions to a crisis is that democratic societies are inherently adversarial. Adversarial actors taking advantage of opportunities are expected in a multiparty system, a capitalist economy, and an independent legal system. Separating opportunism that is hostile to national interests from self-interested opportunism is one of the challenges of crisis management. These categories are illustrated in Table 8.1.

The geographical locus of a crisis is also relevant. Some affect multiple states and require international coordination. Others occur abroad and demand domestic political attention to manage national interests. Governments typically organize institutions to manage national crises, particularly those affecting national power, capabilities, and institutions. Federal structures, rural territories, and capable cities will also be set up to manage local crises, but jurisdictional issues demand vertical and horizontal coordination.

The Uppsala Conflict Data Program recorded 82 violent crises or wars in 2010 and 169 in 2020 (41 one-sided, 72 non-state, and 56 state-based), so the number of these violent conflicts has risen during the last decade (Uppsala Universitet 2021). All such violent conflicts require some measure of military or civilian crisis management. Manipulation or hostile action directed at a government or its ruling party can cause a national security crisis. Democracies cope more successfully with attacks on institutions because their courts, media, ombuds, and civil society organizations have a longer and better developed tradition of independence and professionalism (Bugarič 2015:219). Countries with weak institutions are in danger of drifting toward authoritarianism.

Table 8.1 Types and Levels of Crisis

	Circumstantial crises	*Opportunistic crises*	*Adversarial crises*
Characteristics	Originate in adversity not hostility; there's no one to blame	Originate in adversity but accentuated by hostility	Originate in intentional hostile action, though this may be concealed
International—involving multiple states	European heat wave in 2003; harsh winter	Market manipulation or sabotage of power infrastructure to exacerbate impact	Invasion or attack coinciding with weakness or adversity
Foreign—one state or area abroad	Famine in East Africa; climate-induced survival migration	Channeling or enabling refugees to target countries	Hiding agents in refugee flows; disinformation to accentuate social unrest
National—within the state	Commercial take-over of national asset (e.g. sale of Royal Mail to a CZ magnate)	Foreign ownership, control, and influence of vital assets	Lawfare to obscure source of funds, manipulation of key leaders
Local—affecting only a part of the state, individuals, and communities	A judge's indiscretion on a key political case	Use of a real incident in an information campaign	*kompromat*, false evidence, and removal of alternative players

Source: Elaborated by the Authors and Editors.

Crises may occur under conditions of peace but require civil-military coordination. The tide of migrants entering Southeastern Europe, even when not supported by hostile powers, demands close coordination between civil authorities, coast guards, border patrols, and naval forces. Modern warfare demands coordination between military and civilian crisis management organizations. Russia's use of migration as a hybrid weapon challenges the EU and NATO because populations are already sensitive to migration (Hall, Fleming and Shotter 2021).

Regional security communities have mechanisms to manage conflict that avoid using force. Still, force is a legitimate tool to manage conflict (Acharya 2009:22). Conflicts may escalate into crisis and war (Swanström and Weissmann 2005). In the relational framework for crisis management, pre-crisis management consists of crisis preparedness and prevention, while crisis management consists of crisis incident and post-crisis management (Jaques 2007:150). UN Clusters are an example of how to clarify responsibilities when coordinating emergencies and addressing the gaps and overlaps in assistance. UN Clusters include protection, shelter, health, and logistics sectors for humanitarian action (Stumpenhorst, Stumpenhorst and Razum 2011:589).

Both NATO and the EU find it challenging to develop capacities and strategies to manage international crises in the face of hybrid operations. For example, to manage the southern migration crisis, they must develop a maritime strategy, cybersecurity strategy, public communications, and a broader "global" strategy, synchronize all these aspects, and apply them in a tailored way to evolving situations (Andersson and Tardy 2015:3; Aaronson et al. 2011:113). Crisis management mechanisms must react quickly to situations where lives are on the line, and even minutes may count. Responses to cyberattacks, sabotage, and acts of warfare call for flexibility; training helps people become more aware of the elements of hybrid influence to be better prepared to deter and respond to its effects (European Parliament 2021:104–111).

Crisis management is needed for people and states in crisis areas to cope with the consequences of natural disasters, political turmoil, and war (Hyttinen and Kallonen 2018:71). Communication and cooperation between the diverse personnel of international and local organizations can increase understanding to improve performance of the crisis management mission (Holohan 2019:2). Hybrid warfare has increased both conventional and unconventional forms of conflict affecting crises (Lasconjarias and Larsen 2015).

The European Common Security and Defense Policy is an example of a regional mechanism that can increase collaboration through shared threat assessment, collective decision-making, processes, and operations, which helps promote increased solidarity among member states (Tardy 2015:35). Interoperability challenges are likely to increase in the new security environment. Greater integration strengthens civil and military responses by pooling and sharing resources (Hyttinen, Hario and Österlund 2017:60).

Information sharing and collaboration are important in fighting hybrid influencing and warfare (Cîrdei and Ispas 2017:73–77). Hybrid influence may be kinetic or non-kinetic, targeting physical, information, or psychological consequences (Ducheine 2016:201–202). Hybrid threats increasingly entail cyber threats and disinformation due to the growing role of digital solutions. The targets may be military, civilian administration, critical infrastructure, business, local government, or citizens (Cederberg and Eronen 2015:9). Because it has so many different facets, some aspects of hybrid influencing can be very hard to detect and, without well-coordinated responses, even harder to counter.

Planning crisis management should consider all sectors to promote smoother cooperation between civilian, military, public, private, and civil society organizations. Military authorities have become increasingly involved in non-military missions such as disaster relief (Weijers 2015:7–8). However, the military relies increasingly on the private sector and civil society organizations.

Interaction between civilian actors and military authorities in disaster relief missions can be complex (Tardy 2015:36). Crisis management is becoming more connected. The UN, EU, NATO, and many other networks in the global community work toward joint goals. Coordination on a regional level can better ensure that disaster relief is tailored to victims' needs (Gross and Juncos 2010:146).

Crises, then, include national and international situations demanding extensive cooperation by national assets and coordination in regional and international bodies. Managing crises can be costly and politically risky, even without hostile hybrid operations. Adding hybrid threats raises the potential for damage to national interests and dissipation of national capabilities. One consequence is that small states with limited capacity may be less willing to contribute to international humanitarian efforts for fear that their capacities will be needed at home.

Hybrid Threats, Hybrid War, and Crises

Hybrid threats can exacerbate ongoing crises and generate new forms of crisis. Means include military incursion and sabotage, cyberattacks, disinformation, and proxies. Hybrid threats are intended to be difficult to detect and respond to because they exploit the Grey Zone between peace and war. Hovering on the threshold of war, hybrid warfare employs conventional kinetic military tactics like sabotage and deniable cross-border incursions and non-kinetic means such as cyberattacks, disinformation, economic pressure, and subversion. Hybrid warfare aims to destabilize and coerce a state while avoiding a military response (NATO 2024). Open communication, good relationships, joint plans, and flexibility are required to manage hybrid crises (Paturas et al. 2016). International cooperation, strong social resilience, identifying common vulnerabilities, and preparing for disruptions are key for combating hybrid threats, especially against critical infrastructure (European Commission 2018).

The terms hybrid warfare and terrorism are ambiguous and may have contradictory interpretations (Gasztold and Gasztold 2020). The evolution of viewpoints on non-kinetic warfare influences definitions of hybrid warfare (Monaghan S./MCDC 2019; Lehto and Limnéll 2017). Hybrid warfare includes deliberate and synchronized campaigns to gain political leverage over opponents (Ducaru 2016). These campaigns may seek to precipitate cascading crises that overwhelm a state's ability to respond, or they may seek to avoid crises that would generate an undesirable reaction, such as a heightened security posture. Traditional warfare may be preceded by hybrid warfare (Lehto and Limnéll 2017), including combinations of elements like coercion, extortion, bribery, lying, proxy wars, psychological manipulation, and propaganda (Raitasalo 2017). These tactics can exacerbate crises, creating opportunities for an adversary to exploit traditional military power.

Hybrid approaches weaken the position of the target country by using all possible means, including information management, media, and even traditional military force combined with irregular forces used by state and non-state actors (Lasconjarias and Larsen 2015). The United States sees hybrid war largely from a military perspective. Smaller countries have more holistic perspectives; the response to threats and crises in the Grey Zone should be whole-of-society or at least whole-of-government.

The US-led Multinational Capability Development Campaign (MCDC) provides relevant models for smaller countries, although designed to work within a NATO framework. The three components of the MCDC framework are detect, deter, and respond (Figure 8.1). These can guide phases as appropriate thresholds for strategic countermeasures to combat hybrid aggressors and prevent, suppress, and disrupt attacks (Cîrdei 2016:118).

The first component, *detect*, addresses the issue of being aware of hybrid threats or attacks that require building situation awareness and understanding. Traditional enemy-centric threat analysis may not be enough to understand the complex situations that hybrid threats pose. The

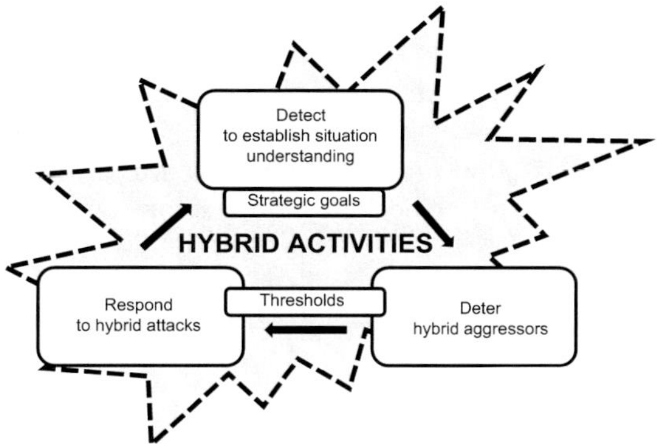

Figure 8.1 Countering Hybrid Warfare Framework.

Source: Adapted by the Authors from Monaghan S./MCDC (2019:22 Figure 2.2).

tabular summary of Grey Zone threats illustrates some of the many ways in which subversive and indirect threats can be generated. All-hazards intelligence supports detection. The *deter* component addresses ways that increase the costs imposed on hostile actors. If efforts to disrupt are less likely to succeed, or if they are known to incur additional costs, then hostile actors may be deterred from undertaking them. The *respond* component includes countermeasures and responses to hybrid attacks. Identifying vulnerabilities and appropriate actions, measures, and thresholds must be considered during all three stages. Holistic situational awareness is essential (Monaghan S./MCDC 2019:23).

Preparedness links internal and external security. Figure 8.2 illustrates the central role of national and international information sharing. EU-wide databases help with general situational awareness. The Schengen Information System and Visa Information System include data on individuals and objects of interest for border security. European Criminal Records Information

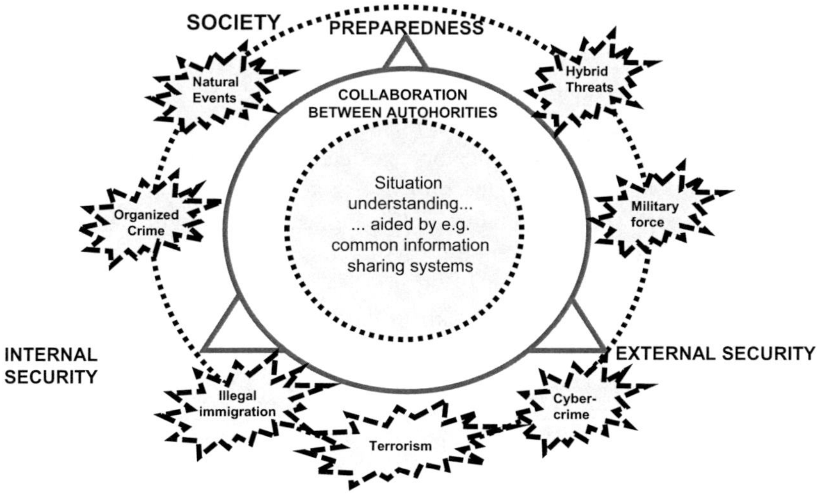

Figure 8.2 Collaboration and Preparedness.

Source: Tikanmäki and Ruoslahti (2021).

System tracks individual criminal records. The Europol Information System includes files on organized crime and terrorism. The European Police Information System and Eurodac for fingerprint data help law enforcement agencies to track suspects and irregular border crossers. Europol's European Cybercrime Centre, the European Union Agency for Cybersecurity, and the European Cybersecurity Organization promote cooperation on cybersecurity. Each European country has its own Cyber Security Incident Response Team, which is networked and shares data and incident information. Rapid and up-to-date information exchange across the full spectrum of policing and security organizations is essential for situational awareness (Tikanmäki and Ruoslahti 2021:429). The EU's Joint Framework on Countering Hybrid Threats contributes to Integrated Political Crisis Response.

Regions outside Europe are not as heavily institutionalized, but there are many networks for sharing and coordinating, and prudent states maximize the use of these networks. The malign use of international organizations should be anticipated. Russia, for example, has used Interpol to issue international arrest warrants for opposition figures like Vladimir Gusinsky (Kasparov 2015:95). Misuse of shared information is a risk depending on the membership and levels of trust in networks.

Future conflicts will likely combine new forms of hybrid warfare, perhaps involving artificial intelligence, deepfakes, space-based assets, and diaspora populations. Security challenges should be approached in multidisciplinary and holistic ways with evidence-based study (Hyttinen, Hario and Österlund 2017:8). In Chapter 7, Gardner, Żakowska and Last introduced DIMEFIL and PESTLE as frameworks for analysis. Here, PMESII-PT will be introduced as a relevant list of factors linking hybrid war to crisis management and contemporary thinking about joint intelligence preparation. These factors are summarized in Table 8.2. PMESII-PT originated with US Army thinking, but Czech defense scientists Vasicek and Hlavizna demonstrate its broad relevance to diverse operations. The third and fourth columns of Table 8.2 provide examples only.

Enhancing Crisis Management and Response

Enhancing crisis management and response demands international as well as whole-of-government and whole-of-society cooperation. Tables 8.1 and 8.2 offer examples of hybrid threats that may exacerbate crises and impede crisis management for small countries, augmenting the list in the Appendix. The detect-deter-respond sequence is continuous and overlapping; this is explicit in the context of the war nexus (Chapter 1) and private sector ideas about crisis management.

Situational awareness develops as indicators are detected, sought out, and confirmed. Decisions are guided by the best estimates and projections available (Hyytiäinen 2021). Situation pictures are always imperfect and especially challenging when faced with hybrid threats (Ducaru 2016). Decisions consider the best current information but also try to anticipate the evolution of the situation (Tikanmäki and Ruoslahti 2019). Democratic politics, market economies, and legal processes in democratic states are adversarial, further blurring the boundary between politics and war (Mälksoo 2018). To build a common and shared understanding among political, civilian, and military organizations and between systems and actors within the operational environment, permanent knowledge development processes that collect, analyze, store, and distribute information are needed, but timely conclusions and early warning are also essential (Thiele 2015:11). Ambiguity challenges awareness (Mumford 2016:42).

Deterring hybrid warfare's mix of soft and hard power measures demands integrated security management systems. Poland's counterterrorism system combines physical and cyberattack measures (Gasztold and Gasztold 2020:11). Intelligence and deterrence are complicated

Table 8.2 PMESII-PT and Sample Threats and Responses

Variables	*Description*	*Hybrid threat*	*Crisis response*
Political	Describes the distribution of responsibility and power at all levels of governance; includes confidence	Subvert politicians or government transition	Protect institutions and politicians
Military	Explores the military, paramilitary, and police capabilities of all relevant actors (enemy, friendly, neutral) in given operational environment; includes regular and irregular elements and military potential of non-military actors	Infiltrate military, police, intelligence, create embarrassing incidents	Buddy-watch systems, reliability checks, public disclosure to retain confidence
Economic	Encompasses individual and group behaviors related to producing, distributing, and consuming resources; includes economic implications of global connectivity	Supply chain or financial attacks; food contamination; and energy infrastructure attack	Macroeconomic planning; consumer protection; and critical infrastructure plans
Social	Describes the cultural, religious, and ethnic makeup within an operational environment and beliefs, values, customs, and behaviors of society members; includes social implications of digital communications	Provoking tensions; *kompromat* for community leaders; discreditation, manipulated conflicts; and targeting weakness in social narratives	Openness, public engagement, rights-based legal regime, reliable public media, and counter-narratives
Information	Describes the nature, scope, characteristics, and effects of individuals, organizations, and systems that collect, process, disseminate, or act on information	Disinformation, media discrediting, deepfakes, triggering events, and data manipulation	Legislation, penalties, fast-track prosecutions, and cyber response teams
Infrastructure	Is composed of the basic facilities, services, and installations needed for the functioning of a community or society; includes space, cyber, and know-how to preserve or repair	Cutting undersea cables, satellite interference, energy blackmail, and funding bribery	Critical infrastructure protection, redundancy, repair capacity, transparency, and whistleblowing
Physical environment	Includes the geography and man-made structures, as well as the climate and weather in areas affecting operations	Damming rivers, sabotaging dams, bridges, and nuclear plants	Critical infrastructure protection, rapid repair capacity, and redundancy
Time	Describes the timing and duration of activities, events, or conditions within operational environment, as well as how the timing and duration are perceived by various actors in the operational environment	Rapid succession of induced crises, overwhelm decision-makers, and cause cascading collapse of systems	Regular rehearsals, surplus capacity for crisis management

Source: Vasicek and Hlavizna (2020: 1–3, inter alia) interpreted by the Authors and Editors.

because threats can emerge from state or non-state sources and may have political, economic, or social motivations, including eco-terrorism or supply-chain disruption (Mumford 2016:7). The value of allies in deterring Grey Zone attacks may be greatest for small states, allowing collective responses and indirect pressure on the originators of hybrid attacks.

Responding to crises can be effective in the United Nations, NATO, and economic and social treaties like the EU or ECOWAS when they include crisis response mechanisms. Treaty mechanisms can come with unwanted obligations or consensus rules that may impede reaction. The advantage Europe enjoys with its many avenues for cooperation is that some will work when others fail. Overlapping competencies in the EU create alternative paths for managing crises, but turf battles prevent full exploitation. This seems to be a generic problem with collaboration to manage crises (Gross and Juncos 2010:11).

The NATO crisis management system is described as a six-step linear process with recurring engagement of the North Atlantic Council (NATO 2010). With political decisions at every stage—reconnaissance, assessment, developing options, planning, execution, and (one hopes) return to stability—there are recurring opportunities for the decision rules of consensus and unanimity to delay or derail action. When members agree, engagement is stronger, but members can hold decisions hostage to national interests. Interorganizational cooperation is also impeded by the dual-consensus rule, which maximizes the autonomy of individual members. It requires consensus within and between organizations to permit cooperative action. This limits organizational agility (Biermann 2015:46–47). If political consensus is elusive, the many working-level options for cooperation provide avenues for security experts to influence their national authorities by sharing information and analysis to develop a common approach.

The European Security Strategy leaves the primary responsibility for combating hybrid threats to member states, responsible for "the full spectrum of action from early detection, analysis, awareness, building resilience, and prevention through to crisis response and consequence management" (European Commission 2020:14). Nevertheless, the strategy envisions collaboration on cybersecurity, security of critical infrastructure, management of illegal online content, and prevention of terrorism. Overcoming the false distinction between physical and virtual worlds and linking security to fundamental values is the common challenge for crisis response (European Commission 2020:26). Other regional organizations have comparable crisis response mechanisms with fewer overlapping agencies and fewer options to bypass roadblocks or delays.

Countering Grey Zone Influence

Threats in the Grey Zone include subversive and clandestine actions that target people, information, and physical assets and influence decision-makers and key policies. The sources and intent of these actions can be difficult and time-consuming to determine. This means that many Grey Zone threats will continue for days, months, or even years undetected and without countermeasures; they are calculated to avoid eliciting countermeasures or to elicit counterproductive countermeasures, like a disproportionate security response.

A fragile society and weak institutions can fragment, fall victim to distrust, or turn on itself in the face of manipulation, deception, disinformation, and targeted attacks. These human reactions produce social and political crises that weaken the state. A resilient society has strong networks of trust within and between communities, between citizens and authorities, and within institutions that enable essential services, including security, transport, communications, health, public information, and public education (Last and Youssofzai 2014:7). Another approach is to focus on engineering critical infrastructure to minimize damage from disruptions like sabotage

Table 8.3 Hybrid Threats to Crisis Management

Crisis management enablers	*What hostile actors do to erode crisis management*	*Prevention and countermeasures*
Stable and determined **leadership** is an essential foundation for crisis management	– Replace local political leaders with hostile operators – Erode confidence in leadership (disinformation, cascading problems) – Political assassinations	– Protection of key personnel – Capacity building for all branches of government – Situation awareness, tracking influencers, following money – Strong legislative framework
International cooperation permits situational awareness, supports national response	– Abuse international law – Disregard humanitarian and legal norms – Use criminal connections to make a pariah state	– International collaboration, liaison, and early detection – Cross-checking criminal, political, and military intelligence – Coordinate prosecutions
Defense capabilities are used for emergency response and to defend against conventional attacks which might follow softening up by hybrid attacks	– Blend conventional and irregular warfare – Flood area with weapons – Arms sales for influence – Separatist militia attacks	– Coordinate national defense and internal security – International cooperation with defense planning, equipment, and training programs
Internal security maintains public confidence and reduces domestic threats	– Coordinated cyberattacks to disrupt – Intimidate police – Violent atrocities erode confidence in security – Intelligence penetration	– Build capacity of interior ministries and police forces to detect and counter unconventional attacks – Domestic criminal and political intelligence, counter-terrorism strategies, and systems in place
Economic order permits people to continue normal lives and provides essential revenue for government	– Degrade the economy (cut exports, supply chains, consumer confidence) to induce or exacerbate crisis – Bribe or influence economic elites	– Macroeconomic planning, triggers to defend critical industries, supply chains – Financial tracking and forecasting – Monitor foreign ownership, control, and influence
Secure **critical infrastructure** permits response to attacks and preservation of normal social order	– Sabotage – Seize or disrupt buildings, ports, and transport hubs – Attack electricity, water supply – Cyberattacks and disruptions	– Critical infrastructure security – Redundancy for critical assets – Local defense, alternative facilities arrangements – Public awareness for maximum warning – Information security for cyber assets
Maintaining **essential services** for the population preserves popular support for government	– Disruption of basic services – Propaganda, disinformation, and misinformation about health, education, and welfare services – Exploitation of social divisions	– Decentralized services, backup for central services (e.g. pension payments) – Trusted sources of information – Liaison with community leaders
Mental resilience allows individuals and communities to withstand shocks and find resources to solve problems in the face of crises and disruption	– Targeted social media – Torture, mutilation, and atrocity – Terror attacks – Disinformation – Hostage-taking – Kidnapping children	– Caution with over-reactions – Trusted communications – Strong national narratives – Due process and legal recourse – Quick response and transparency

Source: Table prepared by Authors and Editors from references cited in the text.

(Thoma et al. 2016). However, leaving people and social relations out of the equation is unlikely to result in a more resilient society in the face of Grey Zone crises.

A hybrid warfare strategy seeks out societal weaknesses, exacerbates them to generate crises, and then takes advantage of crises further to weaken the target in preparation for a conventional attack. Table 8.3 summarizes hybrid threats against crisis management and measures that enhance crisis measurement in the face of these threats.

Countermeasures to support crisis management work at individual, community, and state levels. They target specific threats and general circumstances. Economic vulnerability is a general circumstance that might arise from a region or small state's overdependence on a single port or transport route. Contingency plans for the short term and diversification for the longer term will reduce the impact of economic crises. Specific threats to key economic or political leaders may use extortion, bribery, or manipulation. Personal security, awareness of vulnerability, and countermeasures when threats are detected will help to keep individuals safe. Manipulation of social divisions, disinformation, and propaganda demands a national response supported by civil society organizations to provide trusted and reliable sources of information (Raitasalo 2017:21).

Resilience features heavily in crisis management. NATO identifies seven conditions for increasing resilience: (1) continuity of government and critical government services; (2) resilient energy supplies; (3) ability to deal effectively with the uncontrolled movement of people; (4) resilient food and water resources; (5) ability to deal with mass casualties; (6) resilient communication systems; and (7) resilient civilian transportation systems (Shea 2016). Close cooperation between government, private sector, and civil society organizations within a state, and coordination with neighboring states and international organizations support all seven conditions. Regular preparedness exercises, public discussions, and integration into regional mechanisms help to combat hybrid threats. Crisis management exercises help decision-makers and organizations to prepare for disruption and unanticipated events. Flexible and adaptive command and control are essential for successful military operations in the face of hybrid threats (Got 2020).

Open societies have advantages and disadvantages when dealing with hybrid threats. The advantages stem from strong networks of trusted relationships maintained by an open and honest government and regular, reliable information to citizens and stakeholders (Mälksoo 2018). Defenses against hybrid threats, however, can be detrimental to the core principles of liberal democracy (Ducaru 2016:21).

Conclusion

Crisis management in hybrid warfare is a sophisticated and multidimensional challenge. Crises can evolve quickly or slowly, at home or abroad, because of natural adversity or hostile adversaries. Any crisis can be manipulated and exacerbated by adversaries. Detecting, deterring, and responding to crises is a continuous process demanding comprehensive interagency cooperation and use of all the tools available for coordinating within and between allied states. Preparedness links internal and external security to manage natural crises, organized crime, migration, terrorism, cyberattacks, and hybrid threats (Figure 8.2).

There are many international coordination and information-sharing channels, particularly in Europe. Other regions may rely more on bilateral arrangements. Regularly exercising multilateral channels helps detect and deter hybrid threats and manage crises. Consensus rules, turf battles, and conflicting national interests leave the state with the ultimate responsibility for managing crises.

Hybrid war seeks to weaken a target to reduce threats it poses or to prepare for a conventional attack. Creating cascading crises undermines trust, confidence, and governance. Ultimately, social resilience in the face of adversity is one of the most important assets for managing crises. This can best be fostered by focusing on the basics: good government, public services, secure infrastructure, effective management, and continuous adaptation in a hostile environment.

Note

1 The Authors acknowledge the contributions of the Editors and anonymous Reviewers to this chapter.

References

Aaronson, M., Diessen, S., De Kermabon, Y., Long, M. B. and Miklaucic, M. (2011), "NATO Countering the Hybrid Threat", *Between States,* vol. 2, no. 4, pp. 111–124.

Acharya, A. (2009), *Constructing a Security Community in Southeast Asia: ASEAN and the Problem of Regional Order*, Routledge, London.

Andersson, J. and Tardy, T. (2015), *Hybrid: What's in a Name?*(October 2015), European Union Institute for Security Studies. https://www.iss.europa.eu/sites/default/files/EUISSFiles/Brief_32_Hybrid_warfare.pdf

Biermann, R. (2015), "Designing Cooperation among International Organizations: The Quest for Autonomy, the Dual-consensus Rule, and Cooperation Failure", *Journal of International Organizations Studies,* vol. 6, no. 2, pp. 45–66.

Bugarič, B. (2015), "A Crisis of Constitutional Democracy in Post-Communist Europe: 'Lands in-between' Democracy and Authoritarianism", *International Journal of Constitutional Law,* vol. 13, no. 1, pp. 219–245.

Canyon, D. (2020), "Definitions in Crisis Management and Crisis Leadership", *Security Nexus*, vol. 21 (January 2020), pp. 1–10, Daniel K. Inouye Asia-Pacific Center for Security Studies (DKI APCSS), Honolulu, US. https://apcss.flywheelsites.com/nexus_articles/definitions-in-crisis-management-and-crisis-leadership/?utm_source

Cederberg, A. and Eronen, P. (2015), *How Can Societies Be Defended Against Hybrid Threats?* Strategic Security Analysis no. 9 (September 2015). Geneva Centre for Security Policy (GCSP).

Cîrdei, I. A. (2016), "Countering the Hybrid Threats", *Land Forces Academy Review,* vol. 21, no. 4, pp. 113–119.

Cîrdei, I. A. and Ispas, L. (2017), "A Possible Answer of the European Union to Hybrid Threats", *Scientific Bulletin-Nicolae Balcescu Land Forces Academy,* vol. 22, no. 2, pp. 71–78.

Coombs, W. T. (2007), "Crisis Management and Communications", *Institute for Public Relations,* vol. 4, no. 5, pp. 1–14.

Ducaru, S. D. (2016), *The Cyber Dimension of Modern Hybrid Warfare and Its Relevance for NATO,* Europolity - Continuity and Change in European Governance, vol. 10, no. 1, pp. 7–23.

Ducheine, P. A. L. (2016), "Non-Kinetic Capabilities: Complementing the Kinetic Prevalence to Targeting" Chapter 10, in P. A. L. Ducheine, M. Schmitt and F. P. B. Osinga (eds), *Targeting: The Challenges of Modern Warfare,* T.M.C. Asser Press, The Hague, pp. 201–230.

European Commission (2018), *Increasing Resilience and Bolstering Capabilities to Address Hybrid Threats,* Joint Communication to the European Parliament, The European Council and the Council Join (2018) 16 Final. European Commission, Brussels.

European Commission (2020), *The EU Security Union Strategy,* Communication from the Commission to the European Parliament, the European Council, The Council, The European Economic and Social Committee and the Committee of the Regions 605 Final. European Commission, Brussels.

European Parliament (2021), *Strategic Communications as a Key Factor in Countering Hybrid Threats,* Study No. PE 656.323. Directorate-General for Parliamentary Research Services.

Gasztold, A. and Gasztold, P. (2020), "The Polish Counterterrorism System and Hybrid Warfare Threats", *Terrorism and Political Violence,* vol. 34, no. 6, pp. 1–18.

Got, A. (2020), "NATO Review - NATO Crisis Management Exercises: Preparing for the Unknown", *NATO Review*, 7 February. https://www.nato.int/docu/review/articles/2020/02/07/nato-crisis-management-exercises-preparing-for-the-unknown/index.html

Gross, E. and Juncos, A. E. (eds.) (2010), *EU Conflict Prevention and Crisis Management: Roles, Institutions, and Policies,* Routledge/UACES Contemporary European Studies, London.

Hall, B., Fleming, S. and Shotter, J. (2021), "How Migration Became a Weapon in a Hybrid War", *Financial Times,* 5 December. https://www.ft.com/content/83ece7e4-cc71-45b5-8db7-766066215612

Holohan, A. (2019), "Transformative Training in Soft Skills for Peacekeepers: Gaming for Peace", *International Peacekeeping,* vol. 26, no. 5, pp. 556–578.

Hyttinen, K. and Kallonen, M. (2018), *How to Improve EUs Conflict Prevention Activities to Achieve Longterm Impact?* Prevention. Lahti University of Applied Sciences Series, Vantaa, Finland.

Hyttinen, K., Hario, P. and Österlund, P. (eds.) (2017), *Improving the Effectiveness of Capabilities (IEC) in EU Conflict Prevention,* Laurea Publication no. 86. Laurea University of Applied Sciences, Vantaa, Finland.

Hyytiäinen, M. (2021), *Data Integrity and Situation Picture of Management, in Proceedings of the Scientific Advisory Board for National Defense (MATINE) Research Seminar,* 20 May 2021, Webinar, Helsinki, Finland.

Jaques, T. (2007), "Issue Management and Crisis Management: An Integrated, Non-linear, Relational Construct", *Public Relations Review,* vol. 33, no. 2, 147–157.

Kasparov, G. (2015), *Winter Is Coming: Why Vladimir Putin and the Enemies of the Free World Must be Stopped.* PublicAffairs, New York.

Lasconjarias, G. and Larsen, J. A. (2015), *NATO's Response to Hybrid Threats,* Forum Paper no. 24, NATO Defence College, Rome, Italy.

Last, D. and Youssofzai, F. (2014), "A Framework to Approach Planning for Resilience", *Public Policy and Administration Review September*, vol. 2, no. 3, pp. 01–21.

Lehto, M. and Limnéll, J. (2017), "Kybersodankäynnin kehityksestä ja tulevaisuudesta" (On the Development and Future of Cyber Warfare), *Tiede ja ase 75* (November), pp. 179–212. https://journal.fi/ta/article/view/67730

Mälksoo, M. (2018), "Countering Hybrid Warfare as Ontological Security Management: The Emerging Practices of the EU and NATO", *European Security,* vol. 27, no. 3, pp. 374–392.

Monaghan S./MCDC (2019), *MCDC Countering Hybrid Warfare Project: Countering Hybrid Warfare,* Handbook, The Multinational Capability Development Campaign (MCDC). https://assets.publishing.service.gov.uk/media/5c8141e2e5274a2a51ac0b34/concepts_mcdc_countering_hybrid_warfare.pdf

Moon, J., Sasangohar, F., Son, C. and Peres, S. C. (2020), "Cognition in Crisis Management Teams: an Integrative Analysis of Definitions", *Ergonomics,* vol. 63, no. 10, pp. 1240–1256.

Mumford, A. (2016), *The Role of Counter Terrorism in Hybrid Warfare,* a Report Prepared for NATOs Centre of Excellence for Defence Against Terrorism (COE DAT), University of Nottingham, UK.

NATO (2010), *BI-SC Input to a NEW Capstone Concept for the Military Contribution to Countering Hybrid Threats*, 1500/CPPCAM/FCR/10–270038 and 5000 FXX/0100/TT-0651/SER:NU0040.

NATO (2024), Countering Hybrid Threats. https://www.nato.int/cps/en/natohq/topics_156338.htm

Paturas, J., Smith, S., Albanese, J. and Waite, G. (2016), "Inter-Organisational Response to Disasters", *Journal of Business Continuity & Emergency Planning,* vol. 9, no. 4, pp. 346–358.

Raitasalo, J. (2017), "Getting a Grip on the So-Called 'Hybrid Warfare'", *ASPJ Africa & Francophonie,* vol. 3, pp. 20–39.

Shea, J. (2016), "Resilience: A Core Element of Collective Defence", *NATO Review,* 30 March. https://www.nato.int/docu/review/articles/2016/03/30/resilience-a-core-element-of-collective-defence/index.html

Stumpenhorst, M., Stumpenhorst, R. and Razum, O. (2011), "The UN OCHA Cluster Approach: Gaps between Theory and Practice", *Journal of Public Health*, vol. 19, pp. 587–592.

Swanström, N. L. and Weissmann, M. S. (2005), "Conflict, Conflict Prevention and Conflict Management and Beyond: A Conceptual Exploration", Central Asia–Caucasus Institute & Silk Road Studies Program, Johns Hopkins University-SAIS, Washington DC and Uppsala University, Uppsala, Sweden.

Tardy, T. (2015), *CSDP in Action: What Contribution to International Security?* Chaillot Paper no. 134, European Union Institute for Security Studies. doi:10.2815/634719.

Thiele, R. D. (2015), "Crisis in Ukraine–the Emergence of Hybrid Warfare", I*SPSW Strategy Series: Focus on Defense and International Security*, no. 347 (May 2015), pp. 1–13.

Thoma, K., Scharte, B., Hiller, D. and Leismann, T. (2016), "Resilience Engineering as Part of Security Research: Definitions, Concepts and Science Approaches", *European Journal for Security Research,* vol. 1, no. 1, pp. 3–19.

Tikanmäki, I. and Ruoslahti, H. (2019), "How Are Situation Picture, Situation Awareness, and Situation Understanding Discussed in Recent Scholarly Literature?", in *Proceedings of the 11th International Joint Conference on Knowledge Discovery,* Knowledge Engineering and *Knowledge Management,* 17-19 September 2019, vol. 3, pp. 419–426, SCITEPRESS - Science and Technology Publications, Vienna, Austria.

Tikanmäki, I. and Ruoslahti, H. (2021), "Interdependence of Internal and External Security", in T. Eze, L. Speakman, C. Onwubiko (eds.), *Proceedings of the 20th European Conference on Cyber Warfare and Security*, 24-25 June 2021, pp. 425–432, Academic Conferences and Publishing International Ltd., UK.

Uppsala Universitet (2021), *UCDP - Uppsala Conflict Data Program,* Uppsala Conflict Data Program Department of Peace and Conflict Research. 15 November 2021. https://ucdp.uu.se

Vasicek, R. and Hlavizna, P. (2020), *The Role of Joint Intelligence Preparation of the Operating Environment in Support of Future Military Operations,* Czech University of Defence, Brno.

Weijers, B. (2015), "NATO Disaster Relief Operations: An Analysis of an Underexposed Field of Activity of the Alliance" (Master Thesis), Catholic University of Portugal, Lisbon, Portugal.

9 All-Hazards Intelligence in the Grey Zone and Modern War

David Last[1], *Urszula Jaśkiewicz, and Marzena Żakowska*

Introduction

Intelligence is not a standardized commodity. The word means different things in different contexts and cultures. At its heart, intelligence provides knowledge and foreknowledge to enable decisions. It is set apart from other knowledge by its contribution to security. It is, therefore, inherently competitive (Gill and Pythian 2018:2). Intelligence operations may also attempt to shape how adversaries and allies think and act. What the knowledge is about and who makes the decisions generate many adjectives. Strategic intelligence guides top decision-makers' understanding of risks and opportunities at a national and international level. "Strategic intelligence is the totality of what a government knows (or thinks or hopes it knows) about the outside world" (Davies 2019:3).

Tactical intelligence provides situational awareness to support decisions in specific theatres and engagements (Johnson 2007:1). Intelligence studies in the English-speaking world have been dominated by the intelligence culture of Great Britain and its hegemonic successor, the United States. These cultures are shaped by global engagement and the powerful economic interests that help to preserve major power status. Economic intelligence—about production, resources, capabilities, and flows—is a major field of defense intelligence. Intelligence about social relationships—language, ethnicity, demographics, and population movements—has typically been pursued within strategic and political intelligence. Intelligence about the structure and relationships within a society overlaps with network analysis, which helps to understand who is talking to, influencing, and commanding whom.

All-hazards intelligence implies a holistic approach to understanding all potential sources of harm. Before the 1970s, the term "all hazards" was found in public health and industrial studies. It was common in environmental and disaster management literature by the 1980s. After 9/11, the all-hazards approach was adopted by the US Department of Homeland Security. More recent applications address the intersection of disasters, policing, and security (Vasquez 2021). The evident intersections of critical infrastructure for water, electricity, gas, healthcare, finance, and industry have demanded a more integrated approach to understanding hazards (Soldatos, Praça and Jovanović 2021).

This chapter addresses targets, assets, and organizations for all-hazards intelligence. What is to be known? How is it to be collected? What structures or organizations serve small countries to manage it? Knowledge, assets, and organization are subject to the competing pressures of integration and compartmentalization. A holistic view from an integrated perspective may be more vulnerable to penetration and manipulation, while compartmentalization may fail to identify risks. Specialization and competition may serve a liberal democracy or preserve the power of autocrats by spying on each other.

DOI: 10.4324/9781003428701-14

All-Hazards Intelligence for Small Countries

Histories of intelligence illustrate an evolutionary trajectory and learning between societies. Spies in the Bible date to the 14th century BCE. Sun Tzu's *Art of War* and Kautilya's *Arthashastra* (Art of War) articulated theories of intelligence in the 5th and 3rd centuries BCE (Andrew 2018:54). Overcoming magical thinking and distinguishing between the honesty necessary for cohesive public life and the subterfuge necessary for success in war are perennial lessons. Advances in technique are associated with some eras more than others. Roman emperors learned to collect information systematically through geographically dispersed agents. The Prophet Muhammad tracked trading caravans in his war against Mecca. The Catholic Church perfected organized mass interrogations and record-keeping (Andrew 2018:100). High policing of political dissent emerged in successive European regimes (Liang 1992:4; Brodeur 2010:223). The 19th- and 20th-century colonialism created a paradox. Colonial state structures sought to control indigenous society but often depended on indigenous populations to manage the state. Many countries were established as intelligence states built on security policing, and this apparatus was typically passed on to successor states (Thomas 2005).

Different countries, languages, and political traditions have learned from these historical trajectories, and it would be a mistake to assume that there are common cultures of intelligence, even among close allies. The variety of intelligence approaches and organizations can be a cloak to protect small states from the simplifying assumptions of larger allies and adversaries. English intelligence, Russian разведка (*razvedka*), Chinese 情报 (*qíngbào*) and Arabic dversaries. English i*Al-Istikhbarat Al-Amniya*) concepts have converged with globalization. The history of countries and regimes, including experiences of colonialism, ideological affiliation, and foreign assistance, all influence national approaches. Smaller countries typically have fewer global commitments than larger countries but may be no less globally engaged. Luxembourg is a global finance hub, and Montenegro's ports play an outsized role in European trade networks (Kajtazi 2014).

What information do small states need to support national security decision-making? The Appendix illustrates ways in which security can be undermined. Political, economic, social, technological, legal, and environmental threats are convenient categories but combine in infinite ways. Security leaders of even small states may naturally focus on the military assets of neighboring or rival states. The composition of armed forces, leadership structure, and intentions for employment are combined with information about geography and weather to prepare military plans. Diplomatic intelligence about decision-making combines with knowledge of alliances and timelines for action. However, the variety of Grey Zone threats and the absence of a clear starting line for war make traditional military intelligence inadequate for modern war. The label all-hazards intelligence is applied here to describe a holistic approach to address the full range of harms most countries face in a connected world.

All-Hazards Intelligence Targets

Beginning with political threats, states need intelligence about foreign attempts to influence candidates, political parties, perceptions of the electorate, and electoral processes. These are more threatening to liberal democracies than autocracies controlling election outcomes. From 1945 to 2012, about a third of European elections faced identifiable partisan interventions targeting specific candidates, and more than half saw foreign interference. Small democratic states are frequently targeted by larger, less democratic governments: by China in Taiwan in 2000 and Canada in 2019 and 2021, Russia in Ukraine in 2014, Montenegro in 2016, and Macedonia in

2018 (Karaskova, Berzina-Cerenkova and Nemckova 2023; Shimer 2020). Although specific objectives vary, the overall impact of political interference is to erode autonomy and damage institutions by undermining trust. When common language makes for permeable boundaries, assets from one campaign can be mobilized for another. During the UK referendum on the EU in 2016, 150,000 Russian-language Twitter accounts spread "pro-leave" content on the UK social media (Shimer 2020:152). The same assets were used again to support the Trump campaign in the 2016 US election. Understanding the practices of adversaries and sharing information about the forms and sources of interference are important aspects of political intelligence.

Economic threats combine with political interference. Some countries finance foreign political parties through organized crime. These concerns have been raised concerning the Syriza Party in Greece, the Five Star Movement in Italy, and the National Front in France. Chinese agents infiltrated gambling networks in Taiwan and manipulated them to offer favorable betting odds for China's preferred candidates. This incentivized gamblers and their families to vote as Beijing prefers (Karaskova, Berzina-Cerenkova and Nemckova 2023:25). Russia is estimated to have directed the clandestine transfer of more than $300 million US dollars to political parties and politicians in more than 24 countries between 2014 and 2022 (Karaskova, Berzina-Cerenkova and Nemckova 2023:22). More directly, Chinese intelligence is alleged to have provided more than a million Australian dollars to support a Chinese-Australian candidate for parliament in 2019 (Suri 2020). It is hard to disentangle corporate political donations from foreign engagement with corporations, but all-hazards intelligence seeks awareness of money flows that influence political decisions. Strong electoral finance laws, registration of lobbyists, oversight, and forensic accounting capabilities are essential starting points.

All-hazards intelligence must understand the intersection of political influence, social manipulation, financial irregularities, and sophisticated hacking and cyber-attacks. Defining hostile actors, capabilities, and activities across social media platforms, criminal organizations, economic sectors, and political landscapes is a starting point. The malign influence of allies cannot be ruled out. The rise of intolerance or social schisms may be an early warning of vulnerability to interference. Defensive intelligence efforts may include monitoring social media accounts, criminal organizations, e-mail patterns, and shell corporations. Electoral security and financial reporting may have to be extended to a full-time activity rather than being confined to the margins of electoral campaigns.

The people involved in all-hazards intelligence assessments and other security personnel are potential targets for hostile activity. In 2010, Russian agents targeted Saudi, Belgian, and Italian defense contractors. Hacking and social media operations were used to collect information and influence attitudes and behavior, targeting individuals and corporate accounts. Fake accounts and blogs appeared to be based in Canada (Rid 2020).

Hostile actors can pursue social manipulation over many years using creative social media. Interventions can be linked to electoral cycles. Before the hotly contested and controversial 2020 election in the United States, Facebook's most popular pages for Christian and Black American content were run by troll farms based in Kosovo and Macedonia. An internal Facebook report from 2019 described Facebook's failure to change the algorithms that had promoted and distributed disinformation in the 2016 election. Although the impact is hard to judge, data show that about 75 million Americans interact with these pages monthly (Hao 2021). Interventions can also be tied to protest movements or events in the international media. The Islamic Revolutionary Guard Corps targeted Islamic student unions in Europe using social media accounts, indirect financing, and direct attacks on individuals seen as anti-regime (Mohammadi 2023).

Grey Zone threats in social networks demand a more sophisticated awareness of political and economic motivations and impacts. Key individuals, family influencers, recreational clubs, businesses, professions, religions, and ethnic groups present potential entry points for hostile influence. The targets may be willing, unwitting, or coerced. All-hazards intelligence must consider the hierarchies and cleavages that divide society. Multi-ethnic democratic societies with free markets may form a "vertical mosaic" with hierarchies of social class and power and simultaneous ethnic, religious, and economic cleavages (Porter 2015; Helmes-Hayes and Curtis 1998). Less open societies and economies may be characterized by dominating clans, subordinate classes, and stratified societies. These cleavages are subject to external manipulation, but tensions and conflicts are not necessarily evidence of foreign involvement. Indeed, accusations of foreign involvement may be a tactic to delegitimize them.

Diasporas are often targeted by their country of origin through direct or indirect economic, criminal, or political means to reduce threats to the homeland or seek influence and assets in the country of settlement. In 2014, Chinese consulates in Canada used Chinese students to influence Canadian households where Mandarin was spoken (Karaskova, Berzina-Cerenkova and Nemckova 2023:26). WeChat groups were organized in Chinese diaspora communities and professional business networks to harass or spread information supporting selected candidates in Australian elections in 2017, and in Canada in 2022–2023. Foreign states have used social media groups for fundraising and networking, influencing activists, and encouraging voting with "transportation subsidies" (Karaskova, Berzina-Cerenkova and Nemckova 2023:27–28).

All-hazards intelligence might also need situational awareness about mainstream social groups vulnerable to manipulation by foreign forces. Russian disinformation has fomented and mobilized European anti-immigrant sentiments to support geopolitical goals (Juhász and Szicherle 2017:4). Sunni Islamic communities with ties to Daesh were mobilized for recruitment, facilitation, and attacks in Europe (Brzuszkiewicz 2018).

Key individuals and their close circles present potential threats to the state. Beyond political decision-makers, judges, university leaders, and CEOs of major national companies may be influential assets or targets for foreign operations. The Czech-owned company *Explosia a.s.* is a major supplier of industrial explosives and a manufacturer of 155 mm shells for NATO. In 2024, Sweden and the Czech Republic cooperated to supply shells to Ukraine (Government Offices of Sweden 2024). Efforts to threaten, bribe, or influence the commercial director of *Explosia a.s.* could affect production and distribution operations (EMIS 2024).

Not all vulnerabilities are at the apex of society. Social groupings of technical and professional specialists may also constitute vulnerabilities, particularly when concentrated. A conference of security leaders in a small target country might be a tempting target for a neighbor embarking on war. While security leaders might be protected, other groups may present immediate vulnerabilities. Air traffic controllers in Finland, railway controllers in Slovakia, and doctors in Iceland are all small but important groups who can be easily identified using open sources. Daesh fighters were able to kill 131 and injure 416 across 6 locations without the backing of a state (Boddaert et al. 2017), illustrating the reach of Grey Zone attacks using social analysis of target groups.

Focus on people is important but insufficient for all-hazards intelligence. Financial capital and international markets can cause significant damage. A state made vulnerable by domestic problems, international conflict, political missteps, or loss of confidence in its leadership may be victimized by international speculators and trading algorithms. Currency crises cannot be explained by economic fundamentals alone. Regional contagion and speculative behavior may be important determinants (Kruger, Osakwe and Page 1998). Major powers and corporate interests may mobilize financial power to accentuate destabilization, particularly targeting the US dollar, which has implications for other countries (Rickards 2011). Over a dozen currency crises have been linked to political instability since the turn of the century (Table 9.1). All-hazards

Table 9.1 Recent Financial Crises

Year	*Country*	*Crisis*	*Causal elements*
1992–1994	Yugoslavia	Hyperinflation	The breakup of Yugoslavia and the Bosnian War led to extreme hyperinflation.
1998	Russia	Currency devaluation	Political instability and fiscal deficits led to a severe devaluation of the Russian ruble.
1997–1998	Asia (Thailand, Indonesia, South Korea, Malaysia, Philippines)	Currency depreciation	A lack of confidence in the political leadership in several Asian countries led to a rapid depreciation of their currencies.
1999–2002	Argentina	Devaluation	A series of political crises, including the resignation of two presidents, led to a severe devaluation of the Argentine peso.
2007–2009	Zimbabwe	Hyperinflation	Political instability and poor economic policies led to hyperinflation.
2012	Iran	Currency depreciation	Iran's rial plummeted due to international sanctions related to its nuclear program and political instability.
2009–2018	Greece	Debt crisis	Political instability and a lack of fiscal discipline led to Greece's inability to pay its debts.
2014–2015	Ukraine	Currency depreciation	Political instability following the annexation of Crimea by Russia led to a severe depreciation of the Ukrainian hryvnia.
2016–present	Venezuela	Hyperinflation	The ongoing political and economic crisis in Venezuela has led to severe hyperinflation.
2018	Turkey	Currency depreciation	Political interference in monetary policy and diplomatic spats with the United States led to a severe depreciation of the Turkish lira.
2018	Sudan	Currency depreciation	Political unrest and the overthrow of the government led to hyperinflation and a sharp depreciation of the Sudanese pound.
2019–present	Lebanon	Currency depreciation	A combination of political corruption, poor governance, and civil unrest has led to a significant depreciation of the Lebanese pound.

Source: Compiled by Authors from internet sources.

intelligence spans awareness of major currency fluctuations, supply chain problems, commodity markets, trading instability, links to organized crime, counterfeiting, tax evasion, and credit card fraud. This spectrum from macro- to micro-economic phenomena illustrates vulnerabilities linked to social networks, political leadership, and foreign financial leverage.

The expanding horizons of Grey Zone threats presented in other chapters of this book present additional all-hazards intelligence targets. Climate change affects a growing number of small states. Monitoring its effects on food security, internal migration, and border crossings includes validating satellite reports and data provided by international organizations, including the regular reports of the International Panel on Climate Change, the UN High Commission for Refugees, and the Office for the Coordination of Humanitarian Affairs. Environmental projections on temperature, rainfall, soil salination, harvests, and the activities of neighboring states, like cloud seeding or redirecting watercourses, interact with other all-hazards files. Chemical, biological, radiological, and nuclear environmental damage management informs risk assessments. Commercial arrangements affect access to space imagery and communications services.

All-Hazards Intelligence Assets

Anyone can be involved in all-hazards intelligence collection for small countries. This section concerns mainly the actors or organizations contributing to intelligence rather than the technical means used to collect information. Signals intelligence (SIGINT) includes intercepting and analyzing electronic signals and communications. Increasingly, this means data, and often interpretation of metadata or "big data"—finding patterns in millions of data points. Image intelligence entails analyzing and comparing images. Human intelligence (HUMINT) involves qualitative assessments of information people provide with various motives. Open-source intelligence (OSINT) can draw on anything publicly available. There is a debate about whether OSINT constitutes intelligence (Johnson 2010:229). Is it rather a sophisticated reference library service? The increasing ease of access to data creates an expansive task for OSINT and broadens its application to support security decision-making.

Small countries may be more efficient at transmitting information across organizational boundaries. OSINT removes many of the constraints of security classifications and allows a wide range of contributions. Intelligence agencies, public safety and security organizations, government departments, universities, the private sector, and civil society organizations all have a role to play. Roles vary—from routine engagement to occasional questions and commissions. Historical experience shapes the evolution of national institutions. Minor reorganizations are routine, often driven by resource constraints or internal competition, while major institutional changes usually follow failures, embarrassments, or regime changes. Cataloging the security agencies, related government departments, state institutions, and a sample of private sector assets for a mid-sized country like Poland gives a sense of potential assets for all-hazards intelligence.

Poland is an interesting case because of its location and recent history; a member of the Warsaw Treaty Organization during the Cold War, it is now a NATO member, having undergone significant organizational and cultural change. Its location in Eastern Europe and borders with Russia (the Królewiec enclave), Belarus, Ukraine, and Germany put it in a complex geopolitical neighborhood. State assets include those of the intelligence agencies of military, police, and security intelligence. Paramilitary organizations include territorial defense forces, border police, and customs agencies. International cooperation to investigate and prosecute financial crimes, organized crime, smuggling, counterfeiting, and fraud requires sophisticated intelligence assets.

Table 9.2 summarizes the most important state security assets. It is not an exhaustive list. Almost every government agency has some information collection and analysis functions relevant to Grey Zone threats. In addition to state institutions with a mandate to address commonly understood security threats, information collected for other purposes contributes to the all-hazards intelligence collection effort. In Singapore, intelligence agencies work with the Land Transport Authority and the commissioning centers that implement the national policy requiring all privately owned vehicles to have an in-vehicle unit (IU) to calculate tolls. Policing

Table 9.2 State Security Agencies in Poland with Information Gathering Capacity

ABW	Internal Security Agency (ABW), a national internal security intelligence agency.
AW	Foreign Intelligence Agency (AW), an external security intelligence agency.
BOA	Antiterrorism Operations Bureau (BOA), a specialized anti-terrorist unit concerned with public order.
SOP	State Protection Service (SOP), close protection service of the Polish President, Prime Minister, Deputy Prime Ministers, and Speakers of the Sejm and Senate, which uses intelligence functions, analysis of protective actions, countering cyber threats.
CBA	Central Anti-corruption Bureau (CBA), a special service mandated to protect Poland from corruption throughout public life, the economy, and various levels of government.
GIIF	General Inspectorate of Financial Information (GIIF), a financial intelligence unit within the Polish Ministry of Treasury.
Policja	Policja, the Polish national police service charged with domestic policing.
SCS	Customs and Tax Service (SCS), detects and combats potential irregularities threatening Poland's financial security, protecting and ensuring the customs security of the EU.
SG	Border Guard (SG), an agency charged with maintaining the integrity of Poland's territorial borders, includes internal intelligence components and efforts.
SKW	Military Counterintelligence Service (SKW), an intelligence branch of the Polish Armed Forces that specializes in internal threats to national defense.
SWW	Military Intelligence Service (SWW), the military intelligence branch of the Polish Armed Forces that focuses on external deployments.
WOT	Territorial Defence Forces (WOT), professional and part-time forces dispersed through Poland to support operations.
WS	Special Forces (WS) use more specialized techniques than other types of armed forces in Poland, provide special operations, information support, and direct assault operations.
ZW	Military Gendarmerie (ZW), a specialized military police service.

Source: Table compiled by Authors from Graaf and Nyce (2016); Jager and Daun (2009); Służba Ochrony Państwa (n.d.); Ministerstwo Spraw Wewnętrznych i Administracji (n.d.); Wojsko Polskie (n.d.); Ministerstwo Obrony Narodowej (n.d.).

and intelligence agencies have detailed information on all vehicle movements by tracking IU movements (Shaffer 2022:284). Table 9.3 shows examples from the Polish case.

Market economies leave most social and economic functions and services outside state institutions. These are valuable sources of information for all-hazards intelligence. Private assets include domestic and foreign corporations operating within the state, civil society organizations, educational institutions (which may receive public funds but operate independently), and any institution regularly interacting with people—health services, hospitality, internet service providers and telecommunications companies, charities, and more. Returning to the Polish case, Table 9.4 provides examples of private-sector contributors to all-hazards intelligence.

These assets do not function independently in an intelligence-gathering or analysis role; they have specialized knowledge and expertise, collect data for their own purposes, and organize and manage it to meet the objectives of their stakeholders and directors. Their distance from government intelligence operations will vary. In Singapore, the state-owned Singtel telecommunication company is an integrated resource responsive to the intelligence community. It serves domestically and has expanded beyond Singapore by purchasing neighboring states' telecommunications infrastructure (Shaffer 2020:288).

Table 9.3 Additional Government-Sector Collaborators for All-Hazards Intelligence

KSP	Local police services in Poland, such as the Warsaw Metropolitan Police (KSP)
MI	Ministry of Infrastructure (MI), the Polish ministry that includes responsibility for internal transport, inland and maritime economy, as well as water management
PAŻP	Polish Air Navigation Services Agency (PAŻP), a Polish institution charged with managing the safety of air traffic (similar to the Finnish agency Fintraffic)
PGNiG	Polish state-controlled oil and gas company (PGNiG)
POLSA	Polish Space Agency (POLSA), Poland's space agency that collects climate information
UTK	Office of Rail Transport (UTK), the Polish governmental agency that regulates the rail market, transport safety, and technical issues
WTP, MPK	Various public transportation services that include internal intelligence efforts to collect data on their usage, ticket scammers, and fare evasion, such as the Warsaw Public Transport (WTP) and its counterpart in the Krakow (MPK).

Source: Table compiled by Authors from Internet and government sources.

Table 9.4 Sample of Potential Private Sector Contributors to All-Hazards Intelligence

Polkomtel sp. z.o.o.	Polish telecommunication and media company that operates the country's largest cell network—Plus
National Federation of Polish NGOs (OFOP)	Association of various Polish-based NGOs
University of Warsaw	Largest post-secondary education institution in Poland and home to the largest population of foreign students within the country
STS Gaming Group	Large Polish legal sports betting company
Various niche clubs	Such as the Polish Blockchain Association or the Polish United Travel Club—two online-based social media communities based around the exchange of knowledge on specific subjects

Source: Compiled by Authors from internet sources.

Security leaders seeking full situational awareness and an all-hazards intelligence picture might consider a matrix of possible contributors to organize inputs. Integrating the numerous sources of information is a challenge. Deciding what can be paid for and how, who should be involved, the timeliness of reports, the cyclical process of seeking and reviewing information, and the status of different informers and sources is a challenge for small countries. Table 9.5 simplifies categories of sources without implying organization.

The nature of the international economy and governance means that almost every cell in Table 9.5 extends beyond state boundaries and agencies. Financial intelligence (FINTRAC), for example, is augmented by membership in the Egmont Group of Financial Intelligence Units and the Financial Action Task Force. Police services share intelligence through INTERPOL and regional networks like the Southeast European Law Enforcement Center or the West African Police Information System. Telecommunications agencies share information through the industry association Global System for Mobile Communications Agency, and the UN standard-setting International Telecommunications Union. The Forum of Incident Response and Security Teams is an eclectic global association of computer security response teams, industry experts, government officials, and academics that shares information on cybersecurity incidents, threats, and vulnerabilities. Effective networking can increase capacity, but it also increases vulnerability to

Table 9.5 Simplified Theoretical Table for All-Hazards Intelligence Resources

Intelligence agencies	*Secondary intelligence assets*	*Supplemental government assets*	*Private resources*
Internal security intelligence	Military police	Municipal police services	Telecommunications
Foreign intelligence	Customs and border services	Public infrastructure management	Non-governmental organizations
Military intelligence	Treasury and finance, tax authorities	Air traffic control and airspace monitoring	Universities and academic expertise
Counter-intelligence	Financial intelligence	Rail and road transport agencies	Hospitality industry actors
Anti-corruption task forces	Personnel security and background checking	Space agency	Organizations with technical expertise
Police intelligence	Counterterrorism units	State energy companies	Niche clubs with access to special knowledge
Financial tracking	Territorial defense organizations	Municipal transport services	Civil society groups with special knowledge

Source: Compiled by Authors from discussions, de Graaff and Nyce 2016; Shaffer 2022, 2023

security leaks and penetration by hostile intelligence. Counter-intelligence, personnel screening, and effective information classification help to reduce vulnerability. Organization for effective inter-agency coordination is also essential.

Organization for All-Hazards Intelligence

Intelligence communities take many forms. Some are optimized for control by a regime, some for oversight, and some for efficient operations. Intelligence legislation also takes many forms, and official secrets legislation must be considered alongside laws about privacy and access to information. Many countries have emergency legislation or martial law provisions that expand the powers of intelligence agencies or place new obligations on citizens. The Grey Zone implies that many risks facing small countries arise before such legislation is triggered. Situational awareness, early warning, and effective response may, therefore, depend on initiative and ad hoc responses by professionals in various agencies, including sharing and seeking information that may be on the edge of, or beyond, their mandates. The calculus of repercussions for all-hazards intelligence efforts plays out at every level, from the individual operator and analyst to the head of state.

Decisions about alignment and sharing are continuous, incremental, and sometimes irrevocable. Small states may seek to maximize their capabilities through partnerships and exchange arrangements with larger allies but may never know if they are being manipulated by information they cannot verify. Supporting common objectives and sharing information is the cost of receiving support from allies. Before joining the European Economic Community (EEC) in 1973, Ireland's intelligence community focused narrowly on domestic issues and its relationship with the United Kingdom, particularly the consequences of Irish nationalism. Joining the EU in 1993 opened Ireland to pressures to participate in broader international efforts to support the European project. The capabilities and directions of Dublin's intelligence efforts turned toward partnership arrangements on transnational drug smuggling and human trafficking and alignment with NATO interests in international counterterrorism despite being a neutral state (O'Halpin 2016:189–192).

Internal organization is a function of intelligence culture—values, attitudes, and beliefs that result in ways of organizing behavior. Many states have distinct political cultures, which may cluster for historical reasons. For example, English-speaking parliamentary democracies, French-speaking presidential republics, and successor states of the former Soviet Union have distinct institutions and expectations about the judicial process, the rule of law, and legal oversight of policing and security intelligence. Institutional changes may be undertaken in response to common pressures, like urbanization, privatization, globalization, and digitization. A common pattern is to require individual agencies or clusters of agencies to report to elected officials in legislative or executive bodies (de Graaff and Nyce 2016; Shaffer 2022, 2023). Montenegro's use of the Coordination Bureau of Security Services, with the National Security Agency responsible for monitoring and control, reflects an effort to follow European norms despite the origins of Montenegro's security service in the former Yugoslavia. The structure of Serbia's intelligence services contrasts with those of Montenegro (de Graaff and Nyce 2016).

Maintaining different silos or organizations to address different intelligence problems is a widespread practice. Competition between intelligence organizations is also a common feature of intelligence communities. One scheme for dividing responsibilities relies on a distinction between strategic intelligence to support policy decisions and tactical intelligence to support operations. Another common distinction is between subject areas. Military intelligence concerns kinetic forces and enemy intentions. Political or security intelligence concerns actions related to subversion or the regime's overthrow—sometimes called "high policing" (Brodeur 2010:36–37). Economic, financial, criminal, cyber, and other forms may each have their own agencies, because they focus on different targets, use different assets, engage different parts of the state apparatus, and use different tools. The challenge for the small state is to organize these separate tasks or agencies in ways that permit concerted efforts in the national interest.

The Canadian model includes assets like the Polish model described above, including military intelligence, Canadian Security Intelligence Services (CSIS), the Communications Security Establishment (CSE), criminal intelligence (within the Royal Canadian Mounted Police), FINTRAC, diplomatic relations, including intelligence relationships (Global Affairs Canada—GAC), and the Canada Border Services Agency (CBSA). Organizational distinctions are reflected in the legal mandates of intelligence agencies that restrict their actions and internal policies and shape their overall culture toward intelligence collection. Factors such as the use of force, international deployment, and restrictions on spying on citizens define different organizations.

Tools to bring these agencies and departments together include task forces and centers. The Integrated Terrorism Assessment Centre (ITAC), for example, is an analysis agency focused on terrorist threats to Canadian interests (NSICOP 2018). Task forces are formed to address areas of concern to decision-makers. Each might have a different focus depending on the nature of the concern driving its establishment. Hypothetically, a Middle East Task Force might encompass a wide geographic area and be concerned with both terrorism and support of regional peace processes. A Ukraine Task Force might be mainly concerned with military lessons learned and support to Canada's NATO commitments. An Asia-Pacific task force might focus on China, with trade and investment relations the initial concern, evolving toward greater concern with domestic interference, global influence, and peer military competition. While the task force approach falls short of an all-hazards approach, each task force can tailor its access and composition to address the concerns of its stakeholders.

As Table 9.6 illustrates, many countries have equivalent functional focal points for similar organizations. There is greater variety in the tolerance for using force, covert operations, active deception, spying on or coercing citizens of the home state or allies, oversight by democratic institutions, and violation of domestic or international law. Requiring political oversight

Table 9.6 Polish (right) and Canadian (left) Intelligence Organizations' Focus

Military strategic	*Operational*	*Civilian strategic*	*Criminal*
	Police		
	Internal security (ABW)		
	Anti-corruption (CBA)		
	Foreign Intelligence (AW)		
	Border Guard (SG)		
Military Intelligence (SWW)			
Counter Intelligence (SKW)			
Special Operations (WS)			
	Gendarme (ZW)		Treasury (SCS)
	Protection (SOP)		
			Financial (GIIF)

Military strategic	*Operational*	*Civilian strategic*	*Criminal*
Military Intelligence (CAF)			
	Security Intelligence (CSIS)		
	Federal		
Communications Security (CSE)			
		Border security (CBSA)	
		GAC	
			FINTRAC
	Terrorism Assessment (ITAC)		
Regional and subject-specific task forces bring together experts from multiple agencies.			

Source: Compiled by Authors from de Graaff and Nyce 2016:217, NSICOP 2018, and personal communications. Służba Ochrony Państwa (n.d.); Ministerstwo Spraw Wewnętrznych i Administracji (n.d.); Wojsko Polskie (n.d.); Ministerstwo Obrony Narodowej (n.d.).

of activities verging on the illegal can taint politicians and introduce them to "dirty tricks" they might be tempted to use.

Conclusions

Small states don't have the same resources as their larger counterparts but might have other advantages. All-hazards intelligence focuses on local needs, not global engagements, although some states have wider engagements than others. Governments balance the values of reputation and the moral high ground against the utility of deviousness and illegal actions. Unique local intelligence cultures help cloak these choices. The targets for all-hazards intelligence collection span the full spectrum of intersecting political, economic, social, technological, legal, and environmental subjects, extending to people, relationships, and processes. Assets for gathering and analyzing all-hazards intelligence include the usual military and civil intelligence assets and people and organizations in the private sector and civil society. Organizing these assets effectively demands new thinking about combining assets and information streams and shielding sources and methods.

Note

1 The Authors acknowledge the many contributions of intelligence professionals who prefer not to be named. Others are included in the book's Acknowledgments.

References

Andrew, C. (2018), *The Secret World: A History of Intelligence*, Penguin, London.

Boddaert, G., Mordant, P., Pimpec-Barthes, F., Martinod, E., Aguir, S., Leprince, P., Raux, M., Couëtil, J., Fiore, A., Lescot, T., Malgras, B., Pons, F. and Castier, Y. (2017), "Surgical Management of Penetrating Thoracic Injuries during the Paris Attacks on 13 November 2015", *European Journal of Cardio-Thoracic Surgery,* vol. 51, no. 6, pp. 1195–1202.

Brodeur, J. P. (2010), *The Policing Web*, Oxford University Press, Oxford.

Brzuszkiewicz, S. (2018), "Radicalisation in Europe after the Fall of Islamic State: Trends and Risks", *European View*, vol. 17, no. 2, pp. 145–154.

Davies, P. (2019), *The Authorised History of British Defence Economic Intelligence: A Cold War in Whitehall, 1929–90*, Taylor & Francis, London.

De Graaff, B. and Nyce, J. (2016), *The Handbook of European Intelligence Cultures*, Rowman & Littlefield, New York.

Emerging Markets Information System (EMIS) (2024), *Explosia A.S. (Czech Republic) Company Report.* https://www.emis.com/php/company-profile/CZ/Explosia_AS_en_1420640.html

Fintraffic (2022), *Fintraffic Air Navigation Services Ltd. Annual Report 2022*, Helsinki, Finland.

Gill, P. and Pythian, M. (2018), *Intelligence in an Insecure World*, Polity Press, Cambridge, UK.

Government Offices of Sweden (2024), *Sweden Supports Czech Ammunition Initiative for Artillery Shells to Ukraine*, (18 March), Ministry of Defence. https://www.government.se/press-releases/2024/03/sweden-supports-czech-ammunition-initiative-for-artillery-shells-to-ukraine/

Hao, K. (2021), "Troll Farms Reached 140 Million Americans a Month on Facebook before 2020 Election, Internal Report Shows", *MIT Technology Review*, 16 September. https://www.technologyreview.com/2021/09/16/1035851/facebook-troll-farms-report-us-2020-election/

Helmes-Hayes, R. and Curtis, J. (eds) (1998), *The Vertical Mosaic Revisited*, University of Toronto Press, Toronto, Canada.

Jager, T. and Daun, A. (2009), *Geheimdienste in Europa: Transformation, Kooperation, und Kontrolle (Intelligence Services in Europe: Transformation, Cooperation, and Control)*, VS Verlag für Sozialwissenschaften, Wiesbaden.

Johnson, L. K. (ed.) (2007), *Handbook of Intelligence Studies*, Routledge, New York.

Johnson, L. K. (ed.) (2010), *The Oxford Handbook of National Security Intelligence*, Oxford University Press, Oxford.

Juhász, A. and Szicherle, P. (2017), *The Political Effects of Migration-related Fake News, Disinformation and Conspiracy Theories in Europe*, Friedrich Ebert Stiftung and Political Capital, Budapest.

Kajtazi, R. (2014), "Organized Crime in the Field of Narco-criminality as a Challenge of Western Balkans Countries: A Special Overview of Montenegro", *Mediterranean Journal of Social Sciences*, vol. 5, no.19, pp. 580–591.

Karaskova, I., Berzina-Cerenkova, U. A. and Nemckova, K. (2023), *Foreign Electoral Interference Affecting EU Democratic Processes*, Authority for European Political Parties and European Political Foundations, Brussels.

Kruger M, Osakwe P., Page J. (1998), *Fundamentals, Contagion and Currency Crises: An Empirical Analysis*, International Department Staff, Working Paper 1998-10, Bank of Canada, Ottawa.

Liang, H. H. (1992), *The Rise of Modern Police and the European State System from Metternich to the Second World War*, Cambridge University Press, Cambridge.

Ministerstwo Obrony Narodowej (n.d.), Wojska Specjalne (Special Forces). https://www.gov.pl/web/obrona-narodowa/wojska-specjalne

Ministerstwo Spraw Wewnętrznych i Administracji (MSWiA) (n.d.), Służba Celno-Skarbowa (Customs and Fiscal Service). https://www.gov.pl/web/mswia/sluzba-celno-skarbowa

Mohammadi, M. (2023), "Iran Expands Influence Through Islamic Students Associations in Europe", *Iranian International*, 4 December. https://www.iranintl.com/en/202312043532

National Security and Intelligence Committee of Parliamentarians (NSICOP) (2018), *Chapter 2: A Functional Overview of the Security and Intelligence Community*. Annual Report 2018. Government of Canada.

O'Halpin, E. (2016), "Ireland", in De Graaff, B. and Nyce, J. (eds), *The Handbook of European Intelligence Cultures*, Rowman and Littlefield, Lanham.

Porter, J. (2015), *The Vertical Mosaic: An Analysis of Social Class and Power in Canada*, 50th Anniversary edn. with Introductory Essay by J. Jedwab and V. Satzewich, University of Toronto Press, Toronto.

Rickards, J. (2011), *Currency Wars: The Making of the Next Global Crisis*, Portfolio, New York.

Rid, T. (2020), *Active Measures: The Secret History of Disinformation and Political Warfare*, Farrar, Straus and Giroux, New York.

Shaffer, R. (2022), *The Handbook of Asian Intelligence Cultures*, Rowman & Littlefield, New York.

Shaffer, R. (2023), *The Handbook of African Intelligence Cultures*, Rowman & Littlefield, New York.

Shimer, D. (2020), *Rigged: America, Russia, and One Hundred Years of Covert Electoral Interference*, Alfred A. Knopf, New York.

Służba Ochrony Państwa (n.d.), Przeznaczenie i zadania (Purpose and tasks). https://www.sop.gov.pl/pl/o-sluzbie/struktura-i-dzialanie/przeznaczenie-i-zadania/220,Przeznaczenie-i-zadania.html

Soldatos, J., Praça, I. and Jovanović, A. (2021), *Cyber-Physical Threat Intelligence for Critical Infrastructures Security: Securing Critical Infrastructures in Air Transport, Water, Gas, Healthcare, Finance and Industry,* Now Publishers, Boston-Delft.

Suri, N. (2020), "Australia-China Relations: The Great Unravelling", ORF Issue Brief No. 366 (June 2020), Observer Research Foundation.

Thomas, M. (2005), "Colonial States as Intelligence States: Security Policing and the Limits of Colonial Rule in France's Muslim Territories, 1920–40", *The Journal of Strategic Studies*, vol. 28, no. 6, pp. 1033–1060.

Vasquez, R. (2021), "Law Enforcement Intelligence Process", in M.R. Landahl and T.E. Thornton (eds.), *The Role of Law Enforcement in Emergency Management and Homeland Security*, vol. 24, pp. 131–151, Emerald Publishing Limited, Bingley, UK.

Wojsko Polskie (n.d.), Wojska Specjalne (Special Forces). https://www.wojsko-polskie.pl/wojska-specjalne/

10 Logistics Cooperation for Modern War

Per Skoglund

Introduction

Logistics is about acquiring resources, moving forces, and supporting the forces with supplies, maintenance, and medical support (NATO 2012). These activities are carried out in peace, during Grey Zone conflict, and in war. There is, however, a difference in focus and priorities. In peace, logistics activities are about building capacities and capabilities; in war, the focus is on effectively using the resources. Logisticians often argue that the Grey Zone period is the most important for logistics activities. Demonstrating operational capabilities, including logistics, lowers the risk of war. If preparations do not prevent war, this period gives the country time to reduce imbalances in capabilities and increase the sustainability of their defense forces through procurement, exercises, maintenance of equipment, and increasing stocks of spare parts and consumables. Armed forces must be prepared to conduct comprehensive logistics support for operations in an unpredictable environment (Jałowiec 2017).

We do not know what the next war will look like. We assume it will differ from the past, but we cannot know. We can guess there will be a period before the war breaks out that we call the Grey Zone. Here, we focus on logistics in this period. It is also possible that the Grey Zone is not a prelude to war but a permanent condition like the Cold War, in which states seek advantage by means that fall short of war. There is a risk that these hostile actions erode logistics capacity to respond to war, perhaps making war more likely.

Most countries discuss three levels of logistics: strategic, operational, and tactical (Kress 2016). Strategic logistics is about capacity building, including acquiring facilities, supplies, and services for the armed forces. Operational logistics involves planning and decision-making to use the logistic resources and capabilities for training or combat. Finally, tactical logistics consists of the activities logistics forces conduct to manage operational logistics planning and decisions. Tactical logistics provides the combat forces with supplies and services, including medical support.

In this chapter, small countries are assumed to have one strategic command and one operational command, with subordinate component commands on the tactical level. The logistics command is viewed as a component command at the same level as the army, navy, and air force. The logistics command's tasks are to provide logistics supply and services to the other component commands. While operational and tactical tasks will differ in peace, Grey Zone, and war, the levels will not change. Countries can choose to organize logistics differently, but the logistical principles remain the same, and therefore, national organizational differences are not discussed.

Small countries are vulnerable. They lack the capacity to support and supply their armed forces with domestic production. Small countries understand that they can import and export security but cannot develop the necessary capacity and capability alone (Markowski, Hall and

DOI: 10.4324/9781003428701-15

Wylie 2010). Small countries must decide who to depend upon and to what degree the country should be able to defend itself with its own resources. Ultimately, it is about balancing defense expenditure against other priorities, such as health care. These decisions will influence other fundamental decisions. Will the government invest in infrastructure that supports military operations and territorial defense? Everything from ports, roads, and railways to telecommunications, hospitals, water purification, power stations, and transmission lines can be part of a national logistics support network, which can be vital during wartime. In many small states, these systems may be developed by civilian planners who do not consult military logisticians and engineers.

The war in Ukraine is a typical example of relying on the support of others. The Ukrainian Armed Forces depend on support from other countries supplying equipment and ammunition. The Ukrainian situation is unique but highlights important aspects of logistics (Skoglund, Listou and Ekström 2022). Another example is the Winter War between the Soviet Union and Finland. Finland was small compared to today's Ukraine. However, there are important similarities. Finland depended on supplies from abroad, and Sweden supplied ammunition and weapons (Backström and Ekberg 2000). Discussions between Finland and Sweden were ongoing before the Winter War started, during the period which we now would call the Grey Zone. Similarly, Ukraine put much effort into getting support and building relations with NATO and EU members before the 2022 invasion. A low-intensity war has been going on in eastern Ukraine since 2014. 2014–2022 can thus be viewed as a high-level Grey Zone conflict.

These two examples support the conclusion that few small countries, if any, can independently supply their armed forces with all the necessary supplies for war. The smaller the country, the more dependent it becomes on foreign suppliers to support it, including its armed forces.

Strategic Logistics in Small States

Strategic logistics builds military capability and capacity. To create capabilities and capacities, the armed forces must balance equipment procurement and procurement of the logistics system supporting that equipment (Eccles 1959). Small states tend to prioritize acquisition over logistics (Markowski, Hall and Wylie 2010). A larger number of platforms are assumed to demonstrate force more effectively than a logistic system. However, a good logistics system may be more important for deterrence than the number of weapon system platforms. The problem for deterrence is to convince potential adversaries that a well-functioning logistics system exists. A well-functioning system includes secure external sources, alternative supply chains, stockpiles, warehousing, internal transport, repair facilities, and more. All of this can be eroded in Grey Zone operations by an adversary.

A second explanation for the lower priority of logistics is that taxpayers' expectations limit the national budget for defense. Obvious needs for health, education, and welfare spending eclipse defense. The limited budget is spent on more platforms (Hartley 2011). Defense planners hope to be able to increase spending when tensions rise, increasing the capabilities of the logistics system required to support the main equipment and operations. Grey Zone operations by an adversary may simultaneously target the will to increase spending, erode access to international sources, and prepare sabotage and attacks on internal assets.

A third reason logistics is a lower priority is miscalculation of the extent of logistics resources needed. Small states rarely engage in hostile military activities on any scale and consequently lack some of the experience that larger states have when building a balanced logistics system. This perception can also be manipulated by disinformation and influence operations sponsored by an adversary.

Procuring and building military capabilities for small states is achieved through relations with others—countries and suppliers (Markowski, Hall and Wylie 2010). Most countries depend on international defense markets to supply their forces (Chapter 5). The smaller the country, the greater the dependence on foreign suppliers. Small countries need strategies for supplying their defense forces with equipment and support that differ from those of larger countries. Consequently, these countries must make several long-term strategic decisions: defense capability, alliance and partner strategy, national defense industry strategy, and procurement strategy. Ultimately, the decisions depend on the national economy and the taxpayers' willingness to spend resources on defense during peace and Grey Zone periods (Hartley 2011). Admiral Eccles (1959) summarizes logistics as the bridge between the national economy and the operations of combat forces.

Defense Industry Strategy

The national defense industry in small nations will vary in size and direction. The country's defense ambition is important for the firms' decisions on what markets to act in, how the supply chain will be organized, and what national production capabilities the industry should have in Grey Zone and war. The firms' roles will differ between countries and depend on history and long-term commitment from the state. The Nordic countries can be used to exemplify different strategies. Denmark does not consider national defense industry capacity as critical. Instead, they rely on their supply chains on the international market, NATO, and partner-nation support. Finland focuses its industrial capability on two legs—maintenance and ammunition production. All systems in the Finnish Defense Forces must be maintainable in-house or with national industry capability, and national ammunition producers must have a predefined agreed production capability with long-term contracts. Norway seeks to create interdependent industrial production capabilities, where their industry produces primarily for export, and in exchange, Norway imports other types of defense material. Like Finland, Norway views ammunition as strategically important for national defense capability. Sweden has focused on defense firms that are system integrators with international supply chains. Swedish industry has the ambition to meet the requirements for the most important systems for the Swedish Armed Forces, to maintain and upgrade systems during Grey Zone periods and war. The Swedish government has focused on specific technology areas of strategic security importance, including fighter aircraft and submarine technology.

At the end of the twentieth century, many European countries started to change their national industry strategy due to reduced defense budgets. This affected the larger countries, such as the United Kingdom and France, and smaller ones, such as Sweden and the Netherlands. It resulted in an era of privatization and restructuring in the European defense industry, with mergers, acquisitions, and consolidation to respond to increased competition in the international market. The general trend in business was internationalization and globalization, which also affected the defense industry. Both these factors led to increased interdependencies in supply chains and between countries. These changes and their consequences are, if not irreversible, at least difficult, expensive, and time-consuming to change. However, the Ukraine war will undoubtedly lead to a new situation where small countries review and revise their defense capabilities and defense industry strategies. They must formulate new requirements for their supply chains, excluding suppliers from hostile and untrustworthy nations. Furthermore, they will require new equipment, which the defense industry must develop and produce.

Defense is not a perfect market in economic theory. It consists of few buyers and few suppliers. Suppliers know each other's market power and know how to act strategically in response to their competitors (Markowski, Hall and Wylie 2010:33). The defense market is a monopolistic competition for other products, where the few suppliers have partially differentiated products, and customers balance prices against capabilities and needs. One important question for the future is whether suppliers will find their own path to collaboration or whether states will use funding to force collaboration upon them. Actors like the EU and NATO may also use incentives to encourage nations and industries to collaborate. Since defense material production is strongly related to national defense capability, it can be expected that larger countries and multinational organizations will engage in the industry's future development in small countries. Small and large countries alike have a common interest in preventing the proliferation of new technology to hostile states. Among small countries, none can have the industrial capacity to meet all the requirements of their defense forces. The Nordic countries' different approaches provide examples of how small countries can decide to meet their national needs and make themselves attractive partners in bilateral or multilateral relations. Ultimately, small countries will depend on good relations with other countries and industries to have cost-efficient defense forces.

Sourcing Strategy

A comprehensive perspective on logistics defense capacities and capabilities is a fundamental prerequisite for formulating and subsequently implementing appropriate supply chain strategies, including sourcing strategies (Ekström, Hilletofth and Skoglund 2020a). Defense logistics requires a wide range of services and supplies, for which a small country must allocate resources from all sectors of society, including public authorities and private firms, in addition to relying on partners. Requirements range from delivering and maintaining defense equipment to daily support of forces, including sustenance and medical support, to maintain combat capability. Logistics is carried out in peace, in the Grey Zone, and in war, but priorities and permissible costs differ. This has implications for formulating and implementing supply chain strategies (Tatham 2005). A sourcing strategy for national defense is a two-step strategy (Skoglund 2012). The first step concerns what to make with its own resources and what to procure from suppliers. The second step addresses how to procure—by competition or sole-source from a targeted supplier. Both steps should be based on rules in the sourcing strategy. The choice to make or buy is possibly the most important choice for the defense forces, especially for different services. Most defense forces maintain a lot of their equipment in-house. If suppliers do maintenance, most have maintenance facilities within the country's borders, and only a few pieces of equipment or subsystems are sent abroad for repairs.

During the Cold War, many countries decided on important areas to establish operations in-house or in government-owned firms. Several European countries sold these operations to private investors after the end of the Cold War. However, some have kept control over the firms' decision-making by keeping a shareholder majority or having different varieties of golden share ownership. These stock positions give the nation the right to use a veto against corporate decisions. Through government control, national knowledge and production are retained. Others have legislation that at least partially protects national interests. This way, the armed forces' requirements are protected in peace, Grey Zone periods, and war.

National procurement strategies are connected to a country's ambition for defense capability and its defense industry strategy. Firms and technology areas the government has defined as strategically important require one type of procurement decision, while other areas are open to a larger variety of procurements. Based on a Delphic study of Swedish decision-makers, the segmentation procurement model begins with an understanding of defense needs, market capabilities, and limitations (Ekström, Hilletofth and Skoglund 2021). Figure 10.1 presents a

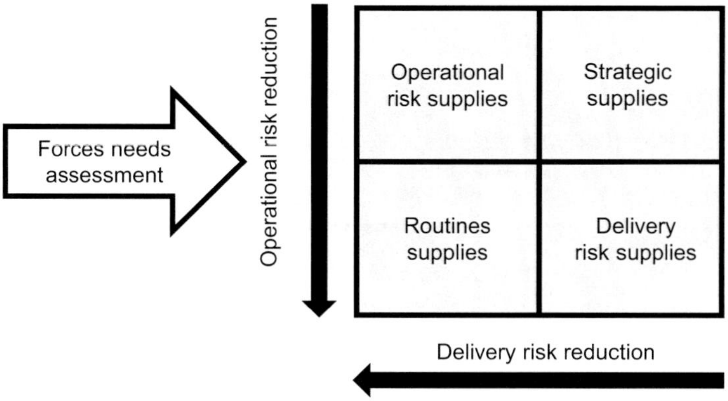

Figure 10.1 Segmentation Model for Procurement.

Source: Adapted from Ekström, Hilletofth and Skoglund (2020).

simplified version. The model begins with a needs assessment of the armed forces' requirements to understand what supply is needed and how important it is for the forces. The second step is analyzing the supply to identify its position in the matrix based on market capabilities and limitations. The third step is to analyze if anything can be done to move the supply toward the lower left corner, which is a more favorable position (Ekström, Hilletofth and Skoglund 2020b).

Applying the segmentation model highlights two procurement strategies: *partnership contracts* for critical materiel and *competitive contracting* for material the market can provide even in war (see Figure 10.2). Materiel may be critical because it is essential for operations or because delivery is time-sensitive.

Partnership contracts are often connected to critical supplies, which require long-term relations for development and support. Competitive contracts are often not critical for military capability. Additionally, supplies provided through competitive procurement can often be substituted with similar supplies. The supply can be critical to defense, but it does not require maintaining a relationship with the supplier.

Procurement has different purposes. Ensuring military operational availability, creating sustainability through stockpiling for future needs, and creating dormant relationships with suppliers expected to provide the forces with resources during Grey Zone periods or in war. Operational availability aims to prepare combat forces for intended tasks without further supply. Sustainability reflects how long the combat forces can remain operationally capable. Procurement for operational availability is directly connected to the number of different military units the nation has and the time these units are expected to operate without additional supplies. Dormant relationships can be both for availability when new units are equipped and for sustainability when existing forces are being resupplied. A small country might have a million or more separate supply items. Consequently, items must be grouped into categories for rational treatment.

The war in Ukraine shows that unexpected operational needs are normal. Small countries cannot build stocks or capabilities for all conceivable threats alone. Dormant relationships with industries and alliances with other nations are required to meet unforeseen contingencies.

A robust and resilient supply system must balance single, dual, and multiple suppliers. Strategic partners generally become single suppliers. They are often positioned at the end of the supply chain, acting as system integrators. Strategic partners can also be upstream in the supply

	Consumables	Major Equipment
Defence Critical Supplies	Short term Partnership Contracts	Long term Partnership Contracts
Non-critical Supplies	Short term Competitive Contracts	Long term Competitive Contracts

Figure 10.2 Different Contract Types.

Source: Author's elaboration.

chain when they produce sub-systems considered critical for the armed forces' capabilities. In industry, it is often argued that the optimal number of strategic partners is few—around five. This number has not been discussed in detail for the armed forces and national defense, but one can argue that the number can be higher because the different branches can have three to five strategic partners each.

Several factors must be considered when deciding on single, dual, or multiple suppliers. The estimated probability of receiving supplies during Grey Zone conflicts and war may be an important factor for small countries. Sourcing decisions are relevant when a supply chain is formed to ensure supply security. It is important not only which firms are engaged but also where the production facilities are located, where the firms are registered, and who owns or influences the firms.

Non-Preferred Suppliers

For 20 years (1994–2014), many Western nations considered the risk of a large conflict in which their own nation was at risk to be negligible. Russia and China were believed to be more interested in international trade than conflict. In line with these assessments, many Western nations increased their international trade, including buying components and subsystems for defense equipment on the open market. They became dependent on global supply chains for availability and sustainability. Starting with the Russian aggression against Georgia in 2008 and Chinese hostile activities in the South China Sea, the view among Western nations gradually shifted back toward the importance of territorial defense. They also realized the inappropriateness of upstream suppliers in global defense supply chains owned by interests-connected nations considered hostile.

There are several good reasons not to do business with some states. They might be viewed as potential enemies or under the influence of potential enemies, which means that in case of a war, all deliveries would stop. There is also a significant risk of technology theft in the supply chains. The potential for sabotage of components or systems also increases with the inclusion of sophisticated software.

Concerns about non-preferred suppliers accelerated with the COVID-19 pandemic and the Ukrainian war. The private sector added its concerns. Industry started to insource or shorten the geographical distance to sub-contractors. Other supply chain risk factors were analyzed. When a small country considers this aspect, there is a need to balance between cost, trustworthiness, and capability to deliver in peace, Grey Zone conflict, or war. One policy response is to procure and hold stock for the Grey Zone and war needs. Monitoring other stockholders should be part of normal security intelligence.

Alliance Strategies

Small nations are, in one way or another, always dependent on alliances. Sweden has proudly declared its 200 years of peace and neutrality, but this is not the only truth. During the Finnish Winter War, Sweden helped Finland with equipment and ammunition. Since World War II, the Nordic countries have cooperated in different ways, such as procurement, exercises, and EU forces. These activities aimed to develop joint thinking and preparedness to assist in times of Grey Zone conflict or war. Staying neutral comes with higher defense costs and the risk of being alone if attacked. Alliances can be bilateral, multilateral, regional, or international. Alliance agreements between nations are often a combination of civil and military arrangements. Figure 10.3 exemplifies some of the Norwegian relations. Norway and the United States have a bilateral agreement on prepositioning Marine Corps equipment and materiel, the Marine Corps Prepositioning Program-Norway (MCPP-N in Figure 10.3). Prepositioning in Norway is one way to reduce logistics needs in a wartime situation. Most of the equipment and supplies needed for a brigade are prepositioned in the planned area of operation. Prepositioning requires Norwegian civilian capacities to support the sustainment of the US materiel. Multilateral alliances can be strictly military, like NATO or Nordic Defense Cooperation (NORDEFCO), or multifunctional, like the UN and the EU. The alliance can furthermore be with or without mutual defense declarations. Norway is a member of NORDEFCO (Figure 10.3), focusing on reducing procurement and logistics costs and equipping the forces with the same type of material to have joint spare parts pools and common ammunition. NORDEFCO may be extended into periods of war now when all members become NATO members.

To continue with the Norwegian example, NATO membership (Figure 10.3) creates congruent logistics strategies between nations. The three most important developments of the NATO concept for logistics are the NATO Support and Procurement Agency (NSPA), the Joint Logistics Support Group (JLSG), and the Host Nation Support concept (HNS). The NSPA focuses on creating economies of scale for the member nations through larger procurement batches and reduced transaction costs for the members. The JLSG focuses on providing logistics support to the nations participating in a wartime operation. The ambition of the JLSG is to reduce the national need to bring logistics resources to the area of operations. In the same way, the HNS concept is an important enabler for the force contributing nations in alliances, where resources can be provided by the host nation. The host nation must plan to provide logistics support and other areas, like legal services and civil-military relations. The backbone of logistics HNS capability is twofold: geographical knowledge, including weather conditions, and access to the local market, in most cases through previously signed contracts. Within NATO, different alignments and organizational arrangements play different roles during peace, Grey Zone conflict, and war, as illustrated in Figure 10.3.

In this example, Norway has a logistics doctrine for peace, the Grey Zone, and war. The solution is dependent on cooperation with other nations. The Norwegian example shows a long-term alliance strategy for building security and logistics capacity with other nations and organizations. Small, neutral countries build their logistics capability and capacity with others. For example, Austria has cooperated with several European nations to procure and maintain defense equipment.

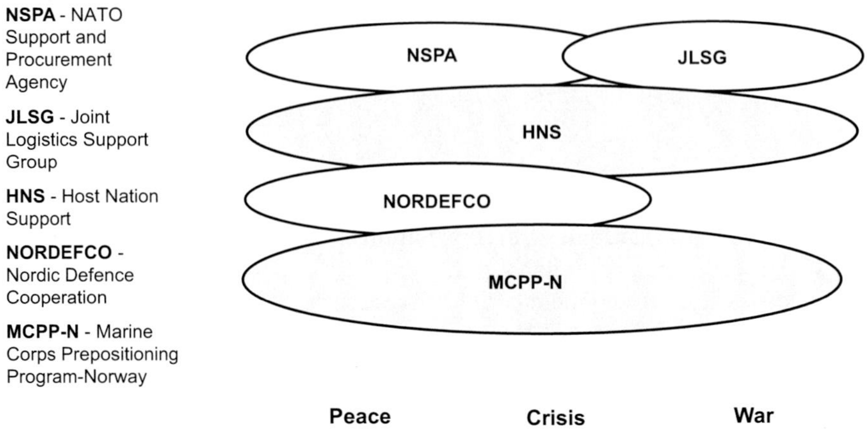

Figure 10.3 Alliances and Organizations for Logistics in the Nordic Region.

Source: Author's elaboration.

Defense Capability from a Logistics Perspective

Implementing the above strategies creates a comprehensive perspective for logistics defense capabilities, where public, private, military, and civilian resources are used. In democracies, the peacetime implementation of strategies depends on the people's willingness to pay taxes to finance the defense sector. The willingness to pay taxes for defense capabilities is strongly connected to the experience of threat levels (Hartley 2011:81). Public perceptions of international expenditures and official risk assessments from the armed forces and the political leadership affect defense expenditure levels. The public must accept estimates and arguments as trustworthy. When defense budget cuts are demanded, countries first reduce general logistic capabilities and subsequently disband high-cost capabilities, like nuclear submarines. When the economy is growing, the same priorities should be used: first, general logistics and, thereafter, new capabilities or an increased number of platforms (Hartley 2011:90).

Logistics, on a strategic level, is about developing and implementing strategies. The strategies are intended to optimize spending during peacetime to produce a defense capability to prevent hostile actions toward the nation. Logistics contribute by maintaining trustworthy relations with national and foreign industries, suitably organized procurement processes, and good bilateral and multilateral defense agreements with other nations. Strategic logistics provide the prerequisites for having available and sustainable armed forces.

Operational Logistics

The military operational level is responsible for planning and conducting larger operations. Kress (2016) discusses operational logistics from a large state and war perspective, nevertheless relevant to small states in Grey Zone conflicts. Kress uses three perspectives on operational logistics—cognitive, functional, and practical. The cognitive perspective explains that operational logistics is the link between long-term planning on the strategic level and practical tasks on the tactical level to support units during training or combat. The functional perspective concerns what operational logistics do and how the decision-making on tactical-level logistics should be prioritized and executed. The practical perspective concerns how planning and decision-making are done.

Small states have continuous operations for national defense. These include exercises and control of national borders during peace, shows of force and infrastructure protection during Grey Zone periods, and full combat during war. The difference between a small and a large state is that the small state has insufficient resources to dedicate to only one task. Because of that, from a cognitive perspective, the logistics capabilities of small states must handle all phases with the understanding that equipment and personnel will be used in subsequent phases, requiring more resources.

From a functional perspective, operational logistics can be viewed as the planning and decision-making process that determines how and where to use the available logistic resources. The principles and processes will be like those of large states' staff work and decision-making, but on a smaller scale and with cultural and traditional variances. Since most small countries have decided to participate in alliances or alignments, they must be interoperable in their staff processes. The planning and decision-making for logistics are based on available logistics resources. These should not be viewed only as the resources the armed forces have on hand. Rather, they include all resources within a small nation's total defense capacity, agreed foreign supplies, and contracted services. The examples of Finland and Ukraine cited above show that more resources are available based on collaboration with other nations and organizations. Mobilizing civilian resources should not be excluded.

From a practical perspective, logistics in peacetime concerns mainly how to supply the defense forces' training needs and only, to a smaller degree, logistics for the protection of national borders. For some countries, border surveillance is mainly done by air force or navy resources and is combined with training objectives. Other countries have specialized border troops with both wartime and peacetime responsibilities. Vehicles and equipment are usually compatible with defense forces, and common logistic support is normal. During a Grey Zone period, three logistics objectives are important. First, peacetime logistics activities continue but with a higher standard to ensure the equipment reaches higher readiness after maintenance. Second, logistics preparations for war are undertaken to speed up maintenance, review storage levels, and, if needed, decide on procurement to increase or update stocks. Related to this second function, there should be heightened attention to threats below the threshold of war, including sabotage, supply disruption, and positioning of hostile assets to impede logistics. This requires close cooperation with all-hazard intelligence (Chapter 9). Third and most importantly, logistics must be ready to support force preparedness to handle incidents and to show force (Jałowiec 2017). As the threshold to war is approached, logistics professionals must provide operational commanders with advice about units, tasks, and priorities in view of the capabilities and limitations of logistics systems. The essential problem of the Grey Zone, or war nexus as Simonyi calls it (Chapter 1), is that these problems may be perennial rather than periodic.

During all phases, the operational logistics staff has the responsibility to develop the logistics to become robust and resilient through contingency planning and foresight. In peacetime, economic efficiency is a dominating factor in the planning and conduct of logistics. During the Grey Zone period, the focus is to transform logistics to become effective, that is to do the right things at the right time and to increase the robustness and resilience capacity of the logistics support. All countries will suffer logistics shortfalls during a war, but the problems for small countries are greater. They continually seek ways to minimize logistics consumption during a Grey Zone period by prioritizing necessary operational training, resource efficiencies, and incident preparedness. Both personnel and materiel must be ready to go from training to full combat within a few minutes for the Air Force, within a few hours for the Navy, and less than a day for the Army. These requirements put high demands on logistics organizations. With the limited resources of a small state, losses of materiel or personnel and supply disruption during the Grey

Zone period can critically impact combat capability. The maintenance organization must repair failures quickly, and logistics must be able to resupply or reallocate.

The principles and processes for operational logistics should be the same during peace, Grey Zone conflict, and war. The tempo increases during Grey Zone conflicts and war, but the processes remain unchanged. The difference in the processes is that input with lower quality may be accepted as urgency increases. For example, only three alternatives might be evaluated instead of the usual five.

Tactical Logistics

Tactical logistics concerns all practical activities to deliver the logistics support to combat forces in line with the operational level's decision. This means that all physical deliveries from stocks to combat forces, maintenance of equipment, and medical support are issues for tactical logistics. These activities are all carried out in peace, Grey Zone conflicts, and war. The prerequisites for tactical logistics are set at the strategic and operational levels, but their conduct is organized at the tactical level. The most important task for a small country is to create robust and resilient supply chains on the tactical level. Proactive activities to avoid disruptions in the supply chains are mainly a responsibility at the tactical level, even though all actors have different roles, all contributing to resilience and robustness.

Resilience and robustness have been discussed in the literature for 20 years. A robust system is hard to break; a resilient system can recover quickly from shocks. Most authors have discussed three concepts related to resilience: systems must absorb shocks, adapt to changes, and restore capabilities. A robust supply chain can be viewed in several ways. With fewer combat units and more advanced equipment, the demands on the supply chains increase. Components cannot be fixed in place. The service staff in the armed forces argue that they should control a larger part of the supply chain to avoid disruptions to their part of it and their operations. The heads of logistics in small states also tend to argue that they are responsible for the supply chains, and the J4 (logistics) within the operational staff discuss their objectives similarly. However, do they fully understand the words robust and resilient, or do they use them interchangeably? Figure 10.4 illustrates the difference between robustness and resilience. The figure

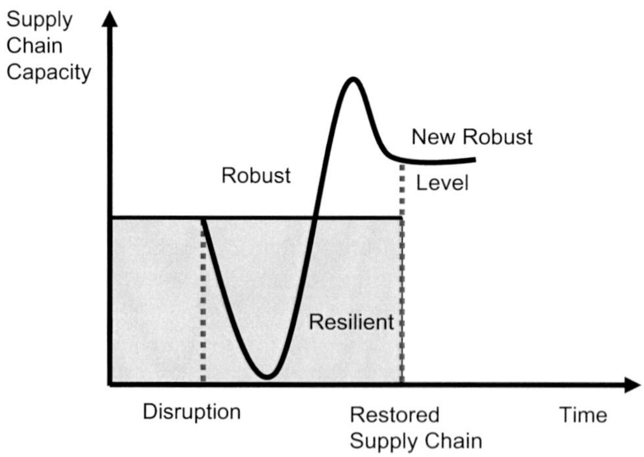

Figure 10.4 Robust and Resilient Supply Chains.

Source: Author's elaboration.

suggests that robust supply chains are not sensitive to disruptions. This can be achieved with overcapacity or redundant solutions. No supply chain can guarantee 100 percent robustness; therefore, a supply chain must also be resilient and have the inherent capacity to cope with disruptions. In this model, the capacity to absorb changes in the supply chain is a robust capability. If a disruption occurs, adaptation describes the supply chain's ability to handle the disruption. Restoration describes how a resilient supply chain recovers after a disruption. The compensation may surpass the original supply chain capacity in a strong system. Often the resulting supply chain may have a more efficient capability after restoration is finalized.

Small states must combine robustness and resilience in their supply chains. Developing a robust solution that can withstand both natural supply chain problems and hostile military actions is important. It is also important to have methods to cope with disruptions and to improve solutions by quickly adapting to new situations and restoring supply chains. The solutions to avoid disruptions and build resilience are often similar but dependent on specific circumstances. Successfully resilient supply chains have routine responses to disruptions.

Conclusions

Grey Zone conflict is seen by many professionals as the most intense and difficult phase during which to conduct logistics. On a strategic level, taxes should prioritize military capability in the ambiguous period preceding war, but this period may now be ubiquitous and perennial, and military capabilities alone cannot secure the state. Increased defense spending should give military decision-makers the funds needed to increase the availability and sustainability of forces through procurement. The difficult choice is to prioritize what should come first. On the operational level, the staff must balance the available resources between training to improve the capabilities of the combat forces and logistics preparedness for war. There are important training requirements, but resources must also be reserved in case of sudden escalation to war. On a tactical level, the focus is to improve the skills and adaptability among logisticians of all categories and to build up the robustness of the logistics system. Small states rely on cooperation with all types of actors to build a robust and resilient logistics system, including private firms, international organizations, and allied partners. For small states, resources will always be limited. Therefore, relations with others are critical for logistics and defense capability.

References

Backström, B. and Ekberg, H. (2000), *Finland i krig 1939–1945*, Schildt, Helsingfors.

Eccles, H. E. (1959), *Logistics in the National Defence*, Greenwood Press, Westport.

Ekström, T., Hilletofth, P. and Skoglund, P. (2020a), "Differentiation Strategies for Defence Supply Chain Design", *Journal of Defense Analytics and Logistics,* vol. 4, no. 2, pp. 183–202.

Ekström, T., Hilletofth, P. and Skoglund, P. (2020b), "Guidance for the Application of a Dynamic Purchasing Portfolio Model for Defence Procurement – A Swedish Perspective", *Necesse,* vol. 5, no. 3, pp. 138–158.

Ekström, T., Hilletofth, P. and Skoglund, P. (2021), "Towards a Purchasing Portfolio Model for Defence Procurement – A Delphi Study of Swedish Defence Authorities", *International Journal of Production Economics*, no. 233, pp. 1–12.

Hartley, K. (2011), *The Economics of Defence Policy: A New Perspective*, Routledge, London.

Jałowiec, T. (2017), *The Effectiveness of Logistics in the Armed Forces*, Publishing House of War Studies University, Warsaw.

Kress, M. (2016), *Operational Logistics: The Art and Science of Sustaining Military Operations*, 2nd edition, Springer Dordrecht, Cham.

Markowski, S., Hall, P. and Wylie, R. (eds.) (2010), *Defence Procurement and Industry Policy: A Small Country Perspective*, Routledge, London.

NATO (2012), *NATO Logistics Handbook.* https://archives.nato.int/nato-logistics-handbook-2

Skoglund, P. (2012), *Sourcing Decisions for Military Logistics in Peace Support Operations; A Case Study of the Swedish Armed Forces*, Jönköping International Business School, Jönköping.

Skoglund, P., Listou, T. and Ekström, T. (2022), "Russian Logistics in the Ukrainian War: Can Operational Failures be Attributed to Logistics?", *Scandinavian Journal of Military Studies,* vol, 5, no. 1, pp. 99–110.

Tatham, P. (2005), "Efficient in Peace but Effective in War: Meeting the Challenge of the Military Supply Chain - An Agenda for Research", *Supply Chain Practice*, vol. 7, no. 2, pp. 42–57.

Design Problem 2

Problem Finding with Convergent Thinking and the Rich Picture

Archipelago of Design (AoD)

Oliver Jones, Kassie Miedema, and Philippe Beaulieu-Brossard

Introduction

Section II of *Modern War and Grey Zones: Design for Small States* explores the connections between policy, strategy, operations, and tactics, focusing on how small states can navigate and leverage their positions. Small states must adapt and innovate in response to the complexities of modern warfare and Grey Zone conflicts. After identifying problems or issues that need to be addressed using a tool like horizon scanning, design tools can be used to hone the focus on the highest priorities. This exercise helps frame a problem using convergent thinking. Convergent thinking focuses on the most promising ideas, chooses the most pertinent information, and focuses on just a few possibilities.

Method

The *rich picture* tool is a creative and comprehensive way to carry out a stakeholder mapping exercise. Stakeholder mapping involves identifying, diagramming, and prioritizing actors by analyzing their influence and interest in a problem space. A rich picture visually represents planners' understanding of a specific situation and the stakeholders tied to that context. A rich picture aims not to produce an organization or process diagram but to develop a metaphor for a complex open system involving cooperation, competition, and conflict between actors. A rich picture activates a robust discussion between participants with different backgrounds, who provide input on actors and elements that impact the situation the participants have been asked to illustrate. This tool aims to create a shared understanding of an existing situation and begin mapping relationships between impacted stakeholders. A small group can produce a rich picture in one to two hours. The rules for producing one can be tailored to the participants' skill sets and knowledge base. The process of developing a rich picture is intended to be open-ended, and participants should negotiate what counts as meaningful to the group and what does not. Participants' can use the outputs of the rich picture tool to identify stakeholders and unseen impacts of the situation under consideration. Thinking about these stakeholders can help identify who a solution is for and how it may impact other actors. See Figure DP2 for an example of a rich picture using the metaphor of a forest fire to illustrate the forces and actors animating the Syrian Civil War.

Figure DP2 is adapted from a rich picture exercise conducted in the Joint Command and Staff Program of the Canadian Forces College. It depicts actors and stakeholders connected to the Syrian Civil War and the Islamic State conflict in the Middle East in 2016. The metaphor of a forest fire qualifies actors and elements that drive, mitigate, or complicate the ongoing conflict. The key states are trees in the forest. The civil war, ISIL conflict, and power struggle in

DOI: 10.4324/9781003428701-16

Figure DP2 A Rich Picture Illustration of Forces Driving the Syrian Civil War.

Source: Archipelago of Design.

Iraq constitute the fire. External jarring impacts, like the Shia-Sunni divide and Russian armed interventions, are depicted as weather events (high winds and lightning). De-radicalization and anti-corruption campaigns in the Muslim world are depicted as rain, reducing the intensity and spread of fire. Humanitarian aid to support economic development and decrease the marginalization of affected groups is depicted as water-bombers. Refugees are depicted as wildlife creatures fleeing the forest fire, with terrorist skunks escaping ahead of them. Syrian Democratic forces are shown as firefighters battling the blaze on the right-hand edge of the fire. Military power is abstracted as a bulldozer posed to bury the fire and bulldoze parts of the forest. Lastly, sparks from the fire depict spreading radicalization float to the edges of the rich picture.

This rich picture consolidates the interactions of many elements in a complex ongoing security challenge as a communicable image. As in the group exercise above, developing the picture provides an opportunity to discuss complex interactions. Design teams can preserve rich pictures for their understanding or use them to communicate alternative perspectives to stakeholders who need to understand the system.

Group Exercise—Develop a Rich Picture

Task. Sketch a rich picture of a situation in modern war and the Grey Zone. Combine elements of competition for space and territory, environmental vulnerability, transnational organized crime, markets, and violent non-state actors.

Consider. How are strategic, operational, and tactical levels of analysis implicated in crisis management? What intelligence requirements support decision-making? To approach this exercise:

- Organize groups of three to six participants from different backgrounds.
- Begin by identifying the enduring structures that inform the situation to be illustrated. These might include organizations, unaffiliated groups, demographic categories, governing bodies, and commercial enterprises. Use visual metaphors to convey these actors' identities.
- Having identified structures and actors, illustrate the ongoing processes, events, and activities that can change the situation. Text explanations can help, but visual metaphors are ideal.
- Now, illustrate relationships between structures, actors, and processes. Identify groups of stakeholders with congruent, competing, or conflicting interests. "Soft" subjective data (feelings, perceptions) matter as much as hard facts, and you should aim to represent both in your rich picture.
- Share the rich picture with others. Try to identify any missing stakeholders or relationships that are important to the situation's dynamics. Discuss various stakeholders' relationships and incorporate any significant changes.

Implications. How can rich pictures be used to hone the focus on the most important or most transformable aspects of a complex problem? The rich picture yields a holistic view of the problem and allows effective communication by metaphor. The links between actors and processes should help to identify leverage points and enable strategic "if…then…" planning cycles with stakeholders engaged in creative thinking.

Using the rich picture developed in the group exercise or the example of the Syrian Civil War, discuss the relationships from the perspective of different actors and stakeholders.

- Should any depictions be rejected, modified, or expanded?
- If the rich picture is a snapshot frozen in time, how do actors or stakeholders want to see it evolve, and what can they do to influence it?
- Can you trace back dependencies or contingencies depicted in the rich picture to causes?
- What can be changed easily, quickly, or with the fewest resources?
- What changes would make the most significant difference to the situation, and how might they be brought about?
- Do new possibilities emerge if we change the timescale (lengthening or shortening the period examined)?

Reference

Phillips, C. (2022), "The International System and the Syrian Civil War", *International Relations*, vol. 36, no. 3, pp. 358–381.

Further Readings

Barabrook-Johnson, P. and Penn, A. (2022), *Systems Mapping: How to Build and Use Causal Models of Systems*, Palgrave Macmillan, Cham.

Carpenter, T. G. (2013), "Tangled Web: The Syrian Civil War and Its Implications", *Mediterranean Quarterly*, vol. 24, no. 1, pp. 1–11.

Ford, R. S. (2019), "The Syrian Civil War: A New Stage, But Is It the Final One?", Middle East Institute, Policy Paper 2019-8, pp. 1–16. https://www.mei.edu/publications/syrian-civil-war-new-stage-it-final-one

Section III

Deterring and Managing Violence

11 International Organizations to Manage Modern War and Grey Zones

Marzena Żakowska and David Last

Introduction

How do small states use international organizations to support their security in the face of Grey Zone challenges? Climate change, technology change, and non-state threats permeate the international environment and transcend borders. In other eras, empires have consolidated small polities to manage transnational threats in their interests. The proliferation of independent states after World War II and again after the Cold War was accompanied by a proliferation of regional and international organizations seeking to manage transnational interests. International organizations generally uphold the fiction of sovereign equality despite differences in state capabilities. This chapter explores small states' use of international organizations to support security interests in the face of Grey Zone threats and emerging war.

Problems of Grey Zone Security and the War Nexus

Persistent problems of violence, population movement, economic instability, and environmental crisis shape the careers of security professionals who may rely on an image of war that no longer prevails in the war nexus. Against this background, international organizations shape the strategic options for small states. States can "bandwagon" with larger powers or seek security alone (Vaicekauskaitė 2017). Medium powers seeking self-sufficiency may have insufficient capabilities for national security demands (Hill 1986; Klein 2019). To bandwagoning and self-sufficiency, we might add hedging and seeking to balance the power of large states (Edström, Gyllensporre and Westberg 2018). International organizations offer a risk-mitigation strategy by adding rules, norms, and procedures for mobilizing support.

Violence may fall below the threshold at which the international community is prepared to take collective action, but international courts can affect norms and influence future decisions. The US mining of Nicaraguan harbors in 1984 led to a case at the International Court of Justice and an ambiguous ruling against the US on humanitarian grounds (Weiss 2016:303). The Colombian government's involvement in the 1997 Mapiripán massacre marked a threshold with the involvement of the Inter-American Court of Human Rights (IACHR 2005). International court cases impose minor costs and no hard constraints on aggressive powers. Coercive force often involves states with veto powers in the UN Security Council. America's invasion of Iraq in 2003, NATO's intervention in Libya in 2010, and Russia's attacks on Georgia in 2008 and Ukraine in 2014 before its full invasion in 2022 all illustrate the impotence of international declarations against force.

Threats and coercive violence can also be met with international investigations, monitoring, observation, and peacekeeping missions. Shedding light on aggression, clarifying murky situations,

DOI: 10.4324/9781003428701-18

and adjudicating competing claims are a basic function of collective security organizations, with the League of Nations foreshadowing much of later UN practice. Regional organizations have monitored conflict since the late 19th century (Wainhouse 1966).

Migration can be induced or encouraged. Russian resources enabled refugees to arrive in Finland through Russia in 2016 and in Poland through Belarus in 2021. Kenya has been suffering from refugee flows related to drought, climate crisis, and the Al-Shabaab insurgency. International organizations like the UN High Commissioner for Refugees (UNHCR) and the International Organization for Migration (IOM) have helped manage these flows and investigate contributing factors but have had limited success in reducing or redirecting migration and have often been critical of host or target countries.

Economic crises are linked to the onset of conflict, and international economic organizations are the front line for managing economic crises. The Bretton Woods institutions established to maintain stability after World War II have survived numerous changes. The International Monetary Fund (IMF), World Bank (WB), regional banks, and exchange mechanisms provide essential tools to manage international trade and finance (Chapter 5). With globalization and economic deregulation of the 1980s, other organizations became salient in managing international economic interactions. The World Economic Forum has provided a platform for rich individuals and corporations to influence international policy since 1971. The International Accounting Standards Board (IASB) is an independent international organization whose power comes from national laws that adopt the International Financial Reporting Standards it recommends. The Society for Worldwide Interbank Financial Telecommunication is an international messaging network that enables rapid and secure international financial transactions. The Basel Committee on Banking Supervision has produced collective responses to under-regulated banks and financial crises. Currency instability, trade balances, inflation, international sanctions, and strategic manipulation of money and trade are best managed collaboratively. International economic institutions may be a vital comparative advantage for small states.

Environmental crises (Chapter 2) are linked to violence, migration, and economic problems. The tragedy of the commons is that each actor overuses common assets out of self-interest while all depend upon them. This is overcome only through the international organization of interests and commitments. The Intergovernmental Panel on Climate Change (IPCC) has helped to articulate a common understanding of climate change and its dangers. The UN Framework Convention on Climate Change is the primary forum for negotiating a common response through the annual Conference of Parties and incentives provided by the Green Climate Fund and the Global Green Growth Institute. Progress to overcome the tragedy of the commons has been slow. Failure to reduce the impact of climate change may lead states to adopt mitigation strategies at the expense of neighbors or weaker actors (Buxton and Hayes 2016).

Violence, migration, economic friction, and climate change all demand security responses that force small states to cooperate within an international community. International organizations are necessary vehicles for managing Grey Zone threats and Modern Wars. What do we know about how they operate?

Characteristics of International Organizations

Security leaders must understand international organizations to work with and within them. Organizations are structured relationships. Institutions are ways of organizing behavior. Norms are standards of behavior. Organizations, institutions, and norms change over time (Archer 2001, Chapter 1). The differences between types of organizations are significant. Governance structures, means of appointing or electing officials, rules for voting and representation, available expertise, reports, and rules for approval all create or remove latitude for bargaining and influencing the organization and other members. Formal intergovernmental organizations, supranational organizations, and private transnational governance are three types typically encountered by security strategies.

Formal intergovernmental organizations are the largest category, including the United Nations, NATO, and OECD. They have formal powers assigned by members and are constituted by treaties under international law. There is typically a tension between effectiveness and legitimacy; technocratic centralization maximizes efficiency, but democratic decision-making confers legitimacy. How an organization is constituted and operates to balance efficiency and legitimacy is a messy political compromise. Norms may change when small states depend more on larger powers for security. Although member states establish international organizations, the organizations have independent legal personalities under domestic and international law. This permits them to hire staff, rent buildings, and make pronouncements for their members. They can also be found liable and forced to pay compensation (Klabbers 2016:133–141).

The powers of formal intergovernmental organizations are conferred either explicitly or by implication by member states. Informal amendment of powers occurs over time through successive statements and members' understanding. NATO's transformation from a defensive alliance during the Cold War to a global police force is cited as an example of evolving powers (Klabbers 2016:149). The Economic Community of West African States became a regional peacekeeping organization with AU and UN support. Organizations that expand into new roles can also retreat from them.

Member states establish *supranational organizations* which then have autonomous regulatory powers over the states that established them (Lindseth 2016:152). The EU is the obvious and most complex example, but WTO, IMF, the International Postal Union, and the United Nations have some supranational regulatory characteristics. Regulatory power is in tension with the democratic self-governance of the constituent states. Collective oversight of regulations is usually built into the way supranational organizations function, and national courts sometimes serve to limit the extent of imposed regulations. Collaborative regulatory regimes established through supranational organizations are important for small states to manage relations with larger states.

Private Transnational governance is a hallmark of the expanding global economy. "Public–public, public–private, and private–private partnerships are particularly salient new transnational organizational forms" (Mattli 2016:171). Standards set by self-regulating industries like accounting and transport, once acknowledged in national law, become part of transnational governance and can shape or constrain the actions of states. For example, the IASB establishes acceptable practices for transfer pricing within companies. These practices permit transferring profits to low-tax jurisdictions, reducing the capacity of states to extract taxes (Palan, Murphy and Chavagneux 2013:246).

Structures and Functions in International Organizations

IOs are important for small-state management of Grey Zone threats because they can open vulnerabilities and opportunities for influence. Strategic engagement with assemblies, executives, and panels can help states shape agendas and perceptions of events. The functions of law-making, monitoring, sanctions, use of force, dispute settlement, and information gathering can support national security objectives in the face of manipulation, deception, and attacks below the threshold of war.

General assemblies, international parliaments, and executive boards can provide platforms for small states to broadcast narratives and amplify bilateral diplomacy. Defending against attacks below the threshold of war may entail highlighting the asymmetry of an attack and mobilizing the power of victimhood. This competitive enterprise can involve mass media, social media, and targeted messaging. Public perceptions will affect government decision-making in many countries. It is surprising, for example, that Russia retains considerable acceptance of its actions outside Western audiences (Economist 2022). Winning support in the international community requires both diplomatic action and public diplomacy, indistinguishable from information war (Chapter 14).

International bodies have internal leadership, including executive heads like the UN Secretary-General, secretariats with key executives like the Director of Peacekeeping Operations, and special posts like Envoys and Chairs of Panels. Executive Heads are usually appointed by assemblies in which all member states have an equal vote, but appointments within secretariats vary widely. The WB and the IMF are exceptions, in which voting shares are allocated per financial commitments to the organization, giving wealthier states greater influence (Chesterman 2016:823). Wealthier states also expand their influence by providing nationally funded staff to augment organizations with insufficient funds to hire experts. This can mean over-representation on important committees or investigative panels. This is both a threat and an opportunity for small states. Under international labor law within the organizations' rules, there may be legal recourse to resist this practice (Herrera 2019).

Executive heads and envoys can serve as "norm entrepreneurs" to influence international behavior. Their secretariats maintain continuity, managing information for decision-makers, deploying resources, and communicating with field operations. This is illustrated in the transition from the Conference on Security and Cooperation in Europe to the Organization for Security and Cooperation in Europe (OSCE). Without its nascent secretariat, the OSCE could not deploy the European Commission Monitoring Mission to the former Yugoslavia in 1991 (Vetschera 2001). The capabilities of secretariats grow over time unless they are starved of resources (Ghebali and Warner 2018).

International organizations may operate on their own behalf. Although comprised of similar professional expertise and representing common aims, the International Military Staff of NATO frequently disagreed with the UN Department of Peacekeeping Operations on operational plans (Last 2000). Panels like ICISS and IPCC issue reports that shift agendas and change political discourse in large states. Both can be seen as victories by smaller states in the face of disinterest or opposing agendas of major powers. The "Iron law of oligarchy" suggests that expertise dominates decision-making (Michels 1915, Chapter 2). This opens doors to small states prepared to commit diplomatic and organizational talent to international organizations. Long service, insider knowledge, and good language skills can confer influence within secretariats.

The rule-making function of international organizations generally favors larger states. International law originates in the codification of rights claimed by the major warring and trading states. Still, there are strong currents of support for natural law and the philosophy of rights not vested in power (Tuck 2002). The sovereign equality of states is embodied in voting principles for international assemblies and their committees, which confer agency on groups of like-minded states.

Monitoring, dispute settlement, and use of force are functions of international organizations related to the generic problem of information and analysis. Multinational engagement was established as early as the Congress of Vienna in 1815 and embedded in the operating rules of the League of Nations in the 1920s and UN Military Observer rules from 1947. Unarmed military observers with diplomatic status were deployed in multinational teams to report to international organizations with a consensus report (Wainhouse 1966). Objective conflict reports help international bodies make informed decisions to manage conflicts about which information is often manipulated. The OSCE pioneered multifunctional monitoring missions that tracked social and economic precursors to conflict, informing conflict management and post-conflict reconstruction (Last 2003).

Decisions about the collective use of force can only be approached based on converging information and agreed analysis. This often entails international organizations working together. States explore options through multiple memberships—the United Nations, regional bodies, and specialized agencies. The secretariats of those organizations and the commanders and staff in field headquarters communicate in the mission area, through diplomatic channels, and in the headquarters and secretariats of international organizations (Last 2000). Collaboration extends beyond states to civil society organizations (NGOs) and the private sector. The United Nations and its agencies have established formal mechanisms for consulting with civil society, and special rapporteurs normally include NGOs in their consultations. These avenues for collaboration open doors to manipulation, deception, and defense against Grey Zone attacks. GONGOs are government-organized NGOs set up to promote national interests and may be used to amplify messages. International organizations can help filter disinformation and respond to it. The OECD established an information hub to counter Russian disinformation about its invasion of Ukraine (OECD 2022). Reporters Without Borders, the Committee to Protect Journalists, and the International Press Institute are NGOs that contribute to awareness of manipulation and deception, amplified through the OECD information hub.

Managing International Organizations

How do small states manage international organizations to respond to Grey Zone attacks and the war nexus? While the norms of the liberal international order are important for small states, it is disproportionately expensive for them to participate (Corbett and Connell 2015). International organizations must be inclusive and member-driven but also effective. It is more difficult to be effective with competing blocs and norms (Corbett, Yi-chong and Weller 2021:9). Norms shape small state strategies to influence decisions about managing international organizations through joining, caucusing, governance, agenda-setting, and instrumental organizational contributions.

Scale forces difficult choices for small states managing international organizations. They must find military and diplomatic resources to build trust and develop defensive alliances. Mongolia committed contingents to UN missions in Sierra Leone, NATO missions in Afghanistan and Kosovo, and the US-led coalition in Iraq. A resource-rich country with a small population bordering Russia and China, Mongolia has been in dialogue with NATO since 2005. NATO has discussed shared security concerns and supported the development of cyber and counter-terrorism capabilities but made no security guarantees (NATO 2024). In classic hedging strategy, Mongolia was also the first observer state in the Shanghai Cooperation Organization (SCO) in 2004 but has remained an observer while other states have become full members (Lkhaajav 2022). According to Freedom House indicators, Mongolia is more democratic than other SCO members. It pursues an independent foreign policy and seeks to avoid regional security alliances that may carry more risk than benefit. Small countries with independent foreign policies can't be everywhere but can compensate for a small professional diplomatic corps by mobilizing friendly nationals, diasporas, and volunteer consular officials.

Decisions about *engaging* and *joining* must balance opportunity and risk. Engaging with NATO has brought technical assistance and military professionalization for the countries in NATO's Partnership for Peace, Mediterranean Dialogue, Istanbul Cooperation Initiative, and partners globally located outside NATOs core Euro-Atlantic area who have NATO observer status.

Engagement offers a path to full alliance membership with security guarantees for some partners. Others found that it put them at increased risk of aggression from Russia, which saw NATO expansion as threatening. Armenia and Azerbaijan have engaged through the OSCE Minsk Group to resolve the disputed territory of Nagorno-Karabakh, but with a growing understanding that co-chairs of the Minsk group—Russia, the United States, and France—were more concerned with stability and their own interests than the resolution of the conflict (Abilov 2018).

Caucusing involves forming a like-minded group for a common purpose. Meetings are often behind closed doors to find common ground and develop strategies. The assemblies of international organizations provide opportunities to caucus, and the margins of the United Nations, OAS, AU, and other international organizations are used to set up meetings. Small state delegations cannot be everywhere, but they may listen more than large states, support initiatives, and extract a price for support. Caucusing allows even the smallest states to influence the agenda of international organizations. Pacific Small Island States succeeded in putting climate and oceans on the UN Security Council Agenda despite the opposition of many larger states. Small island states have succeeded in being norm entrepreneurs: "…they matter on some issues in some IOs, and they have realized some benefits but have not achieved everything they sought…" (Corbett, Yi-chong and Weller 2021:170). Small states together can support the liberal international order on which their survival and prosperity depend.

Agenda-setting requires proactive caucusing and bandwagoning to tilt an international organization toward an outcome, statement, action, or support for actions by members. Active contribution to governance can support agenda-setting through participation in the executive leadership or secretariats. Agenda-setting power includes preventing items from surfacing. Studies of voting patterns in the United Nations indicate that authoritarian regimes use anti-Israel voting to fill the agenda and distract from their own records (Mosler 2021; Haykin, Fox and Mirilovic 2021). Other studies indicate that authoritarian regimes actively seek normative change through strategic voting in bodies like the UN Human Rights Council (Dukalskis 2022; Meyerrose and Nooruddin 2023). In other cases, executive bodies and secretariats have been instrumental in shaping organizational agendas. The EU's response to conflict and crisis management involved collective action driven by the formal role of institutions like the European Commission and informal efforts by many actors to influence the agenda (Dijkstra 2012:455). These informal actions require effective inside knowledge, presence, and connections. Small states may be more sensitive to the range of Grey Zone threats requiring action by international organizations.

Instrumental contributions to international organizations are undertaken with expectations of a quid-pro-quo, which range from amorphous to explicit. Large powers can buy allies through direct payments and political side deals affecting trade, sanctions, and debt relief. Smaller states can be willing partners in exchange for national or regime benefits (Henke 2019:129). The dominant power may extract contributions to a regional international organization as part of a strategy of autocracy (Libman and Obydenkova, 2018:1038). Contributions to peacekeeping forces, permission to use bases, and participation in active legal regimes to manage trafficking, organized crime, and money laundering may be altruistic and instrumental approaches to managing interests through international organizations.

Managing Grey Zone Conflicts with International Institutions

International organizations expect a clear demarcation between war and peace. For states, the transition from peace to war involves shifting resources and the focus of effort. But in Grey Zones and modern wars, there is no clear transition. In the war nexus (Chapter 1), states must reconsider the strategic use of international organizations to manage threats. Peace operations can stabilize and prevent escalation. Humanitarian operations can reduce the consequences of violence. Non-proliferation, disarmament, and confidence agreements can tamp down potential escalation. Finally, economics and social initiatives may help to build norms for positive peace that support small state interests.

In a lexicon from the 1990s, peace operations include peace-making, observation, peacekeeping, wider peacekeeping, peacebuilding, and peace enforcement. Peace-making is the diplomatic process of conflict resolution, distinct from coercive measures to enforce peace imposed on the parties to the conflict. All these measures have the characteristic that violence is the enemy, while parties to the conflict are nominally allies of the third parties attempting to avoid dysfunctional violence. Peacekeeping implies consent to the intervention. Wider peacekeeping implies limits to that consent, and peace enforcement implies superior coercive force (Mays 1996:107–108).

Diplomatic tools for the peaceful settlement of disputes have been extensively catalogued and frequently deployed by states confronted with hostile action short of war. The toolset includes negotiation, inquiry, good offices, mediation, conciliation, arbitration, judicial settlement, resort to regional agencies, and other peaceful means (UN Handbook 1992). This extensive list is continuously refreshed and draws on over 50 years of innovation, incorporating new tools and approaches in the UN and regional organizations (UN Preventive Diplomacy Resource Book 2020). Grey Zone threats can be denied, negating diplomacy as a vehicle for managing them, but third-party engagement can still be useful.

International observation, including unarmed observers with diplomatic status, can help mobilize support against covert attacks. They illuminate ambiguous or covert actions if they are competent, perceptive, and pervasive enough to avoid being manipulated or deceived about events. Observer teams or incident investigators typically operate in multinational teams to avoid the perception of national bias. During the Cold War, teams were either non-aligned or included balanced membership of competing blocs. Non-aligned multinational and multifunctional observer teams can engage the international community to condemn aggression, even when the aggression is concealed or conducted by proxies. OSCE election monitors, NGOs like the International Federation of Electoral Systems, and Intergovernmental organizations like IDEA can help monitor and strengthen political systems that may be under external threat.

Cyber domains can be exploited in the Grey Zone. Cyber peacekeeping was proposed as early as 2003 (Cahill, Rozinov and Mule 2003) and ideas for organizing and managing international cyber peacekeeping have evolved rapidly but have not been deployed and have not had an impact on the offensive use of cyberattacks (Schmitt 2011; Akatyev and James 2015; Robinson et al. 2018; Nabeel 2019, 2020; Almutawa 2020). Active engagement in international organizations can shape the mandates of cyber security organizations. The UN Group of Governmental Experts (UN GGE) on Developments in the Field of Information and Telecommunications in the Context of International Security is an example of norm-building through the United Nations. NATO's Cooperative Cyber Defense Centre of Excellence conducts research and provides resources to member states. The EU Agency for Cybersecurity enhances awareness, cooperation, and capability development across the EU.

Space assets are implicated in Grey Zone conflict (Chapter 3). The UN Office for Outer Space Affairs promotes international cooperation in the peaceful uses of outer space. It supports the development of international space law and norms with the General Assembly's Committee on the Peaceful Uses of Outer Space. This is limited to the exchange of information. However, the voluntary International Code of Conduct for Outer Space Activities is explicit about transparency, confidence-building measures, and peaceful resolution of disputes between nations active in space. As with the cyber domain, a UNGGE on Outer Space recommends space security issues to the UN Secretary-General. These diplomatic avenues are open to small states increasingly active in space.

Humanitarian operations are intended to reduce human suffering. Grey Zone threats can manipulate or exacerbate human suffering, and international humanitarian operations may be a defensive response or a vehicle to advance offensive strategic interests. Authoritarian regimes saw the 2011 Arab Spring and democracy movements as Western-instigated threats to order. The Assad regime's brutal response generated a wave of refugees, which strengthened right-wing movements in EU countries with the help of Russian disinformation (Jankowicz 2020:160). The UNHCR is mandated to identify stateless persons, prevent and reduce statelessness, and protect the rights of stateless people. Focused on the rights of refugees, UNHCR must balance the interests of states protecting borders and internal cohesion. When crises generate refugees and internally displaced people, the Office for the Coordination of Humanitarian Affairs (OCHA) manages emergency aid, which reduces or manages refugee flows by meeting urgent needs close to the point of origin.

Population movements are more visible with the help of data from NGOs like the Internal Displacement Monitoring Centre. Internal displacement, forced refugee movement, and migration to improve life prospects can be hard to disentangle. Supporting and managing refugee and migrant flows on behalf of the international community remains with international organizations that document people, determine their status, and assist or deflect their movement. The 20th-century Nansen Passport has been replaced by refugee travel documents under the Global Compact on Refugees (GCR 2024). The IOM endorses the Migration Governance Framework (IOM, 2016), which promises to regularize and manage migration flows.

Manipulation of population movements has long been an instrument of state policy. From England's highland clearances in Scotland and the expulsion of the Acadians from Nova Scotia in the 18th century to Morocco's resettlement efforts in the Western Sahara and Turkey's use of refugee flows in negotiations with the EU, varying degrees of coercion and persuasion are evident. International organizations bring additional complexity to state policies, both offensive and defensive. Recently, weaponizing population movement has involved instigating, facilitating, manipulating, and sharpening population flows. Strategies of atrocity also bear mention. International organizations play a role in responding to these tactics.

Instigating refugee flows involves violent campaigns of terror and displacement. The Civil War in Syria led to NATO allegations that Russian, Syrian, and ISIL tactics were maximizing civilian damage to create refugees. NATO's air campaign in 1999 was intended to use air attacks on Serbian forces to prevent the expulsion of ethnic Albanians from Kosovo. No comparable campaign attempted to reduce the human impact of the Syrian Civil War. United Nations, EU, NATO, Organization of Islamic Cooperation (OIC), and OSCE were all seized with the Syrian Civil War but unable to prevent its human consequences. The United Nations listed 16 resident and 7 non-resident entities and offices contributing to its efforts in Syria, including Office for the Coordination of Humanitarian Affairs (OCHA), United Nations High Commissioner for Refugees (UNHCR), World Health Organization (WHO), World Food Programme (WFP), Food and Agriculture Organization (FAO), International Labour Organization (ILO), and United Nations Development Programme (UNDP) (UN Syrian Arab Republic 2023). The UNHCR reports that more than five million Syrians left the country during the first decade of the conflict for Türkiye, Lebanon, Jordan, Iraq, and Egypt. Türkiye hosts the largest population of about 3.3 million Syrians (United States for UNHCR 2024). Closing borders and imposing visa requirements reduced refugee flows into Türkiye, the destination of choice.

Facilitating migrant and refugee movements involves opening border crossings or providing transport and route control to channel the flow of refugees. This might involve official complacency about vehicles and small boats smuggling or more aggressively providing official transport to move large numbers close to a border. In facilitation, refugees or migrants don't need to be forced to cross the border; they are helped along a route they have already chosen. Türkiye's failure to stop smuggling from its coast to the Aegean islands and its opening of borders with Bulgaria fall into this category. Facilitation can be disruptive, but manipulation is more sinister and strategic.

Manipulating refugee and migration flows involves forced movement or active state transport of refugees. Russia's active transport of thousands of refugees from Africa and the Middle East and their delivery to the Finnish border in Interior Ministry buses in 2023 is a clear case of manipulation to impose costs and heighten tension with a neighbor. The tactic was introduced as a response to Finland's decision to join NATO. Belarus has transported Afghan refugees to its border with Poland in 2021, having denied them the opportunity to apply for asylum, as did Poland, which blocked their entry due to ongoing concerns about the destabilization of state security, identifying this as the primary objective behind Belarus's orchestration of migration flows at the Polish border. The UNHCR attempted to visit the refugees and urged both sides to allow them to apply for asylum. The European Council on Refugees and Exiles points to refugee rights concerns. International organizations are not yet prepared to recognize state manipulation of refugee flows.

Sharpening refugee flows involves using them precisely for strategic effect. Flows might be increased at a specific point to overwhelm services in conjunction with information operations to embarrass or discredit leaders or parties or as a distraction from other operations. The cover of refugee flows can be used to conceal the entry of special operations forces, saboteurs, or intelligence assets into a target country. Multilateral organizations are less help defending against this action than intelligence cooperation and collective security arrangements.

Strategies of atrocity use human suffering to put pressure on countries and international organizations or to delegitimize actors normally held to a higher standard of accountability. The spectacle of Finland or Estonia resisting inundation by freezing African migrants erodes their claims about the rule of law. The irregular forces of Hamas, Hezbollah, and ISIL incur retaliation in populated areas, where civilian casualties delegitimize their assailants. The international community expects the UNHCR to be humane and balanced, so it is not well-placed to respond critically to strategies of atrocity. Liberal democracies are held to a higher standard. Agencies like United Nations Relief and Works Agency for Palestine Refugees in the Near East (UNRWA) might support one side in a conflict because of their mandates.

Conclusions

Small states can be overwhelmed by Grey Zone threats, from political and economic subversion to demographic attacks, cyberattacks, and denial of essential space-based assets. Careful mobilization of international organizations can support norms and collective responses. Small states can use the structure of international organizations, executives, secretariats, and special panels to build effective coalitions and quid pro quo arrangements with larger states. They also need strategies to prevent them from becoming victims of international action enabled by international organizations.

References

Abilov, S. (2018) "OSCE Minsk Group: Proposals and Failure, the View from Azerbaijan", *Insight Turkey*, vol. 20, no. 1, pp. 143–164.

Akatyev, N. and James, J. I. (2015), "Cyber Peacekeeping", in *Digital Forensics and Cyber Crime: 7th International Conference, ICDF2C 2015, Seoul, South Korea, October 6–8, 2015,* pp. 126–139, Springer International Publishing, Cham.

Almutawa, A. (2020), "Designing the Organisational Structure of the UN Cyber Peacekeeping Team", *Journal of Conflict and Security Law*, vol. 25, no. 1, pp. 117–147.

Archer, C. (2001), *International Organizations*, Routledge, London and New York.

Buxton, N. and Hayes, B. (2016), *The Secure and the Dispossessed: How the Military and Corporations are Shaping a Climate-Changed World*, Pluto Press, London.

Cahill, T. P., Rozinov, K. and Mule, C. (2003), "Cyber Warfare Peacekeeping", in *Proceedings of the IEEE Systems, Man and Cybernetics Society Information Assurance Workshop*, 18-20 June 2003, pp. 100– 105, Military Academy, West Point, NY.

Chesterman, S. (2016), "Executive Heads", in Cogan, J. K., Hurd, I. and Johnston, I. (eds), *The Oxford Handbook of International Organizations*, Oxford University Press, Oxford, pp. 822–838.

Corbett, J. and Connell, J. (2015), "All the World Is a Stage: Global Governance, Human Resources, and the "Problem" of Smallness", *The Pacific Review*, vol. 28, no. 3, pp. 435–459.

Corbett, J., Yi-chong, X. and Weller, P. (2021), *International Organizations and Small States: Participation, Legitimacy and Vulnerability*, Bristol University Press, Bristol, UK.

Dijkstra, H. (2012), "Agenda-setting in the Common Security and Defence Policy: An Institutionalist Perspective", *Cooperation and Conflict*, vol. 47, no. 4, pp. 454–472.

Dukalskis, A. (2022), *A Fox in the Henhouse: China, Normative Change, and the United Nations Human Rights Council*, IGCC Working Paper No 6 (August 2022), University of California, Institute on Global Conflict and Cooperation, La Jolla, CA.

Economist (2022), "Russia is Swaying Twitter Users Outside the West to Its Side", *The Economist*, 14 May. https://www.economist.com/graphic-detail/2022/05/14/russia-is-swaying-twitter-users-outside-the-west-to-its-side

ECRE (2023), *Finland: Partial Re-opening of Borders with Russia Amid Concerns Over Refugee Rights; Border Closure Raises Fears Among Latvia, Lithuania and Estonia; Migrants Stuck in a "Geopolitical Drama"*, 15 December, European Council on Refugees and Exiles.

Edström, H., Gyllensporre, D. and Westberg, J. (2018), *Military Strategy of Small States: Responding to External Shocks of the 21st Century*, Routledge, London.

GCR (2024), "Multistakeholder Pledge: Refugee Travel Documents – The 21st Century Nansen Passport". Global Compact on Refugees.

Ghebali, V. Y. and Warner, D. (2018), *The Operational Role of the OSCE in South-Eastern Europe: Contributing to Regional Stability in the Balkans*, Routledge, London and New York.

Haykin, T., Fox, J. and Mirilovic, N. (2021) "Religious Discrimination, Diaspora, and United Nations Voting on Israel", *Foreign Policy Analysis*, vol. 17, no. 4, pp. 1–20.

Henke, M. E. (2019), "Buying Allies: Payment Practices in Multilateral Military Coalition-building", *International Security*, vol. 43, no. 4, pp. 128–162.

Herrera, M. C. (2019), "Is the UN Violating International Labour Standards?", EJIL: Talk!, *Blog of the European Journal of International Law*, 29 October. https://www.ejiltalk.org/is-the-un-violating-international-labour-standards/

Hill, J. R. (1986), *Maritime Strategy for Medium Powers*, Croom Helm, London.

Inter-American Court of Human Rights (2005), "Case of the "Mapiripán Massacre" v. Colombia, Judgment of September 15, 2005 (Merits, Reparations, and Costs). https://www.corteidh.or.cr/docs/casos/articulos/seriec_134_ing.pdf

International Organization for Migration (IOM) (2016), Migration Governance Framework (MiGOF). https://publications.iom.int/system/files/pdf/migof_brochure_en.pdf.

IPCC (2023), *Climate Change 2023: Synthesis Report. Summary for Policymakers*. https://www.ipcc.ch/report/ar6/syr/

Jankowicz, N. (2020), *How to Lose the Information War*, Bloomsbury, London.

Klabbers, J. (2016), "Formal International Organizations", in Cogan, J. K., Hurd, I. and Johnston, I. (eds), *The Oxford Handbook of International Organizations*, Oxford University Press, Oxford, pp. 133–151.

Klein, J. J. (2019), *Understanding Space Strategy: The Art of War in Space*, Routledge, London.

Last, D. (2000), "Organizing for Effective Peacebuilding", *International Peacekeeping*, vol. 7, no. 1, pp. 80–96.

Last, D. (2003), "Early Warning and Prevention of Violent Conflict: The Role of Multifunctional Observer Missions", in F. Hampson and D. M. Malone (eds.), *Conflict prevention: Path to peace or grand illusion?*, United Nations University Press, Tokyo, pp. 157–182.

Libman, A. and Obydenkova, A. V. (2018), "Regional International Organizations as a Strategy of Autocracy: the Eurasian Economic Union and Russian Foreign Policy", *International Affairs*, vol. 94, no. 5, pp. 1037–1058.

Lindseth, P. L. (2016), "Supranational Organizations", in Cogan, J. K., Hurd, I. and Johnston, I. (eds), *The Oxford Handbook of International Organizations*, Oxford University Press, Oxford, pp. 152–170.

Lkhaajav, B. (2022), "Why Mongolia is Steering Clear of Full Membership in the SC", *The Diplomat*, 29 September. https://thediplomat.com/2022/09/why-mongolia-is-steering-clear-of-full-membership-in-the-sco/

Mattli, W. (2016), "Private Transnational Governance", in Cogan, J. K., Hurd, I. and Johnston, I. (eds), *The Oxford Handbook of International Organizations*, Oxford University Press, Oxford, pp. 171–188.

Mays, T. M. (1996), *Historical Dictionary of Multinational Peacekeeping*, Scarecrow Press, Lanham, MD.

Meyerrose, A. M. and Nooruddin, I. (2023), "Trojan Horses in Liberal International Organizations? How Democratic Backsliders Undermine the UNHRC", *The Review of International Organizations*, vol. 20, pp. 125–126. https://doi.org/10.1007/s11558-023-09511-6

Michels, R. (1915), *Political Parties: A Sociological Study of the Oligarchical Tendencies of Modern Democracy*, Jarrold and Sons, London.

Mosler, M. (2021), *Autocrats in the United Nations General Assembly: A Test of the Decoy Voting Hypothesis*, Working Paper, No. 340, Institute - Leibniz Institute for Economic Research at the University of Munich, Munich.

Nabeel, F. (2019), "Establishment of UN Cyber Peacekeeping Force: Prospects and Challenges", *NUST Journal of International Peace & Stability*, vol. 2, no. 2, pp. 17–31.

Nabeel, F. (2020), "Cyber Peacekeeping: Critical Evaluation of Digital Blue Helmets Program", *NUST Journal of International Peace & Stability*, vol. 3, no. 2, pp. 17–27.

NATO (2024), Relations with Mongolia. https://www.nato.int/cps/en/natohq/topics_85297.htm

OECD (2022), Disinformation and Russia's War of Aggression against Ukraine: Threats and Governance Responses. https://www.oecd.org/en/publications/disinformation-and-russia-s-war-of-aggression-against-ukraine_37186bde-en.html

Palan, R., Murphy, R. and Chavagneux, C. (2013), *Tax Havens: How Globalization Really Works,* Cornell University Press, Ithaca, NY.

Robinson, M., Jones, K., Janicke, H. and Maglaras, L. (2018), "An Introduction to Cyber Peacekeeping", *Journal of Network and Computer Applications*, vol. 114, pp. 70–87. doi:10.1016/j.jnca.2018.04.010

Schmitt, M. N. (2011), "Cyber Operations and the Jus in Bello: Key Issues", *International Law Studies*, vol. 87, pp. 89–118. https://digital-commons.usnwc.edu/cgi/viewcontent.cgi?article=1077&context=ils&httpsredir=1&referer="Cyber Operations and the Jus in Bello: Key Issues.pdf

Tuck, R. (1999, 2002), *The Rights of War and Peace: Political Thought and the International Order from Grotius to Kant*, Oxford University Press, Oxford.

UN Syrian Arab Republic (2023), Monthly Digests of the UN in Syria, December 2023. https://syria.un.org/en/257415-monthly-digest-un-syria-december-2023

UNHCR (2025), Global Appeal. https://reporting.unhcr.org/global-appeal-2025

UNHCR (2024), Global Trends: Refugees. https://www.unhcr.org/global-trends

UNHCR (2023), Global Report. https://www.unhcr.org/what-we-do/reports-and-publications/global-report

United Nations (1992), *Handbook on the Peaceful Settlement of Disputes between States,* UN Office of Legal Affairs, Codification Division, New York. https://www.un.org/law/books/HandbookOnPSD.pdf

UNRCCA (2020), *Preventive Diplomacy Resource Book*. UN Regional Centre for Preventive Diplomacy for Central Asia (UNRCCA) Preventive Diplomacy Academy. https://unrcca.unmissions.org/sites/default/files/preventive_diplomacy_resource_book.pdf

United States for UNCHCR (2024), Syria Refugee Crisis Explained, UNCHR, 13 March. https://www.unrefugees.org/news/syria-refugee-crisis-explained/

Vaicekauskaitė, Ž. M. (2017), "Security Strategies of Small States in a Changing World", *Journal on Baltic Security*, vol. 3, no. 2, pp. 7–15.

Vetschera, H. (2001), "Ten Years of the Conflict Prevention Centre - Origins and Development", in OSCE Yearbook 2001, edited by Institute for Peace Research and Security Policy at the University of Hamburg, Nomos Verlagsgesellschaft, Baden-Baden, pp. 401–420.

Wainhouse, D. W. (1966), *International Peace Observation: A History and Forecast*, Johns Hopkins Press, Baltimore, MD.

Weiss, T. G. (2016), "Humanitarian Action", in Cogan, J. K., Hurd, I. and Johnston, I. (eds), *The Oxford Handbook of International Relations*, Oxford University Press, Oxford, pp. 303–322.

12 Deterrence and Alignment for Modern Wars and the Grey Zone

Hajdi Xhixha, Arian Kadriu, David Last, and Marzena Żakowska

Introduction

Wars are always dirty, and the weak always suffer the most—not just in the Balkans but everywhere. Deterring violence is, therefore, a perennial problem for governments. But most states cannot rely on the uncontested strength of nuclear weapons, and aligning with major powers may create more problems than it solves, particularly when the interests of those powers are insufficient to elicit their commitment or when their interests lie in escalation or perpetuation of conflict. How can deterrence and alignment strategies work for small states in complex neighborhoods? The range of Grey Zone threats further complicates deterrence because the threshold of war is indistinct, and ladders of escalation are inherently unclear.

This chapter is in four parts. First, problems of deterrence and alignment are discussed in the context of Grey Zone threats, which fall below the threshold of war but can damage security. Second, the Western Balkans' specific regional and international dynamics illustrate challenges of deterrence and alignment and suggest new approaches. Third, interpersonal and community dynamics open new possibilities for stable deterrence below the threshold of violence. Finally, the case of the Western Balkans illustrates policy instruments for deterrence and alignment in small states faced with the threat of modern war and Grey Zone conflict.

Deterrence and Alignment

Deterrence consists of efforts to control the behavior of other actors through threats. It is important to note that the contest is not between missile stockpiles or armies but entirely within the minds of decision-makers and in the councils of their advisors. After the blasts at Hiroshima and Nagasaki, nuclear deterrence became the central idea in studies of superpower strategy. If the consequences of even a small nuclear strike are sufficiently appalling, then deterrence is simple, and even North Korea and Iran can achieve it. However, perhaps the deterrence calculations are more complex, requiring escalation ladders and bridges between conventional and nuclear deterrents (Buzan 2018). Deterrence works when decision-makers have a common understanding of costs and benefits. This understanding emerges from interactions between experts with a common body of knowledge, who eventually form an epistemic community, or a mutually accepted way of viewing the world, that can guide decision-making (Haas 2015).

Nuclear deterrence works because the consequences of a nuclear blast are uncontested. Conventional deterrence cannot work the same way because the outcome of conventional conflict is always uncertain (Wirtz 2018). Whatever the correlation of forces, there may be enough optimism about the outcome to encourage parties to go to war. Anything that increases optimism about the outcome of war increases the likelihood of war (Blainey 1988). If nuclear deterrence

DOI: 10.4324/9781003428701-19

is certain, and conventional deterrence is uncertain, how do states deter Grey Zone threats like electoral meddling, cyberattacks, hostage diplomacy, and weaponizing refugees?

Table 12.1 reflects the diminishing potential for deterrence as the certainty of an outcome decreases. The examples in each column are illustrative, and the first column is of marginal interest to nonnuclear states. French strategist André Beaufré argued that even a threat to a few Soviet cities (the capacity to "rip off an arm") would deter Moscow from attacking France. This logic fails when applied to allies or aligned states. It is hard to imagine any French leader risking Paris to protect Pristina from Belgrade. Small states with their own nuclear deterrent are unlikely to extend an umbrella of deterrence to others. Even for superpowers, extended nuclear deterrence requires a tripwire like NATO's troops in the Baltic states.

Deterrence by denial convinces an enemy that the strategy contemplated cannot achieve the objectives. The conventional equivalent of denying an enemy territorial victory might be indigestibility. If the enemy takes land, the land will be untenable. The people will be unruly and uncooperative, and the occupation will be a misery for the occupier. During the Cold War, Yugoslavia developed a "survival and national resistance" strategy modeled on its partisan struggles in World War II. An invader would confront a spider's web of trained community defense organizations with stockpiles of food, fuel, ammunition, and equipment, ready to tie down an invading army until the "spiders" of the Yugoslav National Army attacked to finish it. An unfortunate by-product of the strategy was that upon the dissolution of the federation, the stockpiles became a valuable incentive for each identity group to be the first to seize the assets and escalate to armed conflict.

An alternative to armed "indigestibility" is unarmed civil resistance. The International Centre for Nonviolent Civil Conflict has researched the viability of nonviolent civil resistance both in the face of authoritarian regimes and occupations. Evidence suggests that nonviolent resistance is more effective and more likely to result in democracy than violent alternatives (Chenoweth and Stephan 2011). Nonviolent civilian defense has been proposed as a principle for survival against Russian hybrid warfare in the small Baltic states (Bartkowski 2015). Unarmed civil resistance can bolster national identity against external influence in parallel with armed options (Maszkiewicz 2015).

What is the equivalent denial strategy for the new range of Grey Zone threats? There must be multiple strategies to deny advantage from the many forms Grey Zone threats short of war might take. A counter to hostage diplomacy, for example, might be to demonstrate a willingness to take higher-value hostages. This would have to be weighed against the risk of losing the moral high ground and support from allies. "Weaponized" refugees flooding the border might be countered with border fencing and troop deployments, but processing and welcoming them

Table 12.1 Potential for Deterrence

	Nuclear weapons (certain outcome)	*Conventional (uncertain outcome)*	*Grey Zone Threats (very uncertain)*
Denial	Secure second strike	Indigestibility	Educated population
Punishment or retaliation	"Rip off an arm"	Could be asymmetric	Legal means
Extended/umbrella	Needs trip wire	Requires credibility	Space, sea cables
Proximity	Not a factor	Induces caution	Big tech companies
Containment	Full spectrum, MTCR	Build alliances, isolate enemies	International norms and agreements

Source: Authors' elaboration.

might be another approach to take the high ground. Disrupting and deterring state-based hackers are challenging (Libicki 2021). An educated population might be the most effective defense when we think of cyber threats, disinformation, and distributed physical attacks through the internet (Jankowicz 2020). The effect of an aware and educated population should be to increase the uncertainty of the outcome and thereby reduce the willingness of an actor to invest in Grey Zone threats. The problem is that these threats may be cheap to implement and expensive to counter, making them hard to deter.

Some deterrence value might be ascribed to the proximity of powerful allies. A former Deputy Director of NATO makes the case that Ukraine's proximity to NATO has reduced the intensity of attacks because Russia is fearful that a direct strike in Romania or on NATO assets in the Black Sea would trigger more NATO involvement (Gottemoeller 2023). Proximity deterrence is of little comfort to Ukraine, though it might induce Russian caution. Similarly, some actors might be cautious about Grey Zone attacks on space or cyber assets that affect the interests of many states, or even powerful multinational corporations like Google, Meta, or SpaceX, with the capacity to retaliate if the bad actor becomes known.

How do alignment and alliance affect deterrence? Norms for nuclear deterrence are well-established and effective within epistemic communities like NATO and its nuclear-armed adversaries. National strategists seldom hold direct talks but watch carefully for signals and read doctrine and intelligence reports to understand motivations, capabilities, and intentions. Norms against conventional military aggression are well-established but much more frequently breached. Norms against Grey Zone threats are in their infancy and deterring them so far has been largely ineffective. Most states' politics, economics, law, and social relations have competitive and adversarial qualities that open opportunities for external manipulation.

Military alliances, however, may effectively prevent Grey Zone threats from escalating to conventional military conflict. With hindsight, it seems clear that the cyberattacks, disinformation, and low-level insurgencies Russia sponsored against Ukraine after 2014 were a prelude to the conventional invasion of 2022. Ukraine's alignment with NATO made it a target, but the alliance could not deter the invasion. In contrast with Ukraine, the Baltic states have experienced extensive Grey Zone attacks but have not experienced conventional invasion. The trip-wire deployment of Allied troops in Northern and Eastern Europe is a Cold War tactic with new actors. Eight front-line countries now host token forces from 25 NATO member states, including new members on their first NATO deployments.

NATO is unique, and most regions have no equivalent alliance on which to rely for collective defense against conventional or Grey Zone threats. The African Union's Peace and Security Architecture provides a framework for regional peacekeeping deployment. Judging from internal conflicts and coups, the security architecture does not deter violence.

Regional and International Dynamics

The Western Balkans example is relevant to any country with ethno-linguistic or political interpenetration across boundaries. There are few countries and regions where this is not the case. In a region like the Western Balkans with a recent history of violent conflict, the relative size and power of the states is a baseline for considering potential deterrence and alignment strategies. Table 12.2 draws on World Bank and Yearbook data for population (millions), total area (thousands of square kilometers), and armed forces manpower in thousands. Military strength and direction of change (up, stable, or down) are taken from the US commercial site, Global Firepower, but the counterintuitive strength score is inverted so that larger scores indicate greater

Table 12.2 State Actors in the Western Balkans

Country	*Population (millions)*	*Area (thousands of square km)*	*Armed forces (thousands)*	*Power index*	*Change in power (2023)*
Greece (GR)	10.6	131.9	200	2.164	stable
Bulgaria (BU)	6.8	110.8	25	1.025	up
Serbia (SB)[a]	6.8	88.3	30	1.045	up
Croatia (HR)	3.8	56.5	30	0.823	down
Bosnia-Herzegovina (BH)[a]	3.2	51.2	10.5	0.325	down
Albania (AL)[a]	2.8	28.7	10	0.542	up
Northern Macedonia (MK)[a]	2.0	25.7	8	0.460	up
Kosovo (KO)[a]	1.8	10.8	5	0.304	up
Montenegro (ME)[a]	.6	13.8	2	0.348	up

Source: Table compiled by Authors from World Bank, Yearbook, and Global Firepower sites.

[a]The Western Balkans Six (WB6) seeking EU accession (Zweers et al. 2022).

military power. Population figures are strongly correlated with military strength (.956), military manpower (.936), and total area (.883).

Table 12.2 illustrates the relative conventional power of state actors in the Western Balkans, with Greece, Bulgaria, and Serbia as the dominant actors. A closer look at language and identity politics reveals mutual vulnerabilities arising from interpenetration. Seven language groups are present in the Western Balkans: Hellenic, Slavic, Albanian, Romance, Germanic, Ugric, and Turkic. However, Hellenic, Slavic, and Albanian dominate the region, with distributions not corresponding to national boundaries (Friedman 2007). This generates at least three levels of potential conflict, making deterrence and alignment difficult for each Western Balkan state.

At the International level, Russia has a long history of Slavic solidarity with Serbia in opposition to Western support for other Balkan states (Mourtizen 1997). With British and US leadership, the West helped Royalist forces defeat a communist insurgency in Greece after World War II, supported Yugoslavia after Tito broke with Stalin in 1948, and helped Kosovo break from Serbia in 1999. NATO membership for Romania, Slovenia, and Bulgaria in 2004, Croatia and Albania in 2009, Montenegro in 2017, and North Macedonia in 2020 created a sense of encirclement in Serbia, exacerbated by Russia's growing hostility to Western norms, although Western norms have a constituency in Serbian politics. Russian influence operations in the Balkans emphasize shared cultural ties, trade in key sectors like energy, banking, and real estate, and people-to-people connections that create Russia-friendly local constituencies. Russian efforts to aggravate and prolong political instability and ethnic tensions, to support far-right groups and xenophobia seem aimed at delaying or derailing accession to Western organizations (Stronski and Himes 2019).

Pursuing membership in the Organization for Security and Cooperation in Europe, the European Union, and the Council of Europe has pushed most Western Balkan states toward recognizing minority language rights and norms of open society (Woelk 2012). Western alignment may deter or attract hostile intervention by external powers. Belgrade has reason to perceive Western engagement as hostile after suffering a NATO attack in 1999 and losing territory in Kosovo and Montenegro. Serbia's local rivals Greece and Bulgaria have reason to align with other Western-leaning states of the region, but each is potentially vulnerable to internal threats and instability, particularly in the spectrum of Grey Zone threats.

China has an aggressive program of state-sponsored investment of more than $1 billion annually since 2011, 70 percent of it in Serbia. Albania, Bosnia and Herzegovina, Montenegro, North Macedonia, and Serbia have joined Beijing's Belt and Road initiative to expand Chinese trade with Europe. China preserves good relations with Belgrade by not investing in Kosovo. The US expressed worries about debt trap diplomacy and penetration of the security services by surveillance technology provided by China (US Congress 2022). One alternative open to Western Balkan states is to "play the China card" to gain leverage with the EU and reduce some painful economic costs of accession to Europe (Jaćimović, Deichmann and Tianping 2023). The threat of alignment with China is a widely available strategy for curbing Western demands.

Table 12.2 tells only part of the story about regional power. Greece and Bulgaria are both firmly in NATO, the European Union, and the Council of Europe. They have stable borders, effective rule of law, secure means of transition between governments, and effective oversight of military, police, and security intelligence structures. Serbia, more geographically central, has recently been involved in a protracted civil war, has lost territory as the successor state to Yugoslavia, and has an unresolved conflict with Kosovo. Each state in the Western Balkans seeks extra-regional powers for support. Counting on external support reduced the willingness to compromise and prevented regional powers from reshaping the Western Balkans after the Yugoslav Civil War (Pop 2011). The presence of Russia through its relations with Serbia has galvanized cooperation within the Western-facing states. For example, the long-standing dispute between Greece and North Macedonia was eventually resolved with the Prespa Accord of 2018, providing for recognition with UN and EU guarantees about the teaching of history and claims to common heritage and culture. This does not prevent politicians in Greece, North Macedonia, and Bulgaria from making expansive irredentist claims or using identity and language for local advantage. Such claims can easily escalate to violence without external meddling but also create opportunities for manipulation, disinformation, and destabilization by great powers or regional rivals.

Serbia can exploit mutual vulnerabilities in Bosnia-Herzegovina, Kosovo, Albania, and Croatia to act as a regional spoiler, adopting Russian tactics deployed in Georgia (2008) or Ukraine (2014, 2022) to destabilize its neighbors. As the former heartland of multi-ethnic Yugoslavia, Serbia is home to minorities from each of its surrounding countries, which in turn each have Serb minorities. Bosnia-Herzegovina is fragile, with the Serb-run Srpska Republic and Muslim-Croat Federation forced together by the terms of the General Framework Agreement for Peace in Bosnia and Herzegovina (the Dayton Agreement of 1995). Serbia is not the only potential spoiler. Bulgaria has vetoed North Macedonia's accession to the EU. Both Albania and Greece nurse historical ideas of a greater Albania and a greater Greece, which can be mobilized by political entrepreneurs or external manipulation to disrupt local politics. Democratic institutions in Albania and Bosnia-Herzegovina are weak, and Serbia's border with Kosovo remains problematic following Kosovo's unilateral independence. The EU has been unable to produce stable democracies; rather it has helped entrench autocratic tendencies in the name of stability (Zweers et al. 2022).

Religion also plays a persistent role in the region, helping to define the national and community identities that cut across borders and create opportunities to bolster national identity at the expense of minorities (Halilovich 2012). The international connections and funding of Catholic, Orthodox, and Islamic religious organizations create opportunities for international penetration of national identity and consciousness. Turkish foreign policy since 1990 has increasingly emphasized common history and religious identity in Turkish-Bosnian relations (Rašidagić and Hesova 2020). Russia has used connections with the Orthodox Church to strengthen relations with Serbia (Stronski and Himes 2019). During the COVID epidemic, the Serbian Orthodox Church mobilized almost half the population of Montenegro to defeat the incumbent president, Djukanović (Arandjelovic 2022). Djukanović's authoritarianism and popular dissatisfaction suggest factors beyond external meddling (Šljivić 2021), but the ongoing disputes over Montenegrin identity and common Serbian Orthodox culture suggest that Belgrade is actively pursuing its interests (Banchev 2021).

Mapping religious and other nongovernmental organizations along with identity groups, foreign funding, influence, and control might suggest targets for deterrence by denial or punishment through a carrot-and-stick approach. Targeting the reputation, funding, or status of internationally supported nongovernmental organizations or changing the laws under which they operate might prevent them from interfering in political processes. This targeted deterrence of organizations can support or undermine the autonomy of a state and its democratic process.

Interpersonal and Community Dynamics

The ability of great powers to dictate outcomes within a region is constrained by the interests and actions of the regional actors and the interpersonal and community dynamics that shape the actions of national parties and political leaders. No matter the level of intelligence effort by distant great powers, local actors will likely have a more accurate and nuanced understanding of these dynamics and will often be able to organize effectively to thwart outside interests. A consistent challenge for post-conflict reconstruction, nation-building, and democratization efforts supported by the international community is that it is easy to derail efforts by spreading misinformation, manipulating local actors, and appealing to special interests. Building intercommunal trust, mobilizing support for long-term cooperation, and resisting abusive authorities can be harder. Trust at the community level is the foundation for nonviolent civil resistance as a denial strategy and helps build capacity for democratic governance (Chenoweth and Stephan 2011:Chapter 2).

The memories of those who have suffered in war are a powerful resource that can be mobilized in support of strategies of deterrence by denial. The authors interviewed people affected by the war in Kosovo and recorded narratives of repression and violence motivating individuals to serve in the Kosovo Liberation Army during the struggle for independence. Themes included barbaric occupation, persecution and massacre without reason, imprisonment, expulsion from schools and universities, dismissals from jobs, and the urge to protect land and people. It should be assumed, however, that in the Grey Zone, no memory is sacred, and some actors will seek to manipulate emotional subjects, misrepresent events, and set neighbors against each other in pursuit of a larger strategy.

Full-spectrum deterrence strategies should consider the potential to deter actors from manipulating local grievances to aggravate or prolong violence below the threshold of war. Five kinds of actors may be involved at the subnational level, for which different deterrence or alignment strategies might be contemplated: major power propaganda and information warfare efforts targeting communities, diaspora actors, local political entrepreneurs, local economic actors, and sponsors of intercommunal violence.

Foreign Influence Operations

The EU adopted a defensive and positive counterstrategy to Russian Information Influence Activities in the Western Balkans. It formulated its messages not to contradict basic EU values of democracy and openness (Behbudzade 2020). This approach may not have deterred Russian hostile information operations but may be more effective in achieving EU objectives. This leaves the initiative for individual countries to pursue more active measures to deter both Russian and local sources of hostile propaganda.

Russia ran four types of influence operations in Serbia and the Western Balkans to obstruct Euro-Atlantic integration: disinformation, sponsoring pressure groups, diplomacy, and energy politics. Disinformation, diplomacy, and energy politics were managed with minimum local support, but local pressure groups can expand the reach of an outside power. In Serbia, these groups include hooligans and sports clubs; multiple political parties with little public support; minor political movements and cultural movements; individual artists, journalists, and intellectual fellow travelers; and friendly news portals and media (Životić and Obradović 2022:184). Many of these organizations take the form of civil society organizations. Direct action by the state to repress them would violate democratic norms. However, requiring any agent of foreign influence to register and report funding and activities or face legal and financial consequences might help to identify and contain the extent of foreign influence. Australia's Foreign Influence Transparency Scheme Act (2018) provides an example now contemplated by Canada. An individual acting for a foreign principal can be penalized for failing to register or update information with the Attorney General's Office, misleading or lying to authorities, or destroying relevant records. Penalties for violations range from fines to prison terms of six months to five years. In discussions of electoral interference, registration programs are now widely seen as potential deterrent strategies.

Tit-for-tat cross-border meddling might be considered as a deterrence strategy in some areas with mixed ethnic groups. Serbia might fear that if it meddled in Croatia by mobilizing Serbs, Croatia might meddle in Serbia by mobilizing Croats. In practice, asymmetry might make signaling difficult. Consider the possibility of Belgrade manipulating tensions in Bosnia Herzegovina by rousing Serb minorities. How might Sarajevo or Zagreb respond? Serbia has its own minorities to worry about, but in every minority community, some local factors and relationships might impede or accelerate any response to an external ethnic call to action. The uncertainty attached to any cross-border meddling makes treating a state's own minorities fairly a much more reliable strategy to prevent minority fifth columns or domestic support for insurgents.

Diaspora Actors

The status of diaspora actors may be ambiguous. If they are dual citizens living abroad, are they foreign? What about groups or individuals sharing language, culture, and national identity but opposed to the current political regime? If they are aligned with or employed by hostile powers or corporations with interests deemed hostile to a regime, what measures are appropriate to deter them from intervening in local politics? They may seek to support armed or peaceful opposition groups in their home country, or they may seek to raise funds or lobby their host government in support of one or other faction in a domestic dispute. These activities might be part of normal competitive and adversarial politics or cross the threshold into Grey Zone conflict or hybrid war.

Local Political Entrepreneurs

The same is true of local political entrepreneurs, who may act in good faith or with hostile intent toward established norms of political behavior. Even advocacy of violence and overthrowing political institutions might be interpreted as an acceptable tactic for political transition by some actors and their constituencies. How do states deter escalating political violence that may or may not have external sponsors? The principle of deterrence by denial is to prevent violent actors from achieving their objectives and to impose costs on them for using unacceptable tactics. This requires a strong state security apparatus behaving within the rule of law. It also implies that courts and the legal system are reliable and effective. One of the indicators of culpability is demonstrable links to violent specialists, like hooligan sports clubs or bike gangs identified in Serbia (Životić and Obradović 2022). When these are consistently linked to political actors, prosecutions should follow. Extra-legal action like the death squads of Guatemala, however, is unlikely to support political stability.

Economic Actors

Economic actors can also be involved in escalating violence, repression, intimidation, and erosion of political authority through corruption. Companies and investors may be motivated by immediate profit, seek political change to support future operations, or support political allies who have made bargains. Breaking unions, driving people off their land, intimidating lawyers and judges, and bribing authorities are all economic tactics that fit in the Grey Zone of negative peace but usually fall short of hybrid war. As with political entrepreneurs, deterrence by denial requires a strong state, reliable policing, and a functioning judiciary. Without these assets, retaliation may change the economic calculations. Opposing groups might resort to strikes, boycotts, and divestment, disrupting supply chains, damaging plant and equipment, or symbolic attacks on foreign interests and embassies supporting malign economic actors.

Sponsors of Intercommunal Violence

Foreign information and influence operations, diaspora actors, and political and economic entrepreneurs all have identifiable sources, even if these sources and the levers of influence are carefully obscured. Intercommunal violence can be more diffuse and difficult to deter. State deterrence of intergroup violence must begin with an understanding of the dynamics of violence at the margins of state control. Sociological fieldwork helps to understand the microsociology of intercommunal and intergroup violence. Although intercommunal violence may be the interpersonal level, it must be organized to escalate. This demands some personal experience of violence, often connected to police or military action. Experience brings awareness of the potential to be a victim or perpetrator and the motivation to organize for community security. The first step is the activation of social boundaries between "us" and "them." The recruitment of fighters and mobilization of resources follow. People and resources are organized for specific forms of violence—perhaps pogroms, ambushes, night raids, or riots with punitive rape and looting. Acts of violence establish cycles of reprisal and mutual exhaustion, leading to periodic escalation and de-escalation, or orders of violence (Bakonyi and Bliesemann de Guevara 2014).

Even strong states with effective policing and functioning justice systems may lack the presence and capacity, or sometimes the will, to prevent cycles of intercommunal violence. Pogroms or violence against minorities may serve state policy or may be insufficiently damaging to state legitimacy to warrant a response. Enhancing group solidarity by identifying external enemies or establishing other superordinate goals might redirect the violent energy. Denying resources through disarmament programs, state reprisals by destroying houses or villages on both sides, arresting or killing leaders, taking hostages from leading families, resettling troublesome groups, and deporting ringleaders are some historical examples of policies to deter intercommunal violence, vendettas, or feuds. These tactics include elements of denial, punishment, and extended deterrence. The fundamental problem of unpredictable outcomes makes all efforts to deter intercommunal violence questionable. Economic development programs, inducing young men into long army service in remote locations, education camps, and holistic social programs are more likely to bear fruit.

Policy Instruments for Alignment, Alliance, and Deterrence

Policy instruments are specific actions that governments take to advance their interests. Typical policy instruments include political-diplomatic, cultural-informational, military-policing-security, and economic-commercial-financial. These instruments may be employed directly by the government or indirectly through commercial partnerships, allies, or nongovernmental agencies set up for the purpose. To be sustainable and effective, policy instruments must be mutually supportive, durable, and widely supported. This section addresses instruments for alignment with major powers, regional alignment, and deterrence of Grey Zone threats.

Alignment with Major Powers

Seeking a new alignment or alliance is a serious step for a small state. It may even be an existential decision, opening the door to occupation or long-term subordination by a larger neighbor or overseas power. Asymmetrical power might make for bad bargains, like Armenia's reliance on Russian security guarantees in a region where oil-rich Azerbaijan is in a stronger position. What must states be prepared to offer, and what can they expect to receive in return? Alignments may help to offset dependence on or dominance by a major regional power. Canada's search for a "third way" to balance the United States in the 1970s, Mongolia's Pacific engagement strategy, aimed at managing its challenging geographic position between Russia and China; and the efforts of West African states within Francophonie and the Commonwealth to counterbalance the influence of Nigeria. These strategies, however, tend to offer only marginal security value.

Diplomatic instruments to support alignment with major powers might include expanding consular and ambassadorial services, supportive voting in international organizations, and adopting a helpful fixer role. Information and cultural policy instruments to support alignment might include sharing intelligence, rendering persons of interest, and constraining or shaping local media to portray positive images (or negative images of enemies). Accepting foreign-language broadcasts or cultural centers like the Confucius Institute, the Göethe Institute, or the *Alliance française* might curry favor. These gestures might signal more than influence a major power. Military policy instruments to support alignment include willingness to contribute troops to coalitions and common ventures. Small states may be able to contribute only small contingents without special skills or equipment but sometimes compensate by accepting unglamorous roles: camp guards and sanitation, for example, on the NATO Afghan mission. Offering bases, ports, overflight privileges, and communications facilities at home can be as valuable as contributing to expeditionary coalitions. But inviting troops from one power into the backyard of another can be seen as unnecessarily provocative. Finally, economic instruments supporting alignment might entail concessions to foreign traders and finance, banking arrangements, alignment on trade negotiations, mineral concessions, changes to ownership regulations and tax law, and major borrowing.

Regional Alignment

Policy instruments to enhance alignment with regional neighbors might not require tributary concessions to a major power to which a small state is appealing for protection or extended deterrence. Table 12.1 shows that most of the states in the Western Balkans are similar in size and power and would negotiate more as equals. However, they have difficult shared pasts. Policies of regional trade, cooperative infrastructure projects, and youth and cultural exchanges can build positive regional dynamics. Post-conflict peacebuilding efforts in Western Balkan states started even during the conflict. Training activists and community leaders and extending truth and justice activities or restorative truth-finding from courts into communities can contribute to regional stability. The EU-sponsored Stabilization and Association Process and the REKOM initiative for truth and justice have helped to link parallel communal and national processes. People-to-people initiatives like those of the 1975 Helsinki process and community and transnational peace initiatives can support trade, infrastructure, and other pragmatic initiatives to build a stable regional community (Ağir, Arman and Akçay 2019; Vukosavljevic 2007; Fischer and Petrović-Ziemer 2013). This regional peacebuilding approach is probably preferable to a military defense and deterrence strategy for small states with limited military capacity and pressing economic and social priorities.

Deterrence of Grey Zone Threats

Policy instruments to deter and resist Grey Zone threats are more problematic. Reciprocal threats are unlikely to deter when perpetrators are obscure. Tit-for-tat reprisals risk escalating Grey Zone activities to full-blown hybrid war. Instead, governments might consider how the full range of policy options might influence the behavior of the population targeted by Grey Zone threats, eventually denying the perpetrator success and deterring future investment in threats. What are these options? Governments perform at least 11 activities that might help deter the Grey Zone threats identified. This is an illustrative exercise. Policy initiatives to deter Grey Zone threats would probably be undertaken as iterative experiments—probably many simultaneously. Table 12.3 samples some policy instruments governments might experiment with to address Grey Zone threats.

Conclusions

This chapter has addressed deterrence and alignment strategies and policies for small countries in regional security complexes facing the problems of modern war and Grey Zone threats. The concepts of alliance, alignment, coalition, and deterrence are relevant in any region with historical tensions and hostilities. Governments can develop political, economic, and social strategies to limit the potential for escalating conflict. Interpenetrating minority populations can best be addressed by regional peacebuilding strategies that reduce the potential for violence rather than by conventional deterrence strategies.

Table 12.3 Policy Instruments to Deter Grey Zone Threats

Policy instrument	*Example*	*Grey Zone Threats*
Taxation	Require full beneficiary disclosure in all financial statements	Use of shell companies to mask beneficiaries
Regulation	Require full income and asset disclosure	Elite penetration—prevent hiding assets
Subsidies and grants	Support independent investigative journalism	Combat junk news and misinformation

(*Continued*)

Table 12.3 (Continued)

Policy instrument	*Example*	*Grey Zone Threats*
Provision of services	Investigation of commercial identity and true beneficiary when national interests are involved	Masking activities behind lawyer-client privilege
Budget for agencies	Budget for an agency to trace cyberattacks and hacking; build international connections	Cyberattacks, cybercrime, and international fraud
Structure private rights	Require public participation and government oversight on media boards	Purchase or manipulation of media outlets
Provide frameworks for economic activities	Provide special economic zones to deter smuggling, increase enforcement	Disrupting government revenues by smuggling and grey markets
Education and consultation	Increased focus on awareness of information pollution, junk news	Smears, discrediting, and disinformation from foreign sources.
Bureaucratic and political reforms	Modify mandates of border forces, resettlement assistance,	Addresses weaponized refugee flows

Source: Compiled by Authors from Bardach (2012) Appendix B, and Tabular Summary of Grey Zone Threats.

Timely and reliable intelligence, the interaction of security structures, protection of information technology, robust military capabilities, and defense diplomacy all work best in the context of regional alignment. New approaches to warfighting must consider the risks of adversaries' Grey Zone operations. This requires a sophisticated full-spectrum deterrence strategy using all the policy instruments available to small-state governments. War and deterrence are not just military problems.

References

Ağir, B. S., Arman, M. N. and Akçay, E. Y. (2019), "The Peace-Building Efforts of the European Union in the Western Balkans: The Cases of Macedonia and Kosovo", *Dokuz Eylül Üniversitesi Sosyal Bilimler Enstitüsü Dergisi*, vol. 21, no. 2, pp. 517–531.

Arandjelovic, D. (2022), *The Serbian Orthodox Church as a Transnational Political Actor: A Case Study of Regime Change in Montenegro*, MSc Thesis in Sociology, Anthropology and Criminology, Clemson University, Clemson, SC.

Australia (2018), *Telecommunications and Other Legislation Amendment (Assistance and Access) Act 2018* No 148.

Bakonyi, Jutta and Berit Bliesemann de Guevara, (eds.) (2014), *A Microsociology of Violence: Deciphering Patterns and Dynamics of Collective Violence*, Routledge, London and New York.

Banchev, B. (2021), "The Serbian Orthodox Church in Montenegro and the Conflict of Identities in 2019– 2020", *Balkan Studies/Etudes Balkaniques*, vol. 57, no. 4, pp. 566–591.

Bardach, E. (2012), *A Practical Guide for Policy Analysis: The Eightfold Path to More Effective Problem Solving*, Sage, Los Angeles, Washington DC.

Bartkowski, M. (2015), *Nonviolent Civilian Defence to Counter Russian Hybrid Warfare*, John Hopkins University Center for Advanced Governmental Studies, Washington, D.C.

Behbudzade, Z. (2020), *Analysis of the EU Countermeasures vis-à-vis Russia's Information Influence Activities in the Western Balkans* (Master's Thesis). Lund University, Department of Political Science, Lund, Sweden.

Blainey, G. (1988), The Causes of War, Free Press, New York.

Buzan, B. (2018), "Deterrence", in G. W. Brown, I. McLean, and A. McMillan, (eds.), *Concise Oxford Dictionary of Politics and International Relations*. Oxford University Press, Oxford.

Chenoweth, E. and Stephan, M. (2011), *Why Civil Resistance Works: The Strategic Logic of Nonviolent Conflict*, Columbia University Press, New York.

Fischer, M. and Petrović-Ziemer, L. (2013), "Dealing with the Past in the Western Balkans: Initiatives for Peacebuilding and Transitional Justice in Bosnia-Herzegovina, Serbia and Croatia", *Berghof Report* no. 18, Berghof Foundation, Berlin.

Friedman, V.A. (2007), "Language Politics and Language Policies in Contemporary Western Balkans: Infinitives, Turkisms, and Eurolinguistics", Wilson Centre, Meeting Report 337. https://www.wilsoncenter.org/publication/337-language-politics-and-language-policies-the-contemporary-western-balkans-infinitives

Gottemoeller, R. (2023), "Russian Attacks Would Be Far Worse without NATO's 'Proximity' Deterrence", *Economist*, 16 August. https://www.economist.com/by-invitation/2023/08/16/russian-attacks-would-be-far-worse-without-natos-proximity-deterrence-argues-rose-gottemoeller

Haas, P. M. (2015), "Choosing to Comply: Theorizing from International Relations and Comparative Politics", in P. M. Haas (ed.), *Epistemic Communities, Constructivism, and International Environmental Politics*, Routledge, London, pp. 285–309.

Halilovich, E. (2012), *Potsovjetski geopoliticki prostor i Balkan* (*The Post-Soviet Geopolitical Space and the Balkans*) Dobra knjiga, Sarajevo, Bosnia and Herzegovina.

Jaćimović, D., Deichmann, J. I. and Tianping, K. (2023), "The Western Balkans and Geopolitics: Leveraging the European Union and China", *Journal of Balkan and Near Eastern Studies*, vol. 25, no. 4, pp. 626–643.

Jankowicz, N. (2020), *How to Lose the Information War: Russia, Fake News, and the Future of Conflict*, I. B. Tauris, London and New York.

Libicki, M. (2021), "Can Denial Deter in Cyberspace?", in A. Wilner and A. Wegner (eds.), *Deterrence by Denial: Theory and Practice*, Cambria Press, Amherst, NY, pp. 193–211.

Maszkiewicz, M. (2015), *Civilian Based Defense System: New Approach–Polish, Lithuanian and Ukrainian Experience*, 8 March. https://papers.ssrn.com/sol3/papers.cfm?abstract_id=2575296

Mourtizen, H. (1997), *The Future of International Relations*, Routledge, London.

Pop, A. (2011), "The Independence of Kosovo in the Light of Regional Security Complex Theory", *European Journal of Science and Theology*, Center for Strategic Studies, Eastern Mediterranean University, vol. 9, no. 2, pp. 105–114.

Rašidagić, E. K. and Hesova, Z. (2020), "Development of Turkish Foreign Policy towards the Western Balkans with Focus on Bosnia and Herzegovina", *Croatian International Relations Review*, vol. 26, no. 86, pp. 96–129.

Šljivić, D. (2021), "Montenegro's Canonical Orthodox Church and Transition to Democracy in the Aftermath of the 2020 Parliamentary Elections", *Südosteuropa Mitteilungen*, vol. 61, no. 1, pp. 47–61.

Stronski, P. and Himes, A. (2019), "Russia's Game in the Balkans", *Carnegie Endowment for International Peace,* 6 February, https://carnegieendowment.org/research/2019/02/russias-game-in-the-balkans?lang=en

US Congress (2022), "China Regional Snapshot: Western Balkans", Republican Party Foreign Affairs Committee. https://foreignaffairs.house.gov/chinas-malign-global-influence-regional-snapshots/

Vukosavljevic, N. (2007), "Training for Peacebuilding and Conflict Transformation: Experiences of the "Centre for Nonviolent Action" in the Western Balkans", Berghof Research Centre for Constructive Conflict Management. https://edoc.vifapol.de/opus/volltexte/2011/2585/pdf/vukosavljevic_handbook.pdf

Wirtz, J. J. (2018), "How does Nuclear Deterrence Differ from Conventional Deterrence?", *Strategic Studies Quarterly*, vol. 12, no. 4, pp. 58–75.

Woelk, J. (2012), "Identity-Diversity and the Territorial Dimension in the Western Balkans", *L'Europe en Formation*, 2012/1, no. 363, pp. 189–204.

Životić, I. and Obradović, D. (2022), "Spread of the Russian Propaganda on Western Balkans–Case Study in Serbia". *Presented at the Security Horizons International Scientific Conference: 45 Years Higher Education in the Area of Security – Educational Challenges and Security Perspectives, 26–28 September 2022*, Struga, North Macedonia.

Zweers, W., Cretti, G., de Boon, M., Dafa, A., Subotić, S., Muk, M., Fetahu, A., Abazi Imeri, A., Kujraković, H. and Kuhinja, E. (2022), *The EU as a Promoter of Democracy or 'Stabilitocracy' in the Western Balkans*? Report by the Clingendael Institute and the Think for Europe Network. https://www.clingendael.org/pub/2022/the-eu-as-a-promoter-of-democracy-or-stabilitocracy/

13 Cyberspace Operations, Grey Zone Conflict, and Small States

Mikkel Storm Jensen

Introduction

Offensive cyberspace operations capabilities (OCOCs)—cyber-weapons—are well-suited for Grey Zone conflicts. The different categories of offensive cyberspace operations (OCOs)—cyberattacks—will be explained below, but cyberattacks, in general, can be stealthy, difficult, and time-consuming to attribute. They have potentially severe impacts either directly or as second-order effects, and their digital traces are more easily deniable than the more tangible evidence left by kinetic means. OCOCs are often less suited for achieving very specific effects in time and space, but this disadvantage is more than balanced by the fact that states can use OCOCs to conduct espionage, subversion, and sabotage from the comfort of home through cyberspace with lower risks and on exponentially larger scales than in the analog Grey Zone operations of yesteryear. Modern critical infrastructure and governance depend on IT systems, which can be targeted. Disinformation-dissemination can use the algorithms of social media to optimize impact. Espionage benefits from digitalized data by measuring successful operations in extracted terabytes of information rather than a few documents exfiltrated in a microdot. OCOCs allow all these Grey Zone operations to be conducted from the safety of home without sending vulnerable and, if caught, potentially embarrassing agents abroad. Furthermore, some OCOCs are relatively cheap to acquire. This makes them available even to states with limited resources, allowing them to conduct operations against neighboring states or opponents on the other side of the globe.

Despite popular representations of powerful cyberattacks, they are not a miracle weapon. This chapter briefly introduces the potential and limitations of OCOCs in Grey Zone operations. Other chapters in this book discuss vulnerabilities (Chapter 4), information operations (Chapter 14), crisis management (Chapter 8), society-wide resilience (Chapters 15 and 16), and the role of non-state actors (Chapter 6). This chapter will focus on states' offensive use of OCOCs in Grey Zone conflicts, only briefly discussing defensive use for deterrence (see also Chapter 12). This chapter presents Kilcullen's model of Liminal Warfare as an analytical framework for Grey Zone operations. It then presents types of cyber operations and their roles in Grey Zone conflict, followed by a discussion of offensive and defensive potential. OCOCs are worse than the inconvenience portrayed by skeptics but are not the doomsday weapons of hyperbolic claims. The chapter concludes with a discussion of OCOCs within an alliance framework. OCOCs are well suited to offensive operations in Grey Zone conflicts. There are severe limitations to their use for defense or deterrence, especially by alliances or states dependent on alliances—as many small states are. For such states, societal resilience against cyberattacks and cyber-enhanced disinformation are more immediately useful strategies to diminish the effect of hostile Grey Zone operations and deter their use.

DOI: 10.4324/9781003428701-20

The chapter relies on definitions used in the US or NATO doctrines. The analysis will be limited to states' strategic use of OCOCs—that is, cyber-weapons used to achieve national security ends like weakening an adversary by subversion (see Yarger and Bartholomees 2012; Jakobsen 2022). This includes states' use of non-state actors, like Russia's tacit collaboration with cyber criminals, but excludes non-state actors that engage in OCOs for criminal or ideological purposes. The chapter acknowledges that some criminals present very capable threats, and states must be prepared to defend against them, but these attacks are for criminal purposes, not raison d'état, and hence are inherently not a part of Grey Zone conflict. Ideologically motivated cyber activists might be included if aspiring to be state actors. However, these tend to have limited impact even when extensively engaged, as after the Russian invasion of Ukraine. Neither Russians swearing support for the Motherland nor Ukraine's IT Army have been able to cause significant effects (Schechner 2022; Vu et al. 2024). The Ukraine IT Army is an entirely new concept of national and foreign volunteers pledging to attack Russia in support of Ukraine, but only loosely affiliated with the state and often anonymous. It is too big, recent, and under-examined a topic from strategic and legal perspectives for this chapter to address (see Soesanto 2022, 2023; Vu et al. 2024).

Kilcullen's Model of Liminal Warfare

To provide an analytical framework for the discussion, Kilcullen's model of liminal warfare (Table 13.1) identifies the useful concepts: detection, attribution, and response. Grey Zone conflict is conducted mainly above the threshold of detection. While some operations, like espionage, are intended to remain below the detection threshold, they may also be aspects of Grey Zone conflict due to their objectives or accumulative effects. An example is the consistent Chinese cyber-enabled espionage campaigns to transfer intellectual property from other nations to its own industries (NCSC 2018). Some Grey Zone operations may be designed to be difficult to attribute, while others may be indifferent to this aspect, as in the case of many Russian offensive cyber campaigns. Others even seek attribution as an objective, as in North Korea's Sony

Table 13.1 Grey Zone Operations and Kilcullen's Liminal Warfare Model

Liminal warfare	*Layer*	*Description*	*Zone*
Response threshold	**Over conflict**	Clear and overt conflict, exceeding response threshold	Liminal Zone
Attribution threshold	**Ambiguous action**	Operation and sponsor suspected but unproven, crossing attribution threshold	Liminal/Grey Zone
Detection threshold	**Covert operation**	Operation detected, sponsor unknown, crossing detection threshold	Grey Zone
	Clandestine activity	Existence of operation and identity of sponsor both unknown	Grey Zone

Source: Adopted from Kilcullen (2019:69).

attack (Sharp 2017:911–913). Above the level of response, Grey Zone conflict risks becoming an armed conflict, although response can also be in kind within the Grey Zone.

Grey Zone Cyber-Enabled: Espionage, Disinformation, and Destructive Attacks

While the term "cyber domain" is sometimes used in NATO communications, the term used in NATO doctrines is "cyberspace." NATO defines cyberspace as:

> The global domain consisting of all interconnected communication, information technology and other electronic systems, networks and their data, including those which are separated or independent, which process, store or transmit data.
>
> (NATO 2020:4)

Cyberspace operations are actions in or through cyberspace intended to preserve friendly freedom of action in cyberspace and/or to create effects to achieve commanders' objectives (NATO 2020:18). Cyberspace operations can be divided into defensive cyberspace operations and OCOs. Defensive cyberspace operations and the associated capabilities do not represent coercive means. They are not considered controversial or threatening in international relations and are necessary to protect citizens and corporations against the proliferating and talented criminals in cyberspace. Most states with digital infrastructure have acquired means and set up organizations for defensive cyber operations. These play important roles in societal resilience in cyberspace against hostile state actions, criminal attacks, and cyber-related accidents. Thus, states' defensive cyber capabilities are as essential to deterrence strategies of denial as they are key to resilience (Janczewski and Caelli 2016; Jensen 2018:4). Resilience is the main component in the strategies of the EU and NATO to deter Grey Zone operations conducted through cyberspace (Smeets 2020b, 2021).

When examining interstate relations, the focus should be on cyberattacks (OCO). Current NATO doctrine only distinguishes between defensive and offensive cyber operations, but the US doctrines helpfully distinguish between two subcategories of cyberattacks: cyber-enabled espionage (Cyberspace Exploitation) and destructive attacks (Cyberspace Attacks). Cyberspace exploitations are intrusive but non-destructive operations to collect intelligence, while cyberspace attacks are operations to disrupt, deny, degrade, or destroy information resident in computers and computer networks or the computers and networks themselves (Joint Chiefs of Staff 2018:II-6, II-7; NATO 2020:16–17).

Espionage is not a new tool for states operating in the Grey Zone, but compared to analog means, cyberspace exploitation provides a paradigm shift in states' intelligence collection opportunities by widening the scope and lowering the costs and associated risks (Søilen 2016). Chinese theft of intellectual property, conducted through cyberspace exploitation, rose from a nuisance to being considered a strategic threat by the United States after 2010 (Harold, Libicki and Cevallos 2016). The EU is also becoming more concerned over the threat from China's espionage, conducted largely through cyberspace. The EU has been reluctant to accuse China of cyberattacks and has restrained reactions to recommendations for more resilience against espionage (PricewaterhouseCoopers 2018).

Cyberspace exploitation can support hostile states' disinformation campaigns, which can also be conducted using cyber means—social media and the associated optimizing algorithms to target and reach the most advantageous audience. Propaganda and disinformation are old tools for Grey Zone operations, but cyber means increase their scale exponentially, blur the source, obfuscate attribution, and facilitate targeting the most receptive audiences. A famous and successful example of such cyber-enhanced disinformation operations was the Russian hack of the Democratic National Committee during the 2016 US presidential elections. Russia disseminated stolen material through WikiLeaks to support and substantiate parallel efforts to undermine US cohesion and widen ideological fault lines. Operations were conducted from fake accounts set up by so-called "troll farms" like the Internet Research Agency in St. Petersburg. Russians impersonated American citizens on both sides of ideological divides, fanned the flames by calling for radical action, and disseminated propaganda aimed at radicalizing both sides (US Senate Intelligence Committee 2016:199; Diresta et al. 2019). The effects of the espionage and disinformation campaigns resonated throughout the Trump administration and beyond (Calamur 2019). Cyberspace exploitation and disinformation are potentially powerful tools for states in Grey Zone operations.

Characteristics of OCOs Fit Grey Zone Operations

Cyberattacks provide states with a new tool for sabotage and coercion at or below Kilcullen's attribution threshold and sometimes even below the detection threshold. Cyberattacks normally proceed along a so-called "kill chain." Once a target system is identified, reconnaissance for weaknesses is the next step. A cyber-weapon is chosen when a target is found within the target system. A "tailored access operation" is developed to take advantage of a weakness, either by facilitating the installation of other cyber-weapons inside the system under attack or by delivering an effect like information extraction, data destruction, or hardware damage (Yadav and Rao 2015; Loleski 2019). Effects can be instantaneous if the target is well-understood and the cyber-weapons are well-developed. Deciding and preparing an attack may take months, and there may be a long interval between installing a cyber-weapon and executing an attack. Successful attacks against a vigilant opponent are challenging, as demonstrated by Russia's limited cyber success against Ukraine (Bateman 2022).

Unlike Grey Zone operations with analog means, cyberspace attacks and exploitation often leave the victim uncertain about attribution. Even when an attack is recognized, the extent and nature of the attack may be unclear. The US reaction to the Solar Winds attack, tentatively attributed to Russian intelligence, demonstrates the sense of insecurity that follows the discovery of a deep penetration of critical systems (Fireye 2020; Sanger and Perlroth 2020; Jensen 2022a:15).

The technical and operational attributes of cyberattacks fit the ambiguities of Grey Zone operations. Successful cyberspace exploitation remains below the detection threshold. Detected cyberspace exploitation or successful cyberspace attacks move from the lowest, undetected level in Table 13.1 and into the liminal zone: They may remain unattributed for a time and, even when attributed, may remain below the response threshold. Russian, Chinese, and American examples illustrate this. The 2017 NotPetya attack cost Western companies several billion dollars in losses and damage. It was attributed to Russia by the US, UK, and Danish governments (Greenberg 2018; Demberger 2022). In 2020, the director of the FBI stated that

> The greatest long-term threat to our nation's information and intellectual property, and to our economic vitality, is the counterintelligence and economic espionage threat from China, [...] the people of the United States who are the victims of what amounts to Chinese theft on a scale so massive that it represents one of the largest transfers of wealth in human history.
>
> (Flannery 2020)

Yet arguably, neither Russia nor China has suffered consequences for their actions. The STUXNET cyberspace attacks are a third example of the use of cyber-weapons in a Grey Zone operation: The United States and Israel have never acknowledged the attack, but the cyber-enabled sabotage of the Iranian nuclear program is generally attributed to them (Falco 2012:20). In Kilcullen's terms, the STUXNET cyberspace attacks remained below the detection threshold from 2009 to 2010. Following its discovery and substantial analysis, the attack now hovers around the attribution threshold where the attackers are suspected but not proven. Even so, the attack has remained below the response threshold, something that would have been less likely if the Israelis had opted for an airstrike instead of cyberspace attacks to delay the Iranian program (Steele 2008).

While cyberattacks will leave digital trails, intelligence-based attribution may be time-consuming and initially provide uncertain, low-confidence answers (Lindsay 2015:56). Furthermore, a victim may be reluctant to disclose detailed evidence as it may reveal counterintelligence capabilities and methods. In any case, screenshots or printouts from a log file are less likely to arouse public emotions, particularly with third parties, and can be more easily dismissed as fabrications by the perpetrator than bomb fragments or CCTV images of intelligence agents entering the area where an attack has taken place.

OCOs: Less Than a Revolution, More Than a Nuisance

The technical and tactical properties of cyberattacks challenge the traditional realist understanding of what is possible for large states but impossible for small states based on their available resources (Handel 1990:83). Conventional military means with strategic reach and effect—a capable intercontinental missile force, a blue water navy, or nuclear weapons—require a visible and significant investment from a state with a large military-industrial base to develop. Their costs limit proliferation. Infrastructure makes capabilities visible to external observers. Allies and foes can assess the size and efficacy of conventional and nuclear capabilities and assess intent.

Cyberattacks have changed this situation and, hence, the strategic context. Small states can acquire them. The cost of entry into the group of states capable of cyberattacks is relatively low. Basic capabilities require only commercially available IT equipment and a team of qualified researchers, software developers, and operators (Liff 2012:418; JFQ 2019). This setup can be optimized by coupling it with national intelligence services' collection capabilities. Cyberspace attacks have unlimited geographical reach if the targets are connected to the Internet. As demonstrated by STUXNET, with some extra effort, they may even reach air-gapped targets (Farwell and Rohozinski 2011). Unlike conventional means, cyberattacks do not require large military and industrial investments to develop and deploy. This means that cyberattacks can be developed with little or no recognizable signature. Preparations stay below the detection threshold—useful for Grey Zone operations.

Before the Russian invasion of Ukraine, Russia had conducted nearly a decade of Grey Zone coercive cyber campaigns against Ukraine. Many attacks were above the detection and attribution thresholds but below the response threshold (Demberger 2022). Following the invasion in February 2022, some analysts, including the author of this chapter, expected Russian cyberattacks to significantly affect Ukrainian communications, logistics, air defenses, and perhaps civilian critical infrastructure. While information about the cyber aspects of the war is incomplete, it is apparent that Russia has conducted many potentially destructive attacks against both military targets, like air defense and communications, and civilian infrastructure like governance and banking (Sanger and Barnes 2022). However, Ukraine has been sufficiently resilient to remain functional (Bronk, Collins and Wallach 2022; Kostyuk and Gartzke 2022). There are at least three plausible explanations for this. First, Western analysts may have overestimated Russian competence and ability to conduct and coordinate cyber warfare; Russian conventional operations have surprised analysts with their lack of coordination and skill, even at the tactical level. Second, Russia may have saved its cyber-weapons for other missions. Why temporarily disable a piece of Ukrainian infrastructure with advanced, tailored cyberattacks instead of cheap artillery and rockets? Third, Ukraine has skillfully mitigated the Russian attacks with some support from friendly states and private companies like Microsoft and STARLINK (Nakasone 2022). In doing so, Ukraine has demonstrated a whole-of-society resilience strategy involving extensive public-private partnerships. Likely, all three explanations have played a role (Bateman 2022; Wilde 2022; Hüsch and Jarnecki 2023; Lonergan and Poznansky 2023).

While the war in Ukraine is far above the response threshold in the overt conflict zone and hence outside the realm of Grey Zone operations, it does demonstrate how cyber-weapons, even when freed from the constraints of the Grey Zone, are not miracle weapons. Cyber-weapons do not provide the same ability to conduct power projection as conventional, let alone nuclear weapons. Cyberattacks usually have limited, temporary, and reversible effects; STUXNET and NotPetya are exceptions. Cyberattacks generally do not have the permanent and irreversible destructiveness of conventional weapons.

Unlike conventional weapons, advanced cyberspace attacks cannot breach defenses and deliver their effects by brute force; they depend on identified weaknesses in the target systems (Jensen 2022b:10). Although some systems can be attacked directly and temporarily disabled by simple attacks like overwhelming a website with incoming signals, advanced systems require "tailored access" by cyber-weapons designed specifically for that purpose. Such tools depend on secrecy to achieve their effects; they can only penetrate if defenders are unaware of a technical, procedural, or organizational flaw that can be exploited (Libicki 2009:xiii, 18). Examples are zero-day vulnerabilities in software or items installed on a system with internet access and low-security settings. Physical access to a system might allow electronic or physical tampering or an employee in the opponent's organization might be targeted to click on tailored phishing emails (Taillat 2019:370). Once weaknesses have been identified by intelligence collection and target analysis, tailored development of cyberattacks takes time. Weeks or months may pass between identifying a target and developing a means to take it down. Change to the target system may render the cyber-weapon impotent before it is developed.

Notwithstanding these weaknesses, cyberattacks allow states to reach beyond their borders and seriously damage critical infrastructure. The US strategies acknowledge the theoretical potential for catastrophic damage from cyberspace attacks and include cyberspace attacks in the threats the US nuclear arsenal is tasked to hedge against (Pace 2006:C–1; DOD 2018:38; Schneider 2019:856). The inherent ambiguity of a cyberattack makes it a destabilizing tool with significant potential for crisis escalation (Cimbala 2017). Decision-makers have limited situational awareness (Clausewitz 1918:34). Discovering a cyberattack, the victim may be left in doubt whether its full extent is known. Lack of information on what has happened, who did it, and what will happen next, combined with a lack of international norms and precedents, can lead to unfortunate decisions (Libicki 2012:93–97). These include unintended escalation or counterstrikes against third parties triggering unintended security dilemmas (Libicki 2012:45–49).

Cyber-Weapons and Deterrence

Having established that cyber-weapons attributes are conducive to Grey Zone operations, consider their efficacy if a state crosses the response threshold and responds to hostile operations with cyberattacks or declares this policy to deter such attacks. The cyberattacks would be intended to deter further attacks, but not to escalate the conflict above the threshold of armed conflict. In 2018, the US announced increased emphasis on the role of cyber-weapons as a means of deterrence by punishment. The unclassified version of the 2018 National Cyber Strategy is vague, but a shift toward in-domain deterrence (The White House 2018; Smeets 2020a). The strategy came with classified directions for US Cyber Command, allegedly giving wider scope for cyberattacks and a higher threshold before presidential authorization had to be given (Sanger 2018; Wilner 2019). Mr. Bolton, the National Security Advisor, and General Nakasone, commander of Cyber Command and NSA since 2018, stressed the importance of the US doctrine of "persistent engagement." The persistent engagement doctrine requires the ability to be constantly present in other nations' networks to identify threats as they develop and punish hostile actions (Nakashima 2018a, 2018b; JFQ 2019; Nakasone 2019a, 2019b; Nakasone and Sulmeyer 2020; Jensen 2023b).

It is impossible to assess counterfactually how many state-sponsored cyberattacks the United States would have suffered without the declared strategy based on deterrence by punishment. Russia and China seem undeterred and continue cyberattacks. Theorists on cyber strategy have argued that in-domain deterrence is problematic for cyberattacks in the Grey Zone. Difficulties in attribution, unpredictable responses, and the challenge of response below the level of armed conflict are barriers (Libicki 2009:xix). Deterrence by punishment in the cyber-domain is also difficult for Western powers trying to establish rules-based norms of behavior in the cyber-domain (Fischerkeller and Harknett 2017). The UN Charter prohibits the use or threat of force but permits self-defense in the case of an armed attack. Cyberattacks don't fit neatly in the category of armed attack because they are non-kinetic with unpredictable consequences ranging from inconvenience to death (Nye 2017:47). For a major power, cross-domain deterrence, resilience, and norm-building are reasonable (Nye 2017:55). What about small states?

Cyber-Weapons, Small States, and Alliances

Many national security strategies rest on alliances rather than independent military action. Can these states use cyberattacks as a potential deterrent below the threshold of armed conflict? Most of the literature addresses cyber operations from a large-power perspective (Jensen 2023a:39). In the last decade, more alliance-dependent states have declared the intent to acquire offensive cyber capabilities. The group now includes more than half of NATO's members (Smeets, 2019:7), indicating that cyber-weapons are now seen as attractive assets by medium powers and small states. Offensive cyber operations, however, have limited utility for stand-alone deterrence. Small states should balance their investments in defensive cyber operations, cyber emergency response teams, and public-private partnerships to build cyber resilience for critical infrastructure.

Compared to conventional military forces, cyber-weapons are difficult to use alongside allies. Cyber operations depend on strong operational security and integration with national intelligence, and NATO has refrained from coordinating acquisition and shared planning for cyber-weapons. NATO doctrine allows cyberattacks to be integrated into operations on an ad hoc basis without sharing information about the deployed means (Jensen 2022). Major powers like the United States, Russia, and China and states accustomed to independent military action like Israel might find the threat of cyberattacks in the Grey Zone a useful tool to support deterrence. For states like Denmark, demonstrably willing to use force only in conjunction with allies in a multinational framework, deterrence by punishment using cyberattacks would be a major deviation from traditional national security strategies. It would require a higher appetite for risk than its decision-makers have historically demonstrated. While the United States may follow strategies of nuclear deterrence, persistent engagement, and deterrence by punishment in the cyber domain, most of its allies mitigate threats by deterrence through denial and improved resilience (Smeets 2021).

Conclusion

This discussion of the potential value of cyber-weapons in offensive or defensive Grey Zone operations has demonstrated that cyber-weapons' technical and tactical attributes are highly conducive to offensive operations but less well suited to deter attacks. While cyber-weapons are not wonder weapons for rapid deployment and widespread sabotage, their stealthy nature, remote deployment, and lack of physical footprint make them ideal for Grey Zone operations. Cyberattacks can be designed to remain below attribution and sometimes below the detection threshold. Even when attributable, cyberattacks are often difficult to retaliate against for states attempting to uphold a rules-based international order, thus keeping them below the response threshold. From a defensive perspective, states deploying independent military forces can convincingly use cyberattacks to deter adversaries from Grey Zone operations. However, states that normally only use military force in the framework of alliances should not expect offensive cyber operations to be a credible deterrent. They should rely on resilience and cross-domain deterrence against Grey Zone operations in the cyber domain.

References

Bateman, J. (2022), *Russia's Wartime Cyber Operations in Ukraine: Military Impacts, Influences, and Implications*, Carnegie Endowment for International Peace, Washington, DC.

Bronk, C., Collins, G. and Wallach, D. (2022), *Cyber and Information Warfare in Ukraine: What Do We Know Seven Months In?*, Baker Institute, Houston, Texas.

Calamur, K. (2019), "Trump Still Hasn't Condemned Russia for Meddling in the 2016 Election", *The Atlantic*, 18 April. https://www.theatlantic.com/international/archive/2019/04/trump-russia-meddling-2016-election/587518/

Cimbala, S. J. (2017), "Nuclear Crisis Management and Deterrence: America, Russia, and the Shadow of Cyber War", *The Journal of Slavic Military Studies*, vol. 30, no. 4, pp. 487–505.

Clausewitz, C. (1918), *On War*, Trübner & Co, London.

Demberger, A. (2022), *Merck Awarded $1.4 Billion for NotPetya After 5 Years of Legal Battle*, *Risk & Insurance*, 8 May. https://riskandinsurance.com/merck-awarded-1-4-billion-for-notpetya-after-5-years-of-legal-battle/

Diresta, R., Shaffer K., Ruppel B., Sullivan D., Matney R., Fox R., Albright J. and Johnson B. (2019), *The Tactics and Tropes of the Internet Research Agency*, United States Senate Select Committee on Intelligence (SSCI).

DOD (2018), "2018 Nuclear Posture Review", Department of Defense. https://media.defense.gov/2020/May/18/2002302062/-1/-1/1/2018-NUCLEAR-POSTURE-REVIEW-FINAL-REPORT.PDF

Falco, M. D. (2012), *STUXNET Facts Report*, NATO Cooperative Cyber Defence Centre of Excellence (CCDCOE), Tallinn, Estonia.

Farwell, J. P. and Rohozinski, R. (2011), "Stuxnet and the Future of Cyber War", *Survival*, vol. 53, no. 1, pp. 23–40.

FirEye (2020), *Highly Evasive Attacker Leverages SolarWinds Supply Chain to Compromise Multiple Global Victims with SUNBURST Backdoor*, FireEye, 13 December. https://cloud.google.com/blog/topics/threat-intelligence/evasive-attacker-leverages-solarwinds-supply-chain-compromises-with-sunburst-backdoor

Fischerkeller, M. P. and Harknett, R. J. (2017), "Deterrence Is Not a Credible Strategy for Cyberspace", *Orbis*, vol. 61, no. 3, pp. 381–393.

Flannery, R. (2020), *China Theft of US Information, IP One of Largest Wealth Transfers in History: FBI Chief*, Forbes, 7 July. https://www.forbes.com/sites/russellflannery/2020/07/07/china-theft-of-us-information-ip-one-of-largest-wealth-transfers-in-history-fbi-chief/

Greenberg, A. (2018), *The Untold Story of NotPetya, the Most Devastating Cyberattack in History*, Wired, 22 August. https://www.wired.com/story/notpetya-cyberattack-ukraine-russia-code-crashed-the-world/

Handel, M. (1990), *Weak States in the International System*, 2nd ed., Frank Cass & Co. London.

Harold, S. W., Libicki, M. C. and Cevallos, A. S. (2016), *Getting to Yes with China in Cyberspace*, 22 March, RAND Corporation, Santa Monica, CA.

Hüsch, P. and Jarnecki, J. (2023), *All Quiet on the Cyber Front? Explaining Russia's Limited Cyber Effects*, 1 June, Royal United Services Institute (RUSI), London. https://www.rusi.org/explore-our-research/publications/commentary/all-quiet-cyber-front-explaining-russias-limited-cyber-effects

Jakobsen, P. V. (2022), "Understanding and Teaching Military Strategy as Theories of Success at the Royal Danish Defence College", *Scandinavian Journal of Military Studies*, vol. 5, no 1, pp. 177–191.

Janczewski, L. J. and Caelli, W. (2016), "Security of Small Countries, Summary and Model", in L. J Janczewski and W. Caelli (eds), *Cyber Conflicts and Small States*, 1st ed., Routledge, London.

Jensen, M. S. (2018), "Sector Responsibility or Sector Task? New Cyber Strategy Occasion for Rethinking the Danish Sector Responsibility Principle", *Scandinavian Journal of Military Studies*, vol. 1, no. 1, pp. 1–18.

Jensen, M. S. (2022a), "Denmark's Offensive Cyber Capabilities: Questionable Assets for Prestige, New Risks of Entrapment", *Scandinavian Journal of Military Studies*, vol. 5, no. 1, pp. 111–128.

Jensen, M. S. (2022b), "Five Good Reasons for NATO's Pragmatic Approach to Offensive Cyberspace Operations", *Defence Studies*, vol. 22, no. 3, pp. 464–488.

Jensen, M. S. (2023a), *Offensive Cyber Capabilities and Alliances: Questionable Assets for Prestige, New Risks of Entrapment* (PhD Thesis), University of Southern Denmark, Faculty of Social Sciences. https://doi.org/10.21996/f1k8-7974.

Jensen, M. S. (2023b), "US' Allies' Offensive Cyber: Entrapment Risk or Entanglement Nuisance", *The Cyber Defense Review*, vol. 8, no. 3 (Fall 2023), pp. 67–87.

JFQ (2019), "An Interview with Paul M. Nakasone", *Joint Force Quarterly*, vol. 92, no. 1, pp. 4–9.

Joint Chiefs of Staff (2018), "JP 3–12 Cyberspace Operations (2018)", US Joint Chiefs of Staff.

Kilcullen, D. (2019), "The Evolution of Unconventional Warfare", *Scandinavian Journal of Military Studies*, vol. 2, no. 1, pp. 61–71.

Kostyuk, N. and Gartzke, E. (2022), "Why Cyber Dogs Have Yet to Bark Loudly in Russia's Invasion of Ukraine", *Texas National Security Review*, vol. 5, no. 3, pp. 113–126.

Libicki, M. C. (2009), *Cyberdeterrence and Cyberwar*, RAND Project Air Force, RAND Corporation, Santa Monica, CA.

Libicki, M. C. (2012), *Crisis and Escalation in Cyberspace*, RAND Project Air Force, RAND Coorporation, Santa Monika, CA.

Liff, A. P. (2012), "Cyberwar: A New 'Absolute Weapon'? The Proliferation of Cyberwarfare Capabilities and Interstate War", *The Journal of Strategic Studies*, vol. 35, no. 3, pp. 401–428.

Lindsay, J. R. (2015), "Tipping the Scales: The Attribution Problem and the Feasibility of Deterrence Against Cyberattack", *Journal of Cybersecurity*, vol. 1, no. 1, pp. 53–67.

Loleski, S. (2019), "From Cold to Cyber Warriors: The Origins and Expansion of NSA's Tailored Access Operations (TAO) to Shadow Brokers", *Intelligence and National Security*, vol. 34, no. 1, pp. 112–128.

Lonergan, E. and Poznansky, M. (2023), *Are We Asking Too Much of Cyber?*, *War on the Rocks*, 2 May. https://warontherocks.com/2023/05/are-we-asking-too-much-of-cyber/

Nakashima, E. (2018a), "White House Authorizes 'Offensive Cyber Operations' to Deter Foreign Adversaries", *Washington Post*, 20 September. https://www.washingtonpost.com/world/national-security/trump-authorizes-offensive-cyber-operations-to-deter-foreign-adversaries-bolton-says/2018/09/20/b5880578-bd0b-11e8-b7d2-0773aa1e33da_story.html

Nakashima, E. (2018b), "Trump Gives the Military More Latitude to Use Offensive Cyber Tools Against Adversaries", *Washington Post*, 17 August. https://www.washingtonpost.com/world/national-security/

trump-gives-the-military-more-latitude-to-use-offensive-cyber-tools-against-adversaries/2018/08/16/75f7a100-a160-11e8-8e87-c869fe70a721_story.html

Nakasone, P. M. (2019a), "A Cyber Force for Persistant Operations", *Joint Force Quarterly*, vol. 92, no. 1, pp. 10–14.

Nakasone, P. M. (2019b), "Statement of General Paul M. Nakasone Commander USCYBERCOM before the Senate Committee on Armed Services", Senate Committee on Armed Services. 14 February. https://nsarchive.gwu.edu/document/20133-national-security-archive-167-statement

Nakasone, P. M. (2022), "Posture Statement of Gen. Paul M. Nakasone, Commander", *US Cyber Command before the 117th Congress*: *News*, US Cyber Command. 5 April 2022. https://www.cybercom.mil/Media/News/Article/2989087/posture-statement-of-gen-paul-m-nakasone-commander-us-cyber-command-before-the/

Nakasone, P. M. and Sulmeyer, M. (2020), "How to Compete in Cyberspace", *Foreign Affairs*, 15 August, https://www.foreignaffairs.com/articles/united-states/2020-08-25/cybersecurity

NATO (2020), "AJP-3.20, Allied Joint Doctrine for Cyberspace Operations (Edition A)", NATO Standardization Office (January 2020).

NCSC (2018), "Foreign Economic Espionage in Cyberspace", National Counterintelligence and Security Center. https://www.dni.gov/files/NCSC/documents/supplychain/20190327-economic-espionage-pub02.pdf

Nye, J. S. (2017), "Deterrence and Dissuasion in Cyberspace", *International Security*, vol. 41, no. 3, pp. 44–71.

PricewaterhouseCoopers (2018), *The Scale and Impact of Industrial Espionage and Theft of Trade Secrets through Cyber*, EU Commision. https://op.europa.eu/en/publication-detail/-/publication/4eae21b2-4547-11e9-a8ed-01aa75ed71a1/language-en

Pace P., Chairman of the Joint Chiefs of Staff (2006), "The National Military Strategy for Cyberspace Operations", US Department of Defense, Washington, DC.

Sanger (2018), "Trump Loosens Secretive Restraints on Ordering Cyberattacks", *New York Times*, 20 September. https://www.nytimes.com/2018/09/20/us/politics/trump-cyberattacks-orders.html

Sanger, D. E. and Barnes, J. E. (2022), "Many Russian Cyberattacks Failed in First Months of Ukraine War, Study Says", *The New York Times*, 22 June. https://www.nytimes.com/2022/06/22/us/politics/russia-ukraine-cyberattacks.html

Sanger, D. E. and Perlroth, N. (2020), "More Hacking Attacks Found as Officials Warn of 'Grave Risk' to US Government", *The New York Times*, 17 December. https://www.nytimes.com/2020/12/17/us/politics/russia-cyber-hack-trump.html

Schechner, S. (2022), "Ukraine's 'IT Army' Has Hundreds of Thousands of Hackers, Kyiv Says", *The Wall Street Journal*, 4-6 March. https://doi.org/10.1080/02684527.2020.1729316.

Schneider, J. (2019), "The Capability/Vulnerability Paradox and Military Revolutions: Implications for Computing, Cyber, and the Onset of War", *Journal of Strategic Studies*, vol. 42, no. 6, pp. 841–863.

Sharp, T. (2017), "Theorizing Cyber Coercion: The 2014 North Korean Operation against Sony", *Journal of Strategic Studies*, vol. 40, no. 7, pp. 898–926.

Smeets, M. (2019), "NATO Members' Organizational Path Towards Conducting Offensive Cyber Operations: A Framework for Analysis", in T. Minárik, S. Alatalu, S. Biondi, M. Signoretti, I. Tolga, and G. Visky (eds.), *Proceedings of the 2019 11th International Conference on Cyber Conflict: Silent Battle*, 28–31 May 2019, pp. 163–177, NATO CCD COE Publications, Tallinn, Estonia.

Smeets, M. (2020a), US Cyber Strategy of Persistent Engagement & Defend Forward: Implications for the Alliance and Intelligence Collection", *Intelligence and National Security*, vol. 35, no. 3, pp. 444–453, doi:10.1080/02684527.2020.1729316.

Smeets, M. (2020b), "NATO's Cyber Policy 2002–2019: A Very, Very Brief Overview", Cyber References Project.

Smeets, M. (2021), *NATO Allies' Offensive Cyber Policy: A Growing Divide?*, *The Hague Centre for Strategic Studies,* August 2021. https://hcss.nl/report/nato-allies-offensive-cyber-policy-a-growing-divide/

Soesanto, S. (2022), "The IT Army of Ukraine: Structure, Tasking, and Eco-System", June 2022, Center for Security Studies (CSS), ETH Zürich. https://css.ethz.ch/en/center/CSS-news/2022/06/the-it-army-of-ukraine.html

Soesanto, S. (2023), "Ukraine's IT Army", *Survival*, vol. 65, no. 3, pp. 93–106.

Søilen, K. S. (2016), "Economic and Industrial Espionage at the Start of the 21st Century – Status Quaestionis", *Journal of Intelligence Studies in Business*, vol. 6, no. 3, pp. 51–64.

Steele, J. (2008), "Israel asked US for Green Light to Bomb Nuclear Sites in Iran", *The Guardian*, 25 September. https://www.theguardian.com/world/2008/sep/25/iran.israelandthepalestinians1?mc_cid=8b65f97c0a&mc_eid=fee8fd8a4f

Taillat, S. (2019), "Disrupt and Restraint: The Evolution of Cyber Conflict and the Implications for Collective Security", *Contemporary Security Policy*, vol. 40, no. 3, pp. 368–381.

The White House (2018), *National Cyber Strategy of the United States of America*, Washington DC. https://trumpwhitehouse.archives.gov/wp-content/uploads/2018/09/National-Cyber-Strategy.pdf

US Senate Intelligence Committee (2016), *Report of the Select Committee on Intelligence United States Senate on Russian Active Measures Campaigns and Interference in the 2016 US election Volume 2: Russia's Use of Social Media with Additional Views*, The Select Committee on Intelligence United States Senate, Washington, DC.

Vu, A. V., Thomas, D. R., Collier B., Hutchings A., Clayton R., and Anderson R. (2024), "Getting Bored of Cyberwar: Exploring the Role of the Low-level Cybercrime Actors in the Russia-Ukraine Conflict", arXiv, in *Proceedings of the ACM Web Conference 2024*, 13–17 May 2024, Singapore, ACM, New York, NY. https://arxiv.org/abs/2208.10629.

Wilde, G. (2022), *Cyber Operations in Ukraine: Russia's Unmet Expectations*, Carnegie Endowment for International Peace, Washington DC.

Wilner, A. S. (2019), "US Cyber Deterrence: Practice Guiding Theory", *Journal of Strategic Studies*, vol. 43, no. 2, pp. 245–280.

Yadav, T. and Rao, A. (2015), "Technical Aspects of Cyber Kill Chain", in J.H Abawajy, S. Mukherjea, SM. Thampi, and A. Ruiz-Martínez (eds.), *Proceedings of the Security in Computing and Communications: Third International Symposium,* 10–13 August 2015, Kochi, India, vol. 536, pp. 438–452, Springer International Publishing, Cham.

Yarger, H. R. (2012), *Toward a Theory of Strategy: Art Lykke and the US Army War College Strategy Model*, in Bartholomees J.B. Jr. (ed.), *US Army War College Guide to National Security Issues* (5th ed.) vol. I, pp. 45–51, US Army War College, Strategic Studies Institute, Carlisle, PA.

14 Information War, Lawfare, and Balance of Power

Eryk Habowski and David Last

Introduction

The international law of the 20th century reflected the interests of the dominant powers—the US and its Western allies. Power is three-dimensional—encompassing the ability to do, control agendas, and influence thought (Lukes 1975). Law is central to the second and third dimensions, and information to the third. Rival powers—Russia, China, and authoritarian regimes—use information war to undermine international law, the norms of non-intervention, and restraints on war that were established in the 20th century. Small states rely on these norms for their security and, therefore, have a stake in preserving a balance of power that constrains powerful states, particularly those seeking to undermine norms against violence and intervention. Hostile actors abroad and political actors within states manipulate law and information to attack these norms.

This chapter explores the three-dimensional nature of power in the Grey Zone, the nature of international and domestic law for preserving order, and the use of information war to undermine international law and domestic order. Power rests on norms as much as capabilities. Norms established under domestic and international law are attacked by lawfare supported by information warfare, eroding the trust and the authority of law and norms.

Power in the Grey Zone

The strength or weakness of legal norms reflects the balance of power in international relations and the nature of the international order. When strong powers agree on and enforce norms, there is less scope for operations in the Grey Zone. States violating norms are called out and sanctioned. Rivalry between strong powers over the nature of the international order creates more opportunities for states to break or manipulate norms. They use all three dimensions of power, but changing how people think is particularly important. The Cold War, for example, was a period of intense intellectual competition between the liberal freedoms espoused by the West and the authoritarian governments in Moscow and Beijing. We can generalize from this well-understood stand-off between superpowers to explore the problem of manipulative power short of war. This demands re-examining the nature of power wielded by small states, including hard, soft, smart, and sharp power.

With the 20th-century receding, an honest appraisal of the international order is needed to understand the situation of today's small states. Many countries and peoples prospered under the liberal post-war American world order, based on what American President Franklin Roosevelt described as the four freedoms: freedom of speech, freedom of worship, freedom from want, and freedom from fear. However, some experienced what American linguist Noam Chomsky described as the fifth freedom: "the freedom to rob, to exploit and to dominate, to undertake any

DOI: 10.4324/9781003428701-21

course of action to ensure that existing privilege is protected and advanced" (Chomsky 1989). The fear that a dominant power will exercise this freedom leads small states toward a limited number of security strategies. *Balancing* involves countering the rise of another power by supporting the weaker state or coalition in a conflict. *Bandwagoning* is joining the most powerful state or coalition. *Isolation* requires sufficient self-defense to make an attack too costly. *Hedging* entails pursuing different strategies simultaneously, perhaps appeasing one power and moving closer to security arrangements with another. Balancing, bandwagoning, isolation, and hedging strategies may be pursued within and outside an alliance (Edström, Gyllensporre and Westberg 2020:4–5).

Each strategy calls up images of hard power: coercive force, threat, military measures, and economic sanctions, usually with unambiguous or hard-to-conceal origins. In a contest of hard power, smaller states will lose against larger states, other things being equal. But hard power is not the same as state strength. The strength of a state begins with the idea of the state and concepts of nationhood and community (Buzan 1991; Breede 2017:Chapters 6–8). Strong nationalism and weak social bonds (bonding social capital) leave states struggling with democracy and development. Powerful states like the former Union of Soviet Socialist Republics (USSR) fragmented into weak post-Soviet states. Small states like the Nordic countries and some Commonwealth countries have strong and cohesive state cultures and institutions. Large and powerful states tend to be assemblies of nations and identities, while smaller states are often more coherent. This has implications for comparative advantage in the use of power. Large and powerful states use hard, soft, smart, and sharp power but may be more vulnerable to smart and sharp power wielded by smaller but more cohesive states.

Soft power prioritizes diplomacy and cooperation. These tools to build rules-based orders are most persuasive when led by a hegemonic power, as the United States was able to lead order-building in the last century. But hegemony never lasts. Immanuel Wallerstein and the World systems analysts identified cycles of hegemony and rivalry (Wallerstein 1976). Increasing rivalry may make declining hegemonic powers less cooperative and more war-prone (Choucri and North 1975; Gilpin 1981). New work drawing on large historical datasets suggests common patterns in the decline of states: concentration of wealth, declining life prospects for majorities, and increasing competition among elites (Turchin 2009; Turchin and Nefedov 2023). Faced with the aggressive behavior of large neighbors, small states might resort to smart or sharp power. Gulf states have adeptly used American public relations firms to sway public opinion and elite decision-making in the United States since the first Gulf War in 1990 (Al-Yahsin and Dashti 2009).

Joseph Nye introduced the term "smart power" in 2003 to indicate the effective combination of hard and soft power (Nye 2009:160). Smart power for small states may involve more cooperation than coercion, but even smaller states can generate convincing coercive power at a time and place of their choosing. This may require investment in special skills and equipment tailored to critical missions. Public Affairs, communications, reconnaissance, and special operations are examples of niche capabilities that may be useful for small countries (Rickli 2008).

The National Endowment for Democracy uses the term sharp power to describe manipulative and coercive tactics used by authoritarian regimes to interfere with the domestic politics of other countries or suppress democracy at home. These tactics include disinformation, economic coercion, and technologies to suppress dissent and manipulate public opinion. While the targets of hard, soft, and smart power are foreign decision-makers, the targets of sharp power are citizens and civil society organizations.

Soft, smart, and sharp power go beyond the coercive first dimension of power in attempting to change perceptions and thinking—whether in small groups of decision-makers, specific groups in society, or broad public or international opinion. Small states can be adept at shaping ideas. Between 2014 and 2019 Middle Eastern states donated more than $5 billion US dollars to American universities. Whether this represents soft power or sharp power may depend on perspective. However, the novelty of modern war is that it goes beyond blending hard and soft power to influence states in achieving state aims. Modern wars and Grey Zone conflicts transcend the boundary between war and peace, attacking the norms and rules on which peace rests (Introduction, Chapter 1). This brings us to the problem of law and lawfare in the Grey Zone.

Law and Lawfare in the Grey Zone

International and domestic laws help us explore the connection between law and power in the Grey Zone. International law constructs international order for states and commercial entities, and lawfare can be conducted within the framework of international law and in ways that erode international law. In the same way, domestic laws create domestic order, and lawfare can be used to attack within the framework of domestic law, or it can attack the structure of domestic law. We can describe four distinct lawfare tactics in modern war and Grey Zone conflicts: attacks using domestic or international law, and attacks on domestic and international law. Examples are found in Appendix.

As Rousseau wrote, "The strongest is never strong enough to be always the master unless he transforms strength into right, and obedience into duty." But further, "…force does not create right, and…we are obliged to obey only legitimate powers" (Rousseau 1762). The law reflects and underpins order because it is a form of cooperation. The community in which cooperation arises is relatively homogenous and consensual. Actors are more equal and trusting. Social norms are stronger and the cost of non-cooperation to a defector is high. Free riders are easy to identify and penalize. Individual and collective interests are more closely aligned (Glennon 2015:79–80). Lawfare is less likely in domestic or international communities with these characteristics.

Lawfare uses the law as a substitute for armed force to accomplish international security objectives. Orde Kittrie, a former US Department of State lawyer, developed a typology of lawfare limited to two main types—instrumental lawfare and compliance-leverage lawfare—and addresses mainly the first (Kittrie 2016:11). Kittrie's logic is that law can substitute for war by exploiting the legal system to achieve hostile objectives. More generally, any intent to weaken or undermine state interests or critical actors within a state using the legal system can be described as lawfare. Both Kittrie's forms of lawfare are conducted within the legal system. However, the legal system itself can also be a target. These two forms of attack can be mutually supporting. Attacks through the legal system can undermine a community's trust in and the authority of the legal system.

The eight elements of positive peace introduced in the Introduction provide a checklist of ways a national or international legal system can be attacked through the law. A well-functioning government can be eroded by attacking political parties or politicians in lawsuits or eroding norms of cooperative behavior or compromise. Legal manipulation can make business harder to conduct. Erosion of the tax system and preservation of privilege can undermine the equitable distribution of resources. Litigation against rights can erode acceptance of the rights of others. Vexatious and antagonistic lawsuits can undermine good relations between communities. Corrupting the legal system through bribes and the compromise of public officials attacks trust (see Appendix). Of course, all these forms of attack can come from within a state rather than from a hostile foreign actor. If there is even a hint that they exist, so much the better for a hostile actor looking for cracks to exploit.

International Law and Order

International law has been used over centuries to create order and community. Since the age of discovery, globalization has involved the extension of national to international law (May and McGill 2014). As empires became more extensive, they imposed more coherent regimes of law (Benton and Ford 2016). The great powers of the day used the law to extend order for commercial purposes, and the order they created reflected the interests of the commercial entities with influence in the governments of the day. Under hegemony, the ideology of the day justified the use of force. Spain's conquistadors uttered the *Requirimiento* to requisition land and labor (de las Casas 2014), and colonial powers cited variations of the civilizing mission. Many of the practices of colonial and imperial powers violated peremptory norms for behavior (*jus cogens*—Latin for "coercive law"). These are rules or principles so fundamental that they bind all states. They only achieve this status through recognition by the international community as a whole. "Most authorities agree that the laws prohibiting slavery, genocide, piracy, and acts of aggression or illegal use of force are *jus cogens* laws" (Law 2022:1286).

Reflecting the dominant liberal international order, the International Commission on Intervention and State Sovereignty produced *Responsibility to Protect*. But this may have been the high point for the liberal order and international law. Funding of universities and think tanks in the West by less liberal governments and conservative corporate interests began to erode the liberal consensus (NCRI 2023). Russian President Vladimir Putin's determination to return Russia to the status of a great power involves rejecting external norms and paralyzing the United Nations as an institution (Remler 2020). Expanding the BRICs (Brazil, Russia, India, China) may also provide a new counterweight to Western liberal international norms (Rajput 2019; Hiebert 2023).

Does the decline of Western-led international law necessarily undermine the norms of non-intervention and non-use of force? The limitation of war through legal instruments in the 20th century was undoubtedly a Western innovation (Hathaway and Shapiro 2017) but one that relied on the kind of community described by Glennon rather than a liberal hegemon. The increasing diversity of the international community makes agreement on law more difficult; however, preserving and expanding norms of non-intervention and non-use of force is a project small states might pursue regardless of their differences.

Lawfare within the Framework of International Law

Lawfare within the framework of existing international law can be used to target any state. Many states were concerned that the legal doctrine of *Responsibility to Protect* if enshrined as law, would erode the norm of non-intervention. International law can also be a tool for small states and non-state actors to target larger states. Size is relative, and indicators like population and GDP are imprecise measures. Palestinians and their international allies have used the law to resist settler encroachment in the West Bank and the construction of the separation barrier by Israel. An advisory opinion of the International Court of Justice in 2004 stated that Israel had breached its responsibilities under international law by establishing settlements (ICJ 2004:136). Appeals to Israeli courts have accompanied the strategy of international appeal. Palestinian rights groups filed suit in the International Criminal Court asking for rulings against Israel for apartheid and genocide and calling for the arrest of Israeli Leaders (Al Jazeera 2023). The ICC Office of the Prosecutor had previously opened an investigation into "The Situation in the State of Palestine" (ICC-01/18 2021). South Africa, Bangladesh, Bolivia, Comoros, and Djibouti submitted a referral to the case on 17 November 2023. The lawsuits and referrals to the International Court of Justice (ICJ) and International Criminal Court (ICC) are directly connected to information campaigns and diplomacy targeting foreign policies and international public opinion to increase support for the Palestinian cause and weaken support for Israel.

For Palestinians, Israel is the big power with American support and funding, but for Israel, Iran, and Saudi Arabia are the larger powers it faces, and the money of the oil-rich Arab world is aligned against them. The state of Israel and Israeli family members of victims of Hezbollah attacks have sought to establish third-party liability for attacks by terror groups (Kirchner 2006). At a time when Iran was declaring the intent to destroy the state of Israel, Israel sought to establish the legality of its military options against Iran under international law (Gharagozli 2010). More recently, Israeli families have gone to American courts to establish Iranian culpability for the 7 October 2023 massacres (Times of Israel 2024). As with Palestinian efforts, the legal front is linked to information operations and diplomacy, although the coherence of the links varies.

Attacks on the Framework of International Law

The examples above illustrate lawfare within established international and domestic law frameworks. Lawfare campaigns may also be conducted to erode those frameworks. In March 2023, ICC judges issued arrest warrants for Russian President Vladimir Putin and Commissioner for Children's Rights Maria Alekseyevna Lvova-Belova for the abduction of Ukrainian children to Russia (ICC Press Release 2023). Russia's Ministry of Internal Affairs retaliated by putting ICC President Piotr Hofmański on its wanted list. Russia rejects the authority of the ICC and has opened its own criminal case against the prosecutors and court officials (Chiappa 2023). Like the examples above, this legal action accompanies diplomacy and information operations addressed below. While the ICC's warrants were founded on international law, Russia attempted to delegitimize international law.

Erosion or adaptation of the framework of international law may not result from a hostile or targeted attack. Law is inherently adversarial, a vehicle for resolving disputes and seeking justice. Claims of principle can lead to local applications with damaging consequences, and local decisions can have wider ramifications. The UN Declaration on the Rights of Indigenous Peoples (UNDRIP) is part of a movement toward Indigenous internationalism that has implications for borders and access to settler states (Coates and Holroyd 2014). UNDRIP was not an attack on the framework of international law, but it does change international practice. Indigenous territory spanning the Canada-US border is a potential corridor for smuggling and a field for evolving maritime international law. Hostile interests can seek opportunities to change the law, the perception of the law, or the application of the law. Coat-tailing is a strategy that follows an international ruling with another similar suit but intends to obscure or contradict the record. Winning is less important than the information campaign that accompanies the suit. Trim tabbing is a strategy developed by NGOs to use small but strategic pressure points to bring about large movements, like the trim tabs on the huge rudders of container ships. Russia's indictment of ICC officials combined with information operations and diplomacy might be seen as an effort to bring wholesale discredit to international rulings.

Domestic Law to Create Order

Domestic law creates domestic order in a cooperative and consensual society. Domestic law in most countries includes constitutional, criminal, commercial, and civil law, often with different standards of evidence and quality of jurisprudence. For example, civil law may have lower stakes and lower standards of evidence than criminal law. Constitutional law may attract the highest level of public scrutiny. (See Wacks 2023 for more detail on the roots and branches of law.)

The rule of law in the Western democratic tradition includes three concepts. Regular law predominates, precluding arbitrary authority of the government over citizens. Everyone, including officials, is subjected to ordinary law administered by ordinary courts. Personal freedoms are formulated and protected by ordinary law rather than abstract constitutional declarations (Law 2020:1769).

Preserving the rule of law requires generally accepted laws, competent and impartial judges, prosecutors, and defenders, a corrections system deemed fair by the public, and a police service enjoying public trust. Funding levels must permit smooth and continuous operation without unacceptable delays in the process and adequate remuneration (including pensions) to minimize corruption. When states of exception arise through extraordinary threats or national emergencies, there should be mechanisms to permit an extraordinary response and a swift return to normal. Normalizing exceptional processes—like sealed indictments, closed courts, secret prisons, and coercive policing—corrodes the rule of law (Chapter 1). Non-state terrorist and sometimes hostile-state strategies seek to create the conditions for state overreaction to undermine a government's legitimacy (Sanguinetti 2013; Wilkinson 2011). Lawfare can mount attacks within the framework of domestic law, and it can attack that framework. In both cases, having domestic fellow travelers or anti-government agents is helpful but not essential.

Attacks through Domestic Law

Legal actions that weaken the state or impede its actions to defend the state and society can occur without external sponsorship. It becomes lawfare if hostile actors try to take advantage of these actions. Hostile actors can take many forms because of the blending of state, non-state, and criminal actors in Grey Zone conflicts and modern wars. Caution is warranted about a democratic bias and assumptions of universal values. Governments without all the characteristics of Western liberal democracies may have restrictions on free markets, income redistribution programs, concentration of wealth, or other characteristics but are nevertheless legitimate sovereigns. They may see external efforts to impose political, economic, and social reform as hostile actions. Elites in Russia, Ukraine, Moldova, Kazakhstan, and Georgia have seen democratic movements as external threats (Petrov 2020).

Each form of domestic law is a potential avenue of attack to pursue hostile interests. Constitutional law is a national "operating system" (International IDEA 2024). Changes at the constitutional level, including those that expand or restrict powers of government, can have the greatest impact. Constitutional changes by populist or authoritarian leaders may result in incremental erosion of constitutional democracies from within the system of constitutional law (Roznai and Hostovsky Brandes 2020). Criminal lawsuits can be launched to create reputational damage or derail careers, which they are likely to do even with acquittal. As with other forms of lawfare, information campaigns magnify the impact of lawfare.

National legal codes vary, but civil law is generally limited to settling individual disputes and usually includes commercial cases. Commercial and corporate laws involve both criminal and civil aspects in most countries. Lawfare using civil, commercial, and corporate law has a wide scope for operations. Preliminary hearings can be used to explore an opponent's legal strategy. This may be combined with cyberattacks to steal privileged attorney-client documents to undermine the defense (Carlin and Graff 2018:264). In describing US efforts to detect and deter cyber-thefts, the former Assistant Attorney General for National Security under Barack Obama makes it clear that many small countries will not have the resources to battle state-sponsored cyber-support of legal strategies and, therefore, will be working at a disadvantage. Delays, appeals, and disclosures can exploit the legal system's protections to wear down the targets.

Lawyer-client privilege can be used to mask identity or conceal actors involved in lawsuits or commercial transactions involving national assets or incremental acquisition of control over technology or infrastructure under normal commercial rules. Nefarious activities can be subcontracted through law firms bound by confidentiality provisions.

Attacks through thc legal system can come from within a state, particularly one with intense competition between elites, unscrupulous actors in a competitive market system, or rival identity groups vying for power or recognition. Lawfare within a state will likely be self-limiting because most actors will recognize a mutual interest in preserving the state. Attacks aided or manipulated by hostile external actors—whether state or non-state actors—may be aimed at weakening or overthrowing the state. Elements within the state and civil society can also seek to weaken or overthrow the government. The legal team responsible for defending the former Brazilian President has written about the blending of theories of law and war in lawfare strategies:

> We were facing persecution instigated by agents of the justice system, including police, members of the public ministries and judges, aligned with some of the most influential media, with the objective of manipulating the political situation. These acts were orchestrated with the clear objective of destabilizing and destroying an elected government, as well as with the intention of preventing the politically popular, former President of the Republic, Luiz Inácio Lula da Silva, from remaining in the electoral race.
>
> (Martins, Martins and Valim 2022)

Attacks on the Structures of Domestic Law

The events described by Lula's legal team combine an attack using the legal system with an attack on the legal and electoral systems, or at least activities that weaken trust and confidence in these structures. Escalating the attacks from the starting point described above, we can describe increasing manipulation of the justice system, surveillance, intimidation, false witnesses, and fabricated documents as means of abusing the justice system to achieve results. Each tactic damages the justice system by eroding trust and confidence. However, whether this damage is intentional or collateral damage is important. If it is intentional, it will be accelerated with information campaigns. If only collateral, the impact may be less immediate and wide-ranging.

Escalating further, we can describe the manipulation of courts and tribunals by signaling from powerful actors, bribery, intimidation, family pressures, and information operations. Chinese telecommunications giant Huawei's Chief Financial Officer, Meng Wanzhou, was indicted by the US Department of Justice on wire fraud charges. She was detained for extradition in Canada in December 2018. Canada was subjected to escalating pressures from China, including the arrest of Canadian citizens in China, interference in Canada's 2019 election, and wide-ranging lawsuits to block the extradition. Organized crime trials may require special protection not only of judges, but of the families and extended contacts of judges and juries to prevent bribery and intimidation (Campbell 2014).

Grey Zone Lawfare Threats and Modern War

Consider the problems of power, law, and information together in the context of Grey Zone threats and modern war. Facing an imbalance in coercive power, states and non-state actors may use attacks through law and on domestic and international legal frameworks to undermine governance and the rule of law in their opponents. International and domestic lawfare can achieve specific goals as part of a strategy in states bound by the rule of law. It is less effective or irrelevant when governments or political parties control court rulings. This gives authoritarian states an advantage in waging lawfare against democracies and open societies below the threshold of war.

Strategic industries can be attacked by lawsuits whose origins are masked by lawyer-client confidentiality. National leaders can be attacked by public accusations, compromising information (*kompromat*), or court action. The integrity of the courts and the rule of law can be undermined by bribery, intimidation, attacks on families and friends, or leaked and falsified information. Forgeries, corruption of databases, false witnesses, and evidence tampering can complicate or render the administration of justice impossible. All these tactics can be amplified by information warfare, which can undermine domestic and international norms and distort the values, attitudes, and beliefs that support national cohesion.

Information War—Defense and Offense

Information warfare can support and amplify lawfare. Lawfare is an attack through or on law, and information warfare is an attack through or on information. Law as a basis for cooperation and information as a basis for common understanding are mutually supporting pillars of a cohesive society. Both law and information are adversarial pursuits in an open society, with different voices competing to establish justice under law or truth through information. A brief tour of propaganda, disinformation, and attacks on electoral integrity illustrates the damage of mutually supporting lawfare and information warfare.

Jacques Ellul's classic study of propaganda (1965) remains one of the best descriptions of the psychological and social conditions required for effective propaganda—communication intended to persuade or influence an audience to further an agenda. Ellul emphasizes total propaganda, the use of all media and technologies, and the direct communication between leaders and masses enabled by mass communications. Mass communications serve to divide populations (Ellul 1965:212) and democratic propaganda can serve to mythologize and ultimately destroy democracy (234–241). But Ellul's analysis is positivist—there is an assumption that objective truth exists. Pomerantsev (2019), *This is Not Propaganda*, draws out the postmodern aspects of Russian information operations, now widely dispersed in the information sphere. Ellul missed the growing power of falsehood as a foundation for information war. The explosion of new work on information warfare after 2014 reflected a new way of thinking about the political impact of falsity beyond the "big lie" of propaganda. Jayakumar et al. (2021), Shu et al. (2020), Rid (2020), and Howard (2020) have all explored the power of manipulation, deception, and falsehood. By casting doubt on truth, the authority of law can be undermined. Courts no longer establish the truth but merely reflect the dominant power and can be ignored by those subscribing to alternative facts.

Recall Lukes's three dimensions of power—the power to do, the power to set agendas, and the power to change how people think. The 2007 confrontation over Estonia's Bronze Soldier statue in Tallinn illustrates how information warfare mobilizes all three dimensions of power. Estonia's independence in 1991 left a grateful Estonian majority with a large, potentially resentful, and fearful Russian-speaking minority making up about 20 percent of the population. Symbols like the Bronze Soldier—a monument to Estonia's liberation by the Red Army—became potential flashpoints. Moscow and the Estonian government in Tallinn sought to influence the narrative about events around the statue. Every anniversary was an opportunity to put the statue on the agenda, and physical confrontations between Russian speakers and Estonians were orchestrated by political entrepreneurs seeking advantage. Nina Jankowicz describes the 2007 confrontation from the vantage point of 2017 after a decade in which Russian information operations continued to stir tensions (Jankowicz 2020:Chapter 2).

Propaganda, distortion, rumor, untruth, misinformation, disinformation, smears, and fake news all fell within the Cold War scope of "political warfare" (in US doctrine) or "active measures" in Soviet doctrine. Active measures do not originate with politicians but with intelligence bureaucracies. They include disinformation, forgeries, doctored sources, covert acquisition, cutouts (trusted intermediaries), influence agents, false fronts, and more. Active measures have a purpose, usually weakening or manipulating a targeted adversary. They are covert operations aiming for tangible influence. They use true as well as false or distorted information. They often seek to erode trust in authority (Rid 2020:8–12). The digital revolution fundamentally changed disinformation and active measures. They became not only cheaper, faster, less risky, and easier to scale up and down, but also harder to control and assess once launched (Rid 2020:12): "… by the 2010s, active measures seamlessly overlapped with covert action. Networked computers, their vulnerabilities baked in, meant that information no longer targeted only minds; it could also now target machines" (Rid 2020:13). Electronic active measures with physical consequences blurred the boundaries of intervention in the internal affairs of states. Electronic attacks with physical consequences, like Russia's distributed denial of service attacks on Estonian banks in 2007, call into question the commitment of NATO members to defend against an attack on one of its members.

In contrast to active measures, political lies originate with politicians, parties, and political entrepreneurs. "Lie machines" are the social and technical mechanisms for putting untrue claims into the service of a political ideology (Howard 2020:13). The nature of lie machines within domestic politics makes them useful for hostile actors running active measures to weaken or manipulate an open society. When politicians and governments generate lies, they rely on teams of technical experts with access to demographic and survey data within a state. Electronic distribution includes automating delivery across multiple platforms. Consultants, marketers, and advertisers move messages into the distribution system. This socio-technical system, whether operated by paid contractors or true believers, can be manipulated or purchased by hostile external actors. As internet technology and culture changed in the 21st century, the old Cold War arts of highly skilled, slow-moving, and carefully planned active measures became high-tempo, low-skill, remote, and disjointed (Rid 2020:7). This fourth wave of disinformation has been mobilized to support the lawfare described above to undermine domestic and international legal norms.

Disinformation offensives can be both slow and methodical and fast and chaotic. Networks of accounts and bots can be built years in advance, with hundreds or thousands of ciphers kept ready to mobilize in a campaign. Malware can enslave tens of thousands of computers for use in denial of service, spam attacks, or in response to bots with spurious likes and upvoting of messages (Carlin and Graff 2018:289). Once attacks are launched, viral responses can spiral rapidly out of control in unpredictable ways. These tools are available to any state or non-state actor that cares to invest in them. Carlin describes criminal networks that appear to have partnered with state intelligence agencies, combining commercial exploitation with targeted intelligence gathering, as in the case of Russian businessman and cyber-criminal Evgeniy Bogachev, whose commercial malware attacks on the West concealed data breaches with intelligence implications (Carlin and Graff 2018:296). Poland experienced disinformation offensives during the COVID pandemic, which were associated with Russian centers of influence. When Russia invaded Ukraine, these centers of influence switched quickly to anti-Ukraine and anti-refugee narratives (Zadroga 2023).

Defense against disinformation will inevitably fail if it is reactive and based on individual incidents (Jankowicz 2020:Chapter 7). Interference in elections illustrates the problem. Election interference is now expected, both from foreign and domestic sources. A Freedom House study found that 88 percent of selected countries that held elections or referenda between June 2018 and May 2020 experienced electoral interference. Interference came from politicians, hyper-partisans, and foreign influence agents. The report found three forms of election interference: manipulation of online discussions, technical measures to restrict access to information, and legal measures to punish regime opponents and chill political expression (Shahbaz and Funk 2020). The norm of non-interference is thoroughly broken, but as much from domestic as international interference.

National efforts to counter disinformation are still in the early stages. The Czech Republic instituted the EU's first government-level disinformation task force in 2020, which was moribund within two years (Birch 2022). Canada launched an inquiry into foreign interference in its 2019 and 2021 Federal elections in September 2023 (GOC 2023), but this has already become mired in partisan criticism, as the judge heading the inquiry denied full standing to opposition parties (Globe and Mail Editorial Board 2023).

Conclusions

The international legal order of the last century is under attack. Major powers have eroded the norms against intervention in domestic affairs and the use of war to change boundaries and achieve political objectives. Liberal triumphalism played a part through initiatives like responsibility to protect, globalization, corporate rights eroding borders, and America's pre-emptive wars in the Global War on Terror. Now, resurgent authoritarianism is taking revenge through attacks on and through the law and attacks on and through information. Lawfare and information war erode states' internal cohesion and norms of behavior that make them secure in an international order.

References

Al Jazeera (2023), "Three rights groups file ICC lawsuit against Israel over Gaza 'genocide'", 9 November. https://www.aljazeera.com/news/2023/11/9/three-rights-groups-file-icc-lawsuit-against-israel-over-gaza-genocide

Al-Yahsin, Y. and Dashti, A. A. (2009), "Foreign Countries and US Public Relations Firms: The Case of Three Persian Gulf States", *Journal of Promotion Management*, vol. 14, nos. 3–4, pp. 355–374.

Benton, L. and Ford, L. (2016), *Rage for Order: The British Empire and the Origins of International Law 1800–1850*, Harvard University Press, London.

Birch, K. E. (2022), *Combating Disinformation at the Federal Level: The Czech and American Experiences*, Masaryk Diplomatic Program, Ministry of Foreign Affairs of the Czech Republic and the Embassy of the Czech Republic in Washington, D.C.

Breede, C. (2017), *The Idea of Failed States: Community, Society, Nation, and Patterns of Cohesion*, Routledge, London.

Buzan, B. (1991), *People, States, and Fear: An Agenda for International Security Studies in the Post-Cold War Era*, Harvester Wheatsheaf, Hemel Hempstead.

Campbell, L. (2014), "The Prosecution of Organised Crime: Removing the Jury", *The International Journal of Evidence & Proof*, vol. 18, no. 2, pp. 83–100.

Carlin, J. P. and Graff, G. M. (2018), *Dawn of the Code War: America's Battle Against Russia, China, and the Rising Global Cyber Threat*, PublicAffairs, New York.

Chiappa, C. (2023), "Russia Puts International Court's Top Leadership on Wanted List", *Politico*, 25 September. https://www.politico.eu/article/russia-international-criminal-court-icc-president-piotr-hofmanski-wanted-list/

Chomsky, N. (1989), *Culture of Terrorism*, Haymarket Books, Chicago.

Choucri, N. and North, R. (1975), *Nations in Conflict: National Growth and International Violence*, W.H. Freeman and Company, San Francisco.

Coates, K. and Holroyd, C. (2014), "Indigenous Internationalism and the Emerging Impact of UNDRIP in Aboriginal Affairs in Canada", in T. Mitchell (ed.), *The Internationalization of Indigenous Rights: UNDRIP in the Canadian Context*, Centre for International Governance Innovation, Waterloo, ON, pp. 5–10.

De las Casas, B. (2014), *A Brief Account of the Destruction of the Indies*, Penguin Books, London.

Edström, H., Gyllensporre, D. and Westberg, J. (2020), *Military Strategy of Small States: Responding to External Shocks of the 21st Century*, Routledge, London.

Ellul, J. (1965), *Propaganda. The Formation of Men's Attitude*, Knopf Doubleday Publishing Group, New York.

Gharagozli, B. (2010), "War of Words or a Regional Disaster: The (Il) Legality of Israeli and Iranian Military Options", *Hastings International and Comparative Law Review*, vol. 33, no. 1, pp. 203–237.

Gilpin, R. (1981), *War and Change in World Politics*, Cambridge University Press, Cambridge.

Glennon, M. J. (2015), "The Limitation of Traditional Rules and Institutions Relating to the Use of Force", in M. Weller (ed), *The Oxford Handbook of the Use of Force in International Law*, Oxford University Press, Oxford, pp. 79–95.

Globe and Mail Editorial Board (2023), "Canada's Election Interference Inquiry Starts Off on the Wrong Foot", *Globe and Mail*, 8 December. https://www.theglobeandmail.com/opinion/editorials/article-canadas-election-interference-inquiry-starts-off-on-the-wrong-foot/

Government of Canada (2023), "Government of Canada Launches Public Inquiry into Foreign Interference", 7 September, Democratic Institutions, 7 September. https://www.canada.ca/en/democratic-institutions/news/2023/09/government-of-canada-launches-public-inquiry-into-foreign-interference.html

Hathaway, O. and Shapiro, S. (2017), *The Internationalists: How a Radical Plan to Outlaw War Remade the World*, Simon and Schuster, New York.

Hiebert, M. (2023), "With BRICS Expansion, the Global South Takes Centre Stage", 31 August, Centre for International Governance Innovation (CIGI). https://www.cigionline.org/articles/with-brics-expansion-the-global-south-takes-centre-stage/

Howard, P. N. (2020), *Lie Machines: How to Save Democracy from Troll Armies, Deceitful Robots, Junk News Operations, and Political Operatives*, Yale University Press, New Haven.

ICJ (2004), *Legal Consequences of the Construction of a Wall in the Occupied Palestinian Territory, Advisory Opinion, ICJ Reports*. https://www.icj-cij.org/case/131

International Criminal Court (2021), *Public Redacted Version of "Prosecution's Request for Unsealing Warrants Against Paul Gicheru and Philip Kipkoech Bett*". 8 August 2015. https://www.icc-cpi.int/court-record/icc-01/09-01/15-10-red

International Criminal Court (2023), *Situation in Ukraine: ICC Judges Issue Arrest Warrants against Vladimir Vladimirovich Putin and* Maria *Alekseyevna Lvova-Belova*. Press Release, 17 March 2023. https://www.icc-cpi.int/news/situation-ukraine-icc-judges-issue-arrest-warrants-against-vladimir-vladimirovich-putin-and

International IDEA (2024), *Constitution-Building and the Rule of Law*, International Institute for Democracy and Electoral Assistance, Stockholm. https://www.idea.int/theme/constitution-building

Jankowicz, N. (2020), *How to Lose the Information War: Russia, Fake News, and the Future of Conflict*, I.B. Tauris, London.

Jayakumar, S., Ang, B., and Anwar, N. D. (eds.) (2021), *Disinformation and Fake News,* Palgrave Macmillan, Singapore.

Kirchner, S. (2006), "Third Party Liability for Hezbollah Attacks Against Israel", *German Law Journal*, vol. 7, no. 9, pp. 777–784.

Kittrie, O. (2016), *Lawfare: Law as a Weapon of Warfare*, Oxford University Press, Oxford.
Law, J., (ed.) (2020), *A Dictionary of Law*, Oxford University Press, Oxford.
Law, J. (2022), *A Dictionary of Law (Oxford Quick Reference)*, Oxford University Press, Oxford.
Lukes, S. (1975), *Power: A Radical View*, Bloomsbury Publishing, London.
Martins, C. Z., Martins, V. T. Z. and Valim, R. (2022), *Lawfare: Waging War through Law*, Taylor and Francis, London.
May, L. and McGill, J. (eds.) (2014), *Grotius and Law*, Ashgate, Farnham.
NCRI (2023), "The Corruption of the American Mind: How Foreign Funding in US Higher Education by Authoritarian Regimes, Widely Undisclosed, Predicts Erosion of Democratic Norms and Antisemitic Incidents on Campus", Network Contagion Research Institute, Washington, DC. https://networkcontagion.us/reports/11-6-23-the-corruption-of-the-american-mind/
Nye, J. S. Jr. (2009), "Get Smart: Combining Hard and Soft Power", *Foreign Affairs*, vol. 88, no. 4, pp. 160–163.
Petrov, K. (2020), "Elites and Color Revolutions: The Logic of Russia's Response", *Russian Politics*, vol. 5, no. 4, pp. 426–453.
Pomerantsev, P. (2019), *This Is Not Propaganda: Adventures in the War Against Reality*, Faber & Faber, London.
Rajput, A. (2019), "The BRICS as 'Rising Powers' and the Development of International Law", in H. Krieger, G. Nolte and A. Zimmermann (eds.), *The International Rule of Law: Rise or Decline?* Oxford University Press, Oxford, pp. 105–124.
Remler, P. (2020), *Russia at the United Nations: Law, Sovereignty, and Legitimacy*, Carnegie Endowment for International Peace, Washington, DC. https://carnegieendowment.org/research/2020/01/russia-at-the-united-nations-law-sovereignty-and-legitimacy?lang=en
Rickli, J. M. (2008), "European Small States Military Policies after the Cold War: From Territorial to Niche Strategies", *Cambridge Review of International Affairs*, vol. 21, no. 3, pp. 307–325.
Rid, T. (2020), *Active Measures: The Secret History of Disinformation and Political Warfare*, Farrar, Straus and Giroux, New York.
Rousseau, J. J. (1762), *The Social Contract*, Marc-Michel Rey, Amsterdam [i.e., Geneva].
Roznai, Y. and Hostovsky Brandes, T. (2020), "Democratic Erosion, Populist Constitutionalism, and the Unconstitutional Constitutional Amendments Doctrine", *Law & Ethics of Human Rights*, vol. 14, no. 1, pp. 19–48.
Sanguinetti, G. (2013), *On Terrorism and the State*, Colossal Books, New York.
Shahbaz, A. and Funk, A. (2020), *Digital Election Interference*, Freedom House International. https://freedomhouse.org/report/freedom-on-the-net/2019/the-crisis-of-social-media/digital-election-interference
Shu, K., Bhattacharjee, A., Alatawi, F., Nazer, T. H., Ding, K., Karami, M. and Liu, H. (2020), "Combating Disinformation in a Social Media Age", *Wiley Interdisciplinary Reviews: Data Mining and Knowledge Discovery*, vol. 10, no. 6, e1385, pp. 1–39.
Times of Israel (2024) *American Oct. 7 Victims and Families file $1 billion Lawsuit Blaming Iran for Attack*, 1 https://www.timesofisrael.com/american-oct-7-victims-and-families-file-1-billion-lawsuit-blaming-iran-for-attack/
Turchin, P. (2023), *End Times: Elites, Counter-Elites, and the Path of Political Disintegration*, Penguin Press, New York.
Turchin, P. and Nefedov, S. A. (2009), *Secular Cycles*, Princeton University Press, Princeton.
Wacks, R. (2023), *Law: A Very Short Introduction*, Oxford University Press, Oxford.
Wallerstein, I. (1976), *World-Systems Analysis: An Introduction*, Duke University Press, Durham.
Wilkinson, P. (2011), *Terrorism Versus Democracy: The Liberal State Response*, Routledge, London and New York.
Zadroga, M. (2023), "Disinformation Landscape in Poland", EU Disinfo Lab, 4 December. https://www.disinfo.eu/publications/disinformation-landscape-in-poland/

Design Problem 3

Exploring Solutions with Divergent Thinking and Causal Loops

Archipelago of Design (AoD)

Oliver Jones, Kassie Miedema, and Philippe Beaulieu-Brossard

Introduction

Section III of *Modern War and Grey Zones: Design for Small States* explores deterring and managing violence within Grey Zone conflicts. International organizations, deterrence and alignment, cyberspace operations, and the impact of information warfare and lawfare reflect aspects of small states' efforts to manage their security. Modern conflicts are complex and interconnected. Security planners cannot rely on linear or proforma plans to achieve the many national objectives associated with security. Design thinking helps to find creative solutions to interconnected problems.

Design Problems 1 and 2 went from divergent to convergent thinking to widen and then narrow the focus on a problem and to get that focus right. The next challenge is to develop solutions, first by thinking divergently (Design Problem 3) and then converging on the best possible solution (Design Problem 4). This exercise focuses on *causal loop analysis*. It explores how a system functions by mapping stakeholders, environmental features, and intangibles like time and behavior and their interactions. Visualizing the system as a series of causal loops that diagram a problem space's relevant features allows for targeting specific leverage points and interventions.

A causal loop is an illustration of a system, and those unfamiliar with systems thinking might start with a primer like Braun (2002), Meadows (2008), or Stroh (2018) from which the following summary is drawn. Some basic system concepts (or archetypes) include simple metaphors like the *iceberg* and the *bathtub* (Meadows 2008), and more nuanced archetypes like *shifting the burden*, *limits to growth*, *escalation*, *the tragedy of the commons*, and *fixes that fail* (Braun, 2002). Braun (2002) provides a comprehensive list of system archetypes, whereas Ryan (2017) describes non-archetypal approaches to causal loop analysis. Causal loops, which may depict the interactions of one set of actors or the dynamics of the interactions of several different kinds of actors, can contain aspects of any or all these systems concepts.

Method

The following method is a generic approach to mapping a system and attempting to understand causes.

When mapping a casual loop, it is helpful to think about the role of stakeholders in the system illustration. A stakeholder is any person or group who influences or is affected by the problem space. When identifying stakeholders, it is useful to consider groups who interact with features such as:

- The built environment (infrastructure, buildings, public spaces, etc.)

DOI: 10.4324/9781003428701-22

- The natural environment (forests, deserts, oceans, etc.)
- Related environmental factors (tides, weather, climate, etc.).
- Intangible or non-physical features (time, behavior, economies, society, political features, etc.)

Features and stakeholders should be described in as much detail as is useful to define the problem space. A system involving multiple stakeholders should be defined in more detail than an organizational process. An ecosystem, a city, and a particular industry might be described in layers to identify pressures and points of tension that might be important to understanding potential solutions. In this divergent phase of solution-finding, planners try to look beyond the obvious.

The Israeli-Palestinian conflict provides an example of complex causal loops. Stroh (2018) begins by describing a four-stage cycle of violence: both sides fight for the right to exist, tensions escalate, pressure leads to negotiations, and peace efforts break down. In this example, the observation that the Israeli-Palestinian conflict manifests as a cycle of violence allows an analyst to map the interactions of different kinds of stakeholders in the conflict and the peace process and identify behaviors that are likely to make the cycle of violence worse, or that present barriers to peace. By drawing these relationships in a causal loop diagram, Stroh's analysis moves from the initial observation to additional causal loops affecting the stakeholders.

The following diagram (Figure DP3) illustrates an example of a common system archetype, "shifting the burden," as a causal loop diagram of strategic issues of Irish air defense. "Shifting the burden" describes the tension organizations face between improvising short-term solutions to address visible problems and resisting the implementation of a solution to address the underlying causes. For example, a healthcare system may face rising costs and poor health outcomes. An immediate solution might be to increase spending on healthcare, while the underlying problem might be a lack of prevention, which demands a reallocation of resources. Reliance on short-term solutions allows the underlying problem to persist, which leads to a recurrence of the symptom. More broadly, it illustrates a creative tension between different kinds of problem-solving requiring patience and creativity to manage. Solving underlying causes may require more patience and creativity and may be riskier than short-term fixes. They require deeper understanding and more stakeholder engagement to follow through successfully. In the healthcare example, doctors and hospitals may resist reallocating resources to prevention. In the example below, capability development is a major investment that competes with other public goods and exacerbates competition between service branches.

In this figure of a causal loop illustrating strategic issues of Irish air defense (see Figure DP3), Ireland's potential reliance on external forces, such as the British Royal Air Force (RAF), to deter threats in Irish airspace exemplifies a "Shifting the Burden" dynamic. Although formal defense agreements between Ireland and the United Kingdom remain confidential, anecdotal evidence suggests that RAF intervention may occasionally supplement Ireland's limited defense capabilities. This approach provides a short-term fix, temporarily alleviating the immediate security gap caused by the Defence Forces' lack of air defenses. However, this reliance could carry political costs, as it might lead to questions about Ireland's military neutrality and independence in defense matters. A more sustainable solution could involve developing a domestic combat air capability, directly addressing the root cause of this dependence. However, such a commitment would require significant public and financial support, competing with other national priorities. Mapping this issue as a causal loop clarifies how reliance on external deterrence might

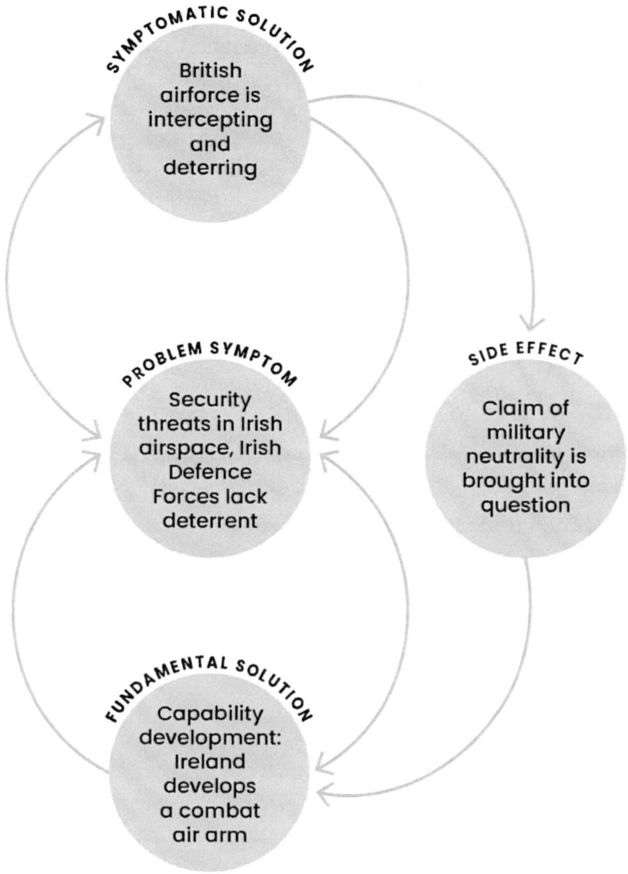

Figure DP3 Causal Loop Diagram Illustrating Strategic Issues of Irish Air Defense—Shifting the Burden.

perpetuate Ireland's defense shortfalls while potentially compromising its neutral stance. Implementing a robust, self-sufficient defense solution could mitigate these tensions but demands long-term investment and careful engagement with the public. Framing this dilemma through the "Shifting the Burden" archetype highlights the need for strategic, long-term thinking to address Ireland's complex defense challenges.

Group Exercise: Causal Loop Analysis

Task. Draw a causal loop diagram for a situation related to section III—international organizations, deterrence and alignment, cyber operations, or information war and lawfare. For example,

- How can international organizations contribute to conflict management?
- Do deterrence systems amongst small states depend upon larger powers, or can they operate independently?
- When does asymmetric retaliation against cyber-attacks serve to deter?
- Can asymmetric information campaigns offset military weakness?

This exercise is best done in groups because it involves brainstorming and iteration. Often, one person will not understand a complex system comprehensively, and different group members will highlight valuable perspectives on how the system works.

- Start by listing all the features relevant to the problem space. Don't question assumptions; write everything down. It often helps to write features down on sticky notes or a whiteboard so they can be modified and moved.
- Next, start drawing arrows connecting the features. Sometimes, labeling the arrows to describe what they represent in a few words is helpful.
- If a feature does not have arrows connecting it to other parts of the system, consider whether it is relevant to your problem space or if a connection is missing.
- Set a time limit for completing the system map (e.g., 20 minutes). This constraint helps focus on the most relevant parts of the system.

Consider. The various system concepts or archetypes are listed above. What does the system look like? What are the system inputs, outputs, and feedback loops? How are the parts of the system interconnected? What dynamics of the system are self-reinforcing? What interventions could help break negative cycles or reinforce positive ones? Refer to the "dependence on retaliation" example (Stroh) to apply some of these questions to the system you are considering:

Cycles of Violence

- Do cycles of violence and retaliation lead to escalating hostilities?
- Have ceasefires, alliances, or partnerships impacted cycles of violence?

Peace and Trust-Building

- How have peace negotiations influenced the system?
- Are there any confidence-building measures that might build trust between communities?
- Do social programs and economic cooperation increase trust and reinforce mutual dependencies?

Economic Factors and Instability

- How do economic hardship impact stakeholders?
- Does economic disparity drive unrest?

Misinformation and Distrust

- Does propaganda increase mistrust or encourage hostile actions?

Implications. Causal loops help to analyze complex systems. A basic characteristic of a system is that changes to one part of it affect other parts. How did diagramming causal loops help to explore this in the problems selected? What interdependencies was the group able to identify? Were there key stakeholders or dominant actors in the system? Can causal loops help to identify sustainability questions or the evolution of a strategic problem over time? How might a team use causal analysis to help with scenario planning? Is there a point of diminishing returns when a group is mapping out a system and its parts? What indicates the system may be too complex to work with, and what can be done to make the analysis more helpful? Causal loop analysis encourages a focus on flows and efficiencies. How can this help optimize the use of limited resources in complex security policy challenges? How does causal loop analysis help to develop and test strategic decisions?

References

Braun, W. (2002), "The Systems Archetypes", *Systems*. https://www.academia.edu/28706184/The_System_Archetypes_The_System_Archetypes

Meadows, D. H. (2008), *Thinking in Systems: A Primer*, edited by Donella Wright. Chelsea Green Publishing, White River Junction, Vermont.

Ryan, A. (2017), "Design in Civvies: The Promise of Creating Degrees of Freedom in Government", *Journal of Military and Strategic Studies*, vol. 17, no. 4, pp. 39–50.

Stroh, D. P. (2018), "A Systemic View of the Israeli-Palestinian Conflict", *System Thinker*. https://thesystemsthinker.com/a-systemic-view-of-the-israeli-palestinian-conflict/

Further Readings

Barabrook-Johnson, P. and Penn, A. (2022), *Systems Mapping: How to Build and Use Causal Models of Systems*. Palgrave Macmillan, Cham.

Bury, P. and Murphy, D. (2023), "No Time to Hide: The Future of Irish Defence and Security", *War on the Rocks*, 3 March. https://warontherocks.com/2023/03/no-time-to-hide-the-future-for-irish-defense-and-security-and-how-our-partners-can-help/

Section IV

Society and People

15 Defending the State in Grey Zone Conflicts and Modern War

Waldemar Kitler, Malwina Ewa Kołodziejczak, and David Last

Introduction

The essence of the Grey Zone is that there is no clear boundary between war and peace. International law sets a high bar for war and aggression. National legislation in liberal democratic countries is framed to limit the powers of the state under normal circumstances and accord additional powers under a state of exception. States must find the right legal tools to operate in the Grey Zone and modern wars without compromising their essential values. Like the response to terrorism, the response to Grey Zone conflict tactics may erode the values democratic states seek to preserve.

This chapter addresses the functions the state performs to preserve security in the Grey Zone, the challenge Grey Zone threats pose to these functions, the interaction of markets and society within the state, and the choice of legal and regulatory responses to Grey Zone threats.

State Functions to Maintain Security

States protect their borders from incursions, their institutions from being overthrown, and their citizens from domestic or foreign depredations. They also cooperate in an anarchical society of states to preserve the conditions necessary for their sovereignty (Bull 2012; Suganami, Carr and Humphries 2017). These three functions reflect the pursuit of human, national, and international security. The Grey Zone and modern wars challenge the separation of international security from national and human security. The three are inseparable, but national legislation and international law do not always recognize this reality. A legal dictionary defines war as "the legal state of affairs that exists when states use force to vindicate rights or settle disputes between themselves." States can engage in hostilities without being at war, and they can be at war without fighting (Law 2022:2024–2025).

Most states have separate national defense and domestic emergency legislation (Table 15.1). Typical domestic emergency legislation addresses unarmed disturbances, armed threats, and natural disasters. It is also common to find published national security strategies, cyber security strategies, banking sector stability strategies, and counter-terrorism strategies. Divided responsibilities between ministries and departments exacerbate the inevitable patchwork of national strategies to address new phenomena. Yet, this patchwork of strategies and jurisdictions is the latticework for national and human security. Neither national defense legislation nor domestic emergency legislation can be adequate defenses against Grey Zone threats, though each is essential for the transition between the Grey Zone and modern wars.

Poland, for example, has legislation to guide the management of the state under external threat and war. But the legislation assumes a clear physical attack. How will the state function

DOI: 10.4324/9781003428701-24

Table 15.1 Examples of National Defence and Emergency Legislation

	National defense legislation	*Domestic emergency legislation*
Botswana	National Security Act	Disaster Preparedness and Relief Act
Canada	National Defence Act	Emergencies Act
Estonia	Defence Forces Act	Law on Emergency Situations
Finland	State of Defence Act (1991)	Emergency Powers Act
Poland	War Powers Act	Act on Crisis Management
Senegal	National Security Law	Civil Defence Law
Sweden	National Security Act	Civil Protection Act
United Arab Emirates	National Security Law	Federal Law on Civil Defence

Source: Compiled by Authors based on the International Committee of the Red Cross, ICRC, International Humanitarian Law Database, and other sources.

under attacks below the threshold of war? Comparing the legislation in Poland, Estonia, and Finland, the key challenge lies in interpreting the connection between external threats below the threshold of war and the national responses that may be legitimate under different circumstances. One of the main difficulties states face in the era of the war nexus is that "external state threats" may be too amorphous to be addressed by existing national legislation.

A further legal problem is that public international law addresses state entities, while Grey Zone conflict and modern war involve actors deliberately unrecognizable in international law, including organized crime, commercial enterprises, and armed groups. The state can also be damaged from within by ideological fellow travelers, "useful fools," or innocents. Citizens' rights to protest and legitimate political activity can disrupt the state at critical times. If states cannot rely on the traditional understanding of hostile state military action, the first requirement to combat external state threats is to identify the state's key functions that must be preserved before and during any hostile action.

In Polish law, the term "external threat to the state" is interpreted narrowly, referring to threats of state aggression defined under international law. However, many non-military situations can evolve quickly into situations requiring a military response. Unresolved internal disputes arising from ethnic, territorial, or sovereignty conflicts, aggressive movements questioning the state's territorial integrity, religious fanaticism, and organized crime crossing borders are examples of external threats requiring a national response. Cyberattacks, theft of state secrets or intellectual property, illegal trade in materials for weapons of mass destruction, and interference in elections or manipulation of public opinion justify some form of state response (see Appendix). In the Polish case, the threat must originate outside the state, and the effect of the threat must have unacceptable political and military impacts.

This raises the central problem of Grey Zone conflicts. States have evolved with a boundary distinction. Armed forces, laws of war, and the protection of the state from external enemies are distinct from police and gendarme actions within the state. Domestic law and order and protection of citizens and state institutions are not structured to withstand erosion from non-attributable hostile actors. Overlaid on this bifurcation of domestic and foreign threats is a framework of international law, which establishes definitions of war and aggression within the anarchical society of states, conditioning the responses deemed to be legitimate.

Domestic emergency legislation typically addresses public safety situations requiring military resources for public welfare (natural disasters and their humanitarian consequences) or public order (support for police and internal security). The functions of states and new forms of Grey Zone threat extend beyond these two circumstances. Threats to state effectiveness are real and serious and have military and political consequences, but they are difficult to address outside the exceptional state of war. The language of legislation does not always recognize this. In Estonian law, a declaration of a state of war is required to mobilize armed forces to meet the state's international obligations. Article 129 of the Constitution of Finland permits the President of the Republic to mobilize the Defense Forces. In Canada, a single Act addresses a range of possibilities. The Emergencies Act describes four types of emergencies: public welfare, public order, international emergency, and war emergency. It can be invoked "to grant temporary additional and necessary powers to the federal government…"

Grey Zone Threats and Essential State Functions

The threats of disorder, armed violence, and vulnerability to external interference can be generated by political, economic, social, ideological/religious, cultural, and environmental factors. The Grey Zone threats listed in Appendix illustrate the extent to which state function can be attacked below the legal threshold for war. The Institute for State Effectiveness (ISE) illustrates the targets of these threats across ten essential state functions: governance, security, rule of law, market engagement, infrastructure, financial management, asset management, human capital, citizen engagement, and disaster resilience (www.effectivestates.org).

The ISE is based in Washington and is enthusiastic about free markets. Still, a stable and effective market framework is essential for human security, regardless of the political economy of the state. Economic threats such as currency or credit-rating attacks, supply chain disruption, and alliances with organized crime can undermine the markets on which citizens depend. Basic infrastructure, including water, electricity, transport, schools, and hospitals, can be eroded directly through sabotage or indirectly through loss of trust. Trust in banking and finance and the government's probity in managing public funds is also necessary for state effectiveness. Attacking these foundations erodes state capacity. Land, water, minerals, and the natural environment are public assets. States that fail to manage these assets lose both trust and capacity.

Effective states invest in their citizens through education, health, and welfare provisions. If the government doesn't provide these investments directly, it must shape the market forces that do so. Citizen engagement and rights that include all sectors of society help build cohesion. Clear frameworks and rules for response to disasters make societies resilient. Disinformation and media manipulation, social nudging away from the common good, deep fakes to erode trust, and evading and undermining regulations can reduce the effectiveness of states.

Public sector governance is the foundation of state effectiveness and supports the legitimacy of the state's monopoly on the use of force. Eroding this monopoly also erodes the rule of law, which includes essential accountability and oversight. These functions can be described as political. Political and legal Grey Zone threats, including corruption strategies, funding extremism, targeted assassinations, *kompromat*, and developing "fellow travelers" hostile to the political order, are Grey Zone tactics to undermine political effectiveness, domestic security, and the rule of law.

Organizing the state for security and defense against this range of Grey Zone threats begins with a legal framework for response. The concept of "armed attack" is one of the barriers to an effective defense. The UN General Assembly's Resolution 3314 (XXIX) of 14 December 1974, Article 3 enumerates specific acts of aggression: invasion, bombardment, blockade, attack by armed forces, or incursion by armed forces. The final category refers to sending armed bands, groups, irregulars, or mercenaries. This non-binding resolution has been reinforced by the Rome Statute of the International Criminal Court, whose 2010 review conference agreed to a binding definition of the crime of aggression in Article 8. This definition refers specifically to the use of *armed force by a state*. The use of armed force is intended to be a broader concept—not every use of force is an attack (Kowalski 2015:180). The role of non-military forces is also problematic. Are cyberattacks armed forces? What about financial attacks, sanctions, or alliances with criminal groups indistinguishable from elements within the state? Regardless of international law, domestic laws can establish a framework that preserves sovereignty, territorial integrity, and political independence. The challenge is to recognize the risk to the state's core values.

Many 20th-century police studies painted political policing as deviant, contrasting the liberal West with authoritarian states. Others took issue with this, distinguishing between high policing and low policing. While low policing addresses common crimes and transgressions, high policing is inherently political, addressing crimes that threaten the stability of the state and the established political order (Brodeur 1983). Tactics for high policing include surveillance, infiltration, manipulation, and deception. These are some of the same tactics that create Grey Zone threats to stability and political order. The Grey Zone is a zone of political and military vulnerability and a field for innovative policing rather than militarization of society.

The crucial change is that modern wars do not begin with the clear starting line theorized in international law. In the wake of 9/11, Western police and intelligence organizations increasingly combined high and low policing tactics, blending surveillance and control of populations in ways that blur any distinction between liberal and authoritarian states (Brodeur 2007:25). This blending has not always been in ways that enhance the essential functions of states or the cohesion of society. Even the most liberal states possess most of the intelligence and policing capacity needed to preserve essential state functions enumerated above. They do not always deploy this capacity in ways that effectively defend those functions without undermining public trust and social cohesion.

Balancing the State, Markets, and Society

Markets and finance play an essential role in the potential destabilization of society (Chapter 5). Unregulated capitalist markets produce economic cycles that benefit some parts of society and are damaging to others. They can destroy livelihoods and generate extreme political responses. Twentieth-century fascism and communism and 21st-century populism are attributed to the failures of market liberalism (Polanyi 1947; Blyth 2002; Frieder 2007; Standing 2016; Crotty 2019). Environmental crises and epidemics can have similarly disruptive effects, setting the conditions for violent protests, extremist movements, and political upheavals. The case for a national security response to these conditions does not rest on preserving control for one political faction over another within the state but on a state's rights to manage these pressures in the public interest without external interference seeking to undermine human and national security.

A protest against vaccine mandates paralyzed Canada's capital in the weeks immediately before Russia invaded Ukraine in February 2022. Canada supported Ukraine with military training and financial assistance over the years, is host to a large Ukrainian Canadian population, and was critical of Russia's seizure of Crimea and occupation of the Donbas region. It might reasonably have been a target of Russian interference. No evidence of Russian involvement was presented to the subsequent public inquiry, and there was adequate hostility to the government to explain the protest (Rouleau 2023). However, there was also clear evidence of foreign financing to sustain and expand the protest. Of $8.4 million US dollars in support, $3.6 million (44 percent) originated in the United States (Thompson, Rocha and Leung 2022). The Canadian government invoked the Emergencies Act (1985). The Public Inquiry into the Public Order Emergency in Canada noted that military support was considered, but appropriate resources—for example, to move large trucks—were not available (Rouleau 2023:176), and using troops against protestors was considered a serious matter with uncertain outcomes (Rouleau 2023:27,212). The Inquiry also noted that the international reaction of allies and trading partners played an important role. The Emergency Measures Regulation banned public assemblies affecting critical infrastructure, impeding movement and trade, and threatening violence (Rouleau 2023:120). The Emergency Economic Measures Order targeted the financial support sustaining the protest and impeding normal government activity in the capital. Bank accounts were frozen, donation conduits were impeded, financial intermediaries were put under surveillance, and financial supporters faced sanctions (Gallant 2023:91).

A similar protest against vaccine mandates obstructed Parliament in New Zealand in February 2022 and was the subject of a publicly funded study of disinformation tactics (Hannah, Hattotuwa and Taylor 2022). It distinguished between misinformation (false information not created with the intent to harm), disinformation (false information created with the intent to harm), and mal-information (true information used with malicious intent). The study found evidence of Russian sources of disinformation and mal-information, linking the platform Telegram to Facebook, but more generally, the information streams were "heavily laden with conspiratorialism, Covid-19 denialism, and other harms, including from QAnon wellsprings in the United States…" (Hannah, Hattotuwa and Taylor 2022:9). As in Canada, identifying external sources did not result in determining a coherent hostile actor. Whether these were attacks or legitimate protests, they impeded normal governance and created vulnerabilities.

Social and economic factors combine. In the Canadian case, Bank executives expressed concern about the impact of the protest on direct foreign investment (Rouleau 2023:120). The protest restricted border crossings, significantly impacting trade and supply chains. The government was concerned about Canadian credibility after formal and informal complaints by foreign governments (Rouleau 2023:107–108). Concern for economic health during social disruption was a factor in the emergency response to a domestic protest, which might have been exacerbated by foreign interference. In turn, some of that foreign influence might have been related to Russia's efforts to support its invasion of Ukraine. There were unintended consequences of the vaccine mandates, which led to the protests resulting in domestic emergency measures: "…the increase in distrust that resulted from the use of mandates in the COVID-19 pandemic may actually lower voluntary vaccine uptake in future pandemics" (Dewar et al. 2024).

A further consequence might be a loss of trust in decisive government action. Although the Public Inquiry into the Public Order Emergency in Canada found the government's action justified, a federal court found that the invocation of the Emergencies Act violated constitutional rights and was unreasonable. A strength of democratic societies is their ability to resolve social conflict through public discourse, including protest and political action. External nudging of existing forces can increase economic costs, exacerbate conflicts, instigate confounding legal action, and erode confidence in governments at critical times for international action.

Finland's response to increasing numbers of migrants after 2015 illustrates two other forms of response to Grey Zone threats. The civil war in Syria and the Da'esh insurgency in Syria and Iraq generated more asylum seekers from Syria and Iraq crossing European borders. According to the European Council, it reached crisis levels in 2015–2016 (Consilium 2024). Migration was combined with Russian disinformation or influence campaigns aimed at Germany, the Czech Republic, and other EU audiences (Jankowicz 2020:Chapter 6). Finland, contemplating NATO membership after Russia's occupation of Crimea in 2014, began to face floods of migrants and asylum seekers bussed to its borders by the Russian authorities. As the number of asylum seekers surged, the percentage of positive decisions declined. Finnish immigration authorities used collective discretion to deny claims under existing guidelines (Vanto et al. 2022).

Finland joined NATO in April 2023. In November 2023, another surge of African migrants arrived from Russia, many under-dressed for winter conditions. Finland's response was to close its border with Russia. The International Organization for Migration urged Finland and Russia to abide by their obligations under international law. The UN High Commission for Refugees condemned the action as "contrary to international law." Finnish Prime Minister Petteri Orpo pointed to a Russian influence operation and claimed the action was a legal and proportionate response (Bryant and O'Carroll 2023). The earlier response—collective discretion by public servants—can be seen as a bottom-up response. The later response—government policy—is top-down. The two cases reflect social and political reactions to a Grey Zone attack on Finnish society.

Legal and Regulatory Responses to Grey Zone Threats

States confronted by external threats like those described above feel that they are victims of aggression. Aggression is not equivalent to an armed attack. Aggression is logically a broader concept that may precede or include an armed attack as one of its components. The challenge for small states and liberal democracies is that an armed attack may be preceded by sufficiently aggressive Grey Zone attacks that the state is weakened or unprepared for the armed attack when it comes. This raises questions about the legislative framework for defense against external interference in the Grey Zone. They are difficult questions because invoking emergencies is a well-established means of increasing executive authority, concentrating power, and eroding democracy. However, international law does not support a broader concept of aggression, and emergency legislation in many countries is consciously framed to separate response to external aggression from domestic conditions. This separation is difficult to sustain in the age of Grey Zone conflict. Indeed, the lines drawn by international law have created the conditions for Grey Zone attacks short of war.

War is a state of exception and justifies extraordinary national efforts. Legal scholars distinguish the use of war as a rhetorical device from its meaning under domestic and international law (Dinstein 2011:5–6). Oppenheim's definition is usually the starting point: "War is a contention between two or more States through their armed forces, for the purpose of overpowering each other and imposing such conditions of peace as the victor pleases" (Dinstein 2011:7 citing Oppenheimer 1952:202). Yet, many states contribute small forces to coalitions for expeditionary wars or peacekeeping without significant impact at home. The forces for these expeditionary operations are typically generated and deployed under the provisions of national defense acts like those listed in Table 15.1, without any anticipation that the homeland will come under threat. These states deploying military forces can still be deemed at war if the coalition seeks to overpower and impose its will on another state. This was the case for America's coalition of the willing in Iraq and for NATO's operations in Afghanistan. Many wars are asymmetrical, with the parties seeking different goals and using different means. War between the regular armed forces of states has been rare in this century. But modern wars may be waged against unwitting victims by non-military means.

Attacks under international law are typically qualified as "armed attacks," which might be made by state or non-state actors (Dinstein 2011:208). With the emergence of new ways to disrupt, damage, and even kill, what constitutes an "armed" attack? The US-Israeli STUXNET attack on Iranian centrifuges in 2005, the repeated Russian cyberattacks on Ukraine's energy grid, and even the incidental damage to the global shipping company Maersk were all conscious attacks (Greenberg 2019). Russia's direction of large numbers of refugees toward Finland and Belorus's similar actions toward Poland are likewise targeted and disruptive attacks. Were they armed attacks? Greenberg can't identify fatalities from cyberattacks. NATO has never invoked Article 5 despite repeated Russian cyberattacks against Estonia. The point about Grey Zone threats is that they are below the threshold of war—attacks but not armed attacks.

If cyberattacks and weaponized refugees are not armed attacks, are they acts of aggression? The crime of aggression under international law has been framed in the context of aggressive war. "An act of aggression—to qualify as a crime subject to the jurisdiction of the International Criminal Court—must, "by its character, gravity and scale," constitute a "manifest violation" of the Charter of the United Nations" (Dinstein 2011:143). Aggressive use of force short of war would fail to qualify as criminal. The US invasion of Iraq (2003), NATO bombing of Libya (2010), and Russia's invasions of Georgia (2008) and Ukraine (2022) have all been described as crimes of aggression, but none has been prosecuted (McDougall 2023). The bar for international law is set high, and practice has raised it. Even the most egregious cases of international aggression and aggressive war have not been prosecuted since Nuremberg.

Civil wars have sometimes been used as an excuse for intervention. The justice of an insurgent cause might be cited as a reason to intervene against the government they are fighting. International law holds this to be inadmissible in all cases. Support for the government, however, is permissible to the extent agreed by the government (Dinstein 2011:126). The peaceful "color revolutions" saw Western pro-democracy non-governmental organizations train and support anti-authoritarian movements in many countries. Western governments supported these efforts. Being non-violent, the support for color revolutions could not be deemed illegal intervention in civil wars, but it was certainly threatening to authoritarian regimes. They developed effective defensive strategies to prevent regime change (Finkel and Brudny 2014). The subsequent development of disinformation, electoral interference, and cyberattacks might even be seen as retaliation for the success of democratic interference. International law did not help the authoritarian states against democratic interference and will not help democratic states against foreign interference short of war or armed attack.

States may feel that they have been victims of aggression and unfair treatment. The high bar set by international law for war and armed attack forces them to find other tools for self-defense in the Grey Zone.

Defending the State

Early in this century, democratic pressures began to threaten authoritarian states. Authoritarian states not only developed defense mechanisms, but they pushed back with threats of their own. Democratic states have been developing new mechanisms to defend the state, often outside the scope of national defense and emergency legislation.

A report by the UK-based Institute for Global Change on the authoritarian threat to liberal democracies sets out six challenges: military threats, political and economic intimidation, influence to reshape the intellectual climate, disruption (by propaganda, cyberattack, and subversion), changing norms, and threats to national cohesion (Joshi 2018). The list applies equally to challenges experienced by religious or authoritarian regimes, which might respond differently. Thinking about both democratic and authoritarian cases helps us envision small states' challenges in the Grey Zone.

By setting a high bar for war, armed attack, and aggression, international law expands the Grey Zone below the threshold of war. Grey Zone democratization offensives and color revolutions removed authoritarian regimes in Serbia (2000), Georgia (2003), Ukraine (2004), and Kyrgyzstan (2005). But after that, removing authoritarian regimes got harder. Belarus (2006) and Khazakstan (2022) successfully repressed popular uprisings. When the Arab Spring blossomed in 2011, many affected regimes experienced only minor political changes (Carbonnier 2013). Authoritarian regimes learned to manage Grey Zone threats.

Authoritarian rulers mix five strategies to resist democratization. They *isolate* democratic movements and individual activists from support. They *marginalize* individuals and movements from national politics, controlling media. They *distribute* rewards to reinforce loyalty to the regime. They *repress* or punish by withholding benefits such as jobs, housing, and schooling. Finally, they *persuade* citizens that democratic movements are alien and unwanted (Finkel and Brudny 2014:5). These strategies are achievable when regimes have strong parties, revolutionary ideologies or theologies, are well-funded, have cohesive coercive apparatus, and have state control over the economy (Way 2008). Authoritarian states have become more resistant to popular unrest; they have also found new ways to threaten democratic neighbors.

Democratic regimes have tools to manage authoritarian Grey Zone threats like intimidation, influence, and disruption. Five strategies for defending democratic states can be proposed based on an understanding of these threats. Russian intervention in Ukraine from 2014 to the 2019 election stands as an example of the combination of threats and potential responses, as reported by the Atlantic Council's Ukrainian Election Task Force Report (2019).

The mechanism for democratic succession must be protected. For democracies, this means protecting free and fair electoral competition. Apolitical electoral commissions, transparency, and campaign finance laws are essential foundations. Non-partisan supervision of media to identify foreign influence removes claims from the political realm and improves confidence in electoral results. Canadian legislation takes an "electoral ecosystem" approach. Elections comprise interconnected actors and institutions, including governments, political parties, voters, third parties, online platforms, and election management bodies. Multiple defense strategies address campaign finance, purchase of advertising, disinformation on social media, and strengthening cybersecurity (Dawood 2021). Australia mixes digital and analogue processes to reduce cyber vulnerability but acknowledges that disinformation and cyberattacks permit interference and erosion of social cohesion (Dowling 2022).

Second, *economic openness creates political vulnerabilities, which should be monitored.* Foreign interests can buy media, advertising, and state assets that confer political power (Joshi 2018). National monitoring of foreign ownership, control, and influence requires an understanding of markets, investment, forensic accounting, and money laundering. Political corruption, organized crime, concealed foreign influence, and overt foreign direct investment are all vectors that can erode national security below the threshold of war or domestic emergency. A Rand Europe study for the Netherlands identified seven vectors of economic risk: foreign ownership, government intervention, supplier dependence, natural resource dependence, espionage and access to sensitive information, corruption and fraud, and socioeconomic inequality (Retter et al. 2019:95). Foreign land ownership is a special case for two reasons. It allows foreign actors to control territory within a state, and it can exacerbate nationalist divisions, as it has in Hungary and Poland (Tesser 2004).

Third, *an educated public and critical journalism are vital defenses* against disinformation and manipulation (Joshi 2018). Strengthening national culture and education in civics and critical thinking helps to make populations more resistant to disinformation and manipulation. Almost half of Mongolia's population is nomadic, sparsely distributed across open lands between China and Russia. Reforming national security education began with a survey of perceptions in schools (Ulziihuyag 2021). National education programs help to reduce receptivity to disinformation and manipulation. Support for journalism and information is necessary to improve the supply of reliable and credible information. Blocking access to some channels is more controversial in free societies.

Fourth, *a discerning and independent judiciary enjoying the public's confidence is a safeguard against economic intimidation and lawfare.* Chapter 15 explored the potential to erode the state through legal means and to undermine the rule of law through attacks on the judiciary (see also Appendix).

Finally, *states need friends in the international community*. One prescription to protect electoral mechanisms includes clearly articulated multinational retaliatory mechanisms spelled out in advance (Joshi 2018). However, setting international norms and dominating the discourse of international institutions like the United Nations is no longer the prerogative of liberal democracies alone. Sovereign democracy and illiberal democracy compete with liberal democratic values. New constellations of trade and military alliance are emerging.

We return to legislation for national defense and emergencies. International law enables the Grey Zone by setting a high bar for war, armed attacks, and aggression. National legislation can identify a broader range of aggressive behaviors, including cyberattacks, disinformation, intimidation, proxy attacks through corruption, and organized crime—the full range of Grey Zone threats (see Appendix). Relevant legislation is not limited to national defense and emergency legislation sampled in Table 15.1. Finland's closure of its border with Russia did not rely on emergency legislation but on a 2022 amendment to the Finnish Border Guard Act permitting closures when "necessary to avert a serious threat to public order, national security or public health" (Library of Congress 2023). This is a subject of intense debate in many countries, given commitments to freedom of movement and international commitments to accept asylum seekers (ECRE 2024).

Militarizing the response to Grey Zone threats risks weakening democratic states' capacity to use all available tools to resist intimidation, disruption, and subversion. As a substitute for or precursor to modern war, Grey Zone threats are designed to fall below the threshold where military means would be useful. Even emergency legislation intended to offer a flexible response to any domestic situation requiring extraordinary measures may be inadequate in the face of legal challenges and provisions consciously included to preserve civil liberties.

Conclusion

Defending the state must acknowledge that Grey Zone threats may be a substitute or precursor to modern war. Ukraine was at war for eight years before Russian tanks crossed its borders. Other states are experiencing Grey Zone conflicts now. The experience is not limited to liberal democracies. Illiberal democracies and authoritarian states can similarly experience Grey Zone threats. While international law sets a high threshold for war, armed attacks, and aggression, different regimes have different mechanisms for preserving their preferred order and different comparative advantages when projecting Grey Zone threats. The essential functions described by the ISE indicate the range of domestic legislation that can be brought to bear to support the state against Grey Zone threats. Declaring war and putting the military in charge is unlikely to be an effective response.

References

Blyth, M. (2002), *Great Transformations: Economic Ideas and Institutional Change in the Twentieth Century*, Cambridge University Press, Cambridge.

Brodeur, J. P. (1983), "High Policing and Low Policing: Remarks About the Policing of Political Activities", *Social Problems*, vol. 30, no. 5, pp. 507–520.

Brodeur, J. P. (2007), "High and Low Policing in Post-9/11 Times", *Policing: A Journal of Policy and Practice*, vol. 1, no. 1, pp. 25–37.

Bryant, M. and O'Carroll, L. (2023), "Finland Close Entire Border with Russia after Tensions over Asylum Seekers", *The Guardian*, 28 November. https://www.theguardian.com/world/2023/nov/28/finland-closes-entire-border-with-russia-after-tensions-over-asylum-seekers

Bull, H. (2012), *The Anarchical Society: A Study of Order in World Politics*, Columbia University Press, New York.

Canada (1985), Canada's Emergencies Act. R.S.C., 1985, c. 22 (4th Supp.), Government of Canada. https://laws-lois.justice.gc.ca/eng/acts/e-4.5/

Carbonnier, J. (2013), "The Arab Spring and Its Different Outcomes: Explaining the Variation in the State of Democratization" (Master's Thesis), Lund University, Department of Political Science.

Consilium (2024), *Response to the migration crisis and force majeure situations*. https://www.consilium.europa.eu/en/policies/migration-crisis/.

Crotty, J. (2019), *Keynes Against Capitalism: His Economic Case for Liberal Socialism*, Routledge, London and New York.

Dawood, Y. (2021), "Combatting Foreign Election Interference: Canada's Electoral Ecosystem Approach to Disinformation and Cyber Threats", *Election Law Journal: Rules, Politics, and Policy*, vol. 20, no. 1, pp. 10–31.

Dewar, J., Wilson, D., Pacheco, G. and Meehan, L. (2024), "Unintended Consequences of NZ's COVID Vaccine Mandates Must Inform Future Pandemic Policy – New Research", *The Conversation*, 27 February. https://theconversation.com/unintended-consequences-of-nzs-covid-vaccine-mandates-must-inform-future-pandemic-policy-new-research-222989

Dinstein, Y. (2011), *War, Aggression and Self-Defence*, Cambridge University Press, Cambridge.

Dowling, M. E. (2022), "Foreign Interference and Australian Electoral Security in the Digital Era", *Australian Journal of International Affairs*, vol. 76, no. 1, pp. 40–56.

ECRE (2024), "European Council on Refugees and Exiles: EU Eastern Borders", *ECRE Weekly*, 19 April 2024.

European Council (2024), Timeline - EU Migration and Asylum Policy. https://www.consilium.europa.eu/en/policies/eu-migration-policy/migration-timeline/

Finkel, E. and Brudny, Y. M. (2014), "No More Colour! Authoritarian Regimes and Colour Revolutions in Eurasia", in E. Finkel and Y. M. Brudny (eds.), *Coloured Revolutions and Authoritarian Reactions*, Routledge, London, pp. 1–16.

Frieder, J. (2007), *Global Capitalism: Its Rise and Fall in the Twentieth Century*, Norton and Company, New York.

Gallant, M. (2023), "Modern Finance-Centric Governance: The 2022 Emergency Measures, Property and Financial Powers", *Manitoba Law Journal*, vol. 46, no. 1, pp. 91–100.

Greenberg, A. (2019), *Sandworm: A New Era of Cyberwar and the Hunt for the Kremlin's Most Dangerous Hackers*, Anchor Books, New York.

Hannah, K., Hattotuwa, S. and Taylor, K. (2022), "The Murmuration of Information Disorders: Aotearoa New Zealand's Mis-and disinformation Ecologies and the Parliament Protest", *The Disinformation Project. Pacific Journalism Review*, vol. 28, no. 1&2, pp. 138–161.

ICRC (International Committee of the Red Cross) (n.d.), State Defence Act 1991.

Jankowicz, N. (2020), *How to Lose the Information War: Russia, Fake News, and the Future of Conflict*, I.B. Tauris, London.

Joshi, S. (2018), *Authoritarian Challenges to the Liberal Order*, Tony Blair Institute for Global Change. 21 June. https://institute.global/insights/geopolitics-and-security/authoritarian-challenges-liberal-order

Kowalski, M. (2015), Ius ad bellum a Systemowy Charakter Prawa Międzynarodowego, in R. Kwiecień (ed.), *Państwo a Prawo Międzynarodowe jako System Prawa*, Wydawnictwo UMCS, Lublin, pp. 167–202.

Law, J. (2022), *A Dictionary of Law (Oxford Quick Reference)*, Oxford University Press, Oxford.

Library of Congress (2023), "Finland: Government Closes Most Border Crossings to Russia", *Law Library of Congress, Global Legal Monitor*, 28 November. https://www.loc.gov/item/global-legal-monitor/2023-11-28/finland-government-closes-entire-land-border-with-russia/

McDougall, C. (2023), "The Imperative of Prosecuting Crimes of Aggression Committed against Ukraine", *Journal of Conflict and Security Law*, vol. 28, no. 2, pp. 203–230.

Polanyi, K. (1947), *The Great Transformation: The Political and Economics Origins of Our Times*. Beacon Press, Boston.

Retter, L., Frinking, E., Hoorens, S., Lynch, A., Nederveen, F., and Phillips, W. (2019), *Relationships between the Economy and National Security*, RAND Europe. RAND Corporation, Santa Monica, CA and Cambridge, UK.

Rouleau, P. S. (2023), *Report of the Public Inquiry into the 2022 Public Order Emergency*, Public Order Emergency Commission. https://publicorderemergencycommission.ca/final-report/

Standing, G. (2016), *The Corruption of Capitalism: Why Rentiers Thrive and Work Does Not Pay*, Biteback Publishing, London.

Suganami, H., Carr, M. and Humphries, A. (eds.) (2017), *The Anarchical Society at Forty: Contemporary Challenges and Prospects*, Oxford University Press, Oxford.

Tesser, L. M. (2004), "East-Central Europe's New Security Concern: Foreign Land Ownership", *Communist and Post-Communist Studies*, vol. 37, no. 2, pp. 213–239.

Thompson, E., Rocha, R. and Leung, A. (2022), "Hacked Convoy Data Shows More Than Half of Donations Came from U.S.", *CBC News*, 14 February. https://www.cbc.ca/news/politics/convoy-protest-donations-data-1.6351292

Ukrainian Election Task Force Report (2019), *Foreign Interference in Ukraine's Democracy*. Atlantic Council, May. https://www.atlanticcouncil.org/in-depth-research-reports/report/foreign-interference-in-ukraine-s-election/

Ulziihuyag, B., Soronzonbold, T., Mendee, J. and Orkhonchimeg, K. (2021), "The Issue Development of a New Defense Education System in Mongolia" *Presented at the 13th Annual Conference of the International Society of Military Sciences (ISMS), Progress in Military Sciences for Resilience and Cohesion in the Face of New Forms of Disruption*, 11–14 October 2021, Royal Military College of Canada, Kingston, Ontario.

UN General Assembly (1974), *Resolution 3314 (XXIX) of 14 December 1974.*

Vanto, J., Saarikkomäki, E., Alvesalo-Kuusi, A., Lepinkäinen, N., Pirjatanniemi, E. and Lavapuro, J. (2022), "Collectivized Discretion: Seeking Explanations for Decreased Asylum Recognition Rates in Finland after Europe's 2015 'Refugee Crisis", *International Migration Review*, vol. 56, no. 3, pp. 754–779.

Way, L. (2008), "The Real Causes of the Color Revolutions", *Journal of Democracy*, vol. 19, no. 3, pp. 55–69.

16 Social Cohesion and Grey Zone Conflict

Danuta Kaźmierczak, Mirosław Laskowski, David Last, and Marzena Żakowska

Introduction

Polarization during the Cold War created clear divisions and lines of support and allegiance. Post-Cold War globalization, especially digital communications, softened those boundaries. The return to polarization with the rise of authoritarian nationalism and a backlash against globalization, even in liberal democracies, has changed the security environment for most small states. This chapter addresses how Grey Zone threats can be mobilized to undermine social cohesion in small states, rendering them less able to defend their citizens' well-being and the institutions that enable their autonomy.

Social cohesion is the solidarity or connectedness among groups in society. It is a necessary foundation for collective well-being and national security efforts directed by the government. Having explored these connections, this chapter considers political, economic, social, technological, legal, and environmental forms of interference that can erode social cohesion below the threshold of war. The erosion of well-being and weakening of the state can be an objective of modern war that is more important than the defeat of the state. It is the essence of the war nexus described by Simonyi (Chapter 1). This chapter concludes by discussing factors increasing and mitigating vulnerability to externally induced fragmentation.

Social Cohesion

The war nexus encompasses all domains of society. For states pursuing Gray Zone conflicts, the objective may be destabilization of a target rather than victory over it. Preserving social cohesion, then, is the essence of defense in the war nexus. Social cohesion is a concept from sociology that has often intruded into national politics. It includes two main dimensions: an individual sense of belonging or common identity and relationships among members and groups in society. Social cohesion is a social process that brings a plurality of citizens together by reducing inequality, socioeconomic disparities, and fractures. It reflects the need for personal development and a sense of belonging, linking individual freedom with social justice, economic efficiency, fair sharing of resources, pluralism, and common rules for resolving conflicts (Manca 2014). Social cohesion in liberal democracies must reconcile market forces, freedom of opportunity, and enterprise with a commitment to solidarity and mutual support. Without open access to benefits and protection for all members of society, dissatisfaction risks eroding cohesion. This liberal-democratic definition of social cohesion within states is mirrored in the objectives of the European Union. Article 2 of the Treaty of the European Union states that the tasks of the Union include: "Maintaining economic and social cohesion and solidarity between all Member States of the Union."

DOI: 10.4324/9781003428701-25

Cohesion can be built on other foundations. Orwell's dystopian novel *Nineteen Eighty-Four* portrays the cohesion generated by the hunt for internal enemies, the resistance to external enemies, and the euphoria of nationalist victory. Orwell's novel was modeled on the show trials and propaganda of Stalin's Soviet Union. Community and belonging can also evolve in hierarchical and inegalitarian societies, where people "know their place" and belong to coherent groups with a common ideology or narrative that justifies their position. The Indian caste system, for example, creates strong communities and social bonds that resist efforts to create more egalitarian societies (Patnaik 2017). The experience of violent disruption of traditional orders is a motivation to find more durable forms of social cohesion. New authoritarianism and "sovereign democracy" are spurred by the liberal disruption of revolutionary regimes that failed to deliver solidarity and equality. Sovereign democracy in Russia is a response to the threat of democratization. This ideology legitimizes Vladimir Putin's personal rule, articulated by Vladislav Surkov, Putin's deputy Chief of Staff, in 2006. It includes sovereignty as resistance to Western interference, collectivist democracy, and a social contract of prosperity for those supporting the regime (Kortukov 2020).

Social cohesion can also be built on religious foundations.

> The power of social cohesion depends on the extent [to which] members of the society realize the principles and values associated with the public interest of a society… cohesion guarantees individuals' well-being through equal opportunities and freedom from all forms of marginalization.
>
> (Ibrahim 2021)

Islam is the religion of submission. Religion plays a powerful role in political socialization and serves the same function as ideologies in justifying authority. Submitting to religiously or ideologically based authority is no less rational than submitting to the authority of market forces.

Several decades after the concept entered national policy documents, systematic reviews of social cohesion suggest ways of developing it (Fonseca, Lukosch and Brazier 2019). If there is a recipe for social cohesion in small states, it probably includes at least seven ingredients cooked over generations. The first is socialization and education about national symbols, shared cultural practices, public ceremonies, and common identity. These cannot be grown quickly but emerge over generations, reinforced at successive stages in individual and family lives. Good governance, strong institutions, transparency, and accountability form a second ingredient that makes a regime resilient to disruption. The third ingredient is economic stability, sound policies, and reduced (or justified) inequality. Fourth, small states need peer groups. Diplomacy and alliances can reinforce norms and standards, validate national leaders, and ground national society in a larger community. A fifth ingredient is social inclusion and equal rights. Scapegoating and the politics of exclusion can also build cohesion of an in-group in opposition to an out-group. However, that can be a dangerous shortcut that leaves enemies within and a weaker society. A sixth ingredient is a civil society—the sum of voices not controlled by the government. The cacophony of these voices is the price of mediating interests for the collective good. However, suppressing social conflicts may also be part of a strategy for cohesion—not too much or too little public discord. A final essential ingredient is public communication about the present. If socialization and education draw heavily on an imagined past and a hoped-for future, public communication is in the present. It reveals threats, manages anxiety, and reassures the public, preserving confidence in government and institutions.

Political Threats

Tactics for a political assault on social cohesion include elite penetration, family strategies, insider threats, corruption, funding political parties and activists, hostage-taking, intimidation, targeting leaders through their constituents, *kompromat*, fellow travelers, astroturfing, targeted assassination, electoral manipulation, and support for extreme political factions. These are summarized in Appendix. The tactics can be combined in different ways depending on the strategic objectives. Is the aim to weaken a target state before an attack, to cause a reaction in other states, to produce opportunities for the originating state, or to produce a response in other states through indirect effects? Each tactic carries risks for the originator, but many are minimal if a state can distance itself from attribution. Between deniability and acceptance of rogue status in some circles, states have wide latitude in using these tactics against neighbors.

The state's sovereignty is conditioned by internal stability and its external environment. The distinction between allies and enemies is often not clear. Individual citizens and groups within states have ties that cross boundaries. They do not always identify strongly with the state. Events like elections and referendums can exacerbate competing identities. The state may be unable to prevent external interference, disinformation, or active measures. Interference may take many forms, often difficult to attribute to specific actors, but with the general effect of disrupting processes and eroding trust in the state (Chapters 14 and 15). Some political actors may benefit from foreign interference, although the proof of benefit does not prove either complicity or hostile intent; deniability is one of the principles in organizing active measures and disinformation.

The political goal of disinformation is "to exacerbate existing tensions and contradictions within the adversary's body politic by leveraging facts, fakes, and ideally a disorienting mix of both" (Rid 2020:7). Routine untruths uttered by politicians are not disinformation campaigns; disinformation campaigns are the systematic output of bureaucracies assigned to the task. They are professionally organized and managed, continually adjusted to achieve objectives, and targeted against specific adversaries who may be domestic or foreign. The content of disinformation may be fabricated or forged. The sources may be misrepresented. Accurate information may be acquired covertly. Influence agents may misrepresent themselves. Online accounts may be inauthentic or artificially amplified by bots—programs that mimic live users. Disinformation does not have to be coherent or consistent; it is often enough to throw a mix of confusing and discrediting information into the mix. The Russian response to the downing of a Malaysian Airways flight over Donbas by Moscow-aligned militias was a barrage of bizarre alternative explanations (Pomerantsev 2019:90). This campaign could be seen as defensive, deflecting blame from Russia, but it also provided fodder for conspiracy theorists in Ukraine.

Disinformation is the main tool for election disruption. Cyberattacks can support it by disrupting actors and agencies involved in elections. Election mechanisms are critical infrastructure for democracies. This includes the information available to voters and the public discourse surrounding national decisions. A 2018 report describes five cases of cyber and information attacks affecting elections in the Netherlands, France, the United Kingdom, Germany, and Sweden in 2017 and 2018. The impact was minimized by vulnerability tests and contingency plans, inclusive legal measures, public statements to deter threats, training, education for political actors, public information to warn voters, government dialogue with media, and international sharing of best practices (Brattberg and Maurer 2018).

Disinformation campaigns can be integrated with political processes. Political clients engage chief architects of the campaign, who hire anonymous digital influencers and some visible key opinion leaders, whose views are then amplified by community-level fake account operators whose posts create an illusion of engagement, with some grassroots intermediaries to impact the public (Ong and Cabañes 2018:8). Entrepreneurs and contractors can replicate this structure in a market society like the Philippines, or state bureaucracies do so under autocratic governments. The more autocratic, the more careful an authority is to deny these tools to the opposition. Unsurprisingly, the techniques have been pioneered in rough-and-tumble democracies with wealthy and powerful parties like the Philippines, Mexico, and the US.

Political unity in a state can be enhanced by making an enemy of its neighbor and mobilizing resources to defeat it (Morris 2014; Turchin 2016). Indeed, that may have been the original function of the state—making war to extract resources (Tilly 1985). Eroding the cohesion and political capacity of a neighbor can have similar benefits. Domestic chaos makes other states less willing and able to ally with a neighbor and weakens it for an eventual attack. Putin's attack on Ukraine has created "a suppurating sore that will prevent Ukraine from ever joining NATO or any other alliance" (Zygar 2023:308).

International organizations are an important source of support for small states facing political destabilization, but the organization's rules bear careful examination (Chapter 11). Consensus rules can paralyze decision-making if one member can be persuaded to defect. Cosmopolitan political mobilization, drawing on international organizations and like-minded states, can help small states resist criminality or the pressure of large neighbors (Kaldor 2012:183–191). A cosmopolitan military—more broadly, cosmopolitan security—provides protections for all, not just insiders of a particular society. We can envision global community policing but acknowledge its implausibility in a world of sovereign states (Gilmore 2015:4–5). Cosmopolitan ideals have faced a backlash that political entrepreneurs can mobilize to reduce the capacity of international organizations and the cohesion of groups within a society. As market liberalism came to be associated with globalization, which left many behind, illiberal nationalism began to reject international organizations (Kubicek 2022). This created an opening for rival states to attack the liberal international order directly through rule changes and indirectly through illiberal nationalist movements (Ikenberry 2018).

Economic Threats to Social Cohesion

Tactics for an economic assault on social cohesion benefit from the preconditions of income inequality, unemployment, recession, globalization, housing shortages, and personal debt crises. These existing conditions create tensions and contradictions that can be leveraged not only by disinformation and active measures but also by direct economic action. Ideologies of personal responsibility and market fundamentalism create fertile ground for manipulation and erosion of trust in government. Small countries with a narrow economic base may be vulnerable to erosion of foreign reserves, domination or disruption of supply chains or supplies of key commodities, and dominance of an economic sector. Manipulation of the energy sector has been a Russian tactic. Siphoning government excise revenues from alcohol, tobacco, and gasoline to support anti-government forces or organized crime was endemic in the Western Balkans in the 1990s.

Mergers and acquisitions of corporations may remove the levers of national economic control from important industries, with consequences for employment and self-sufficiency. This can be aided by intimidation of executives and shareholder manipulation. Some of these techniques have been articulated as the technology of the corporate raid (Nemno 2015:331–335). The use of legal screens and shell companies makes attribution difficult. When companies have been taken over, asset stripping can reduce their value or utility as national economic champions, with consequences for employment and national pride.

Money laundering and the introduction of counterfeit currency can distort local markets. Credit card fraud and economic scams targeting the vulnerable not only help finance other aspects of organized crime but also generate a sense of persecution, injustice, and government impotence to protect the vulnerable.

Organized crime is best understood as commerce by illegal means (Chapter 6). It typically deals with four categories of commodities and services: prohibited goods such as drugs, guns, and commercial sex; regulated commodities such as antiquities, fauna, or flora; differentially taxed commodities such as alcohol and cigarettes; and stolen goods such as cars (Williams 2001:107). A fifth category is illegal financial services, including instructions on methods to conceal funds, move money internationally, acquire fraudulent loans, and conduct real estate transactions for money laundering purposes. The Vancouver model involved international financial transfers and laundering drug money through casinos and real estate (Schneider 2020:118). Two years after the Vancouver model had been exposed, little progress had been made toward controlling these illegal economies and others that erode national security and the trust of allies (Chrustie and Luna 2023). Small states hosting organized crime can be labeled targets for action by larger neighbors, and they can perceive themselves as victims of collusion by other states, effectively caught between other actors unless they can contribute to the effective policing of transnational economic activity.

Cybercrime, state intelligence, and economic warfare converge. By 2010, Russia accounted for about a third of cybercrime revenue—between 2.5 and 3.7 billion US dollars. Russian law enforcement tolerated known Russian hackers if their targets were outside Russia. Sometimes, hackers have permission and protection from law enforcement. The Russian attack on NASDAQ servers in 2010 appeared to be an effort to steal the software to set up a Russian exchange—an intellectual property theft. Other Russian-based hacks in 2012 targeted banks, draining accounts and disrupting online operations with denial-of-service attacks. Ransomware was another income stream. American criminal investigators found that the infected computers used to spread banking and ransomware vulnerabilities were also routinely searching for emails of intelligence and security officials in target countries, including Turkey and Georgia. This was a combined criminal and espionage operation (Carlin and Graff 2018:280–296).

Direct Erosion of Social Cohesion

The social effects of the economic attacks described above are indirect. They include economic costs, rising real estate prices, unemployment, frustration, and distrust of government. Many of these effects can occur without any hostile foreign action. Social cohesion can also be attacked directly. Addictions, disinformation, and secessionism are three tools that can be deployed to erode social cohesion directly.

Addictions and substance abuse are personal responses to dislocation; fueling them can be a profitable and effective strategy. The first and second Opium Wars provide a historical example. In the economic order of the 19th century, Britain needed to balance its trade deficit with China and hit upon opium exports. There followed a rising crisis of government ineffectiveness and addiction among Chinese elites (Hanes and Sanello 2002:75–166). The narrative of the "century of humiliation" instigated by rapacious Western capitalism is not universally accepted. China's opium problem preceded Western traders, and social unrest was already widespread. There is a parallel to today's opioid epidemic in the United States. There are rising social costs, and China provides a significant proportion of the synthetic opioid fentanyl (Realuyo 2019:136). Is the damage done to American society by drugs self-inflicted or part of a strategic attack? The evidence leaves room for doubt. Trafficking in fentanyl is a global criminal enterprise with China as its main source. The Chinese government offers little cooperation to prevent it, but the evidence that it actively supports criminal enterprises is weak (Greenwood and Fashola 2021). Manufacturing and smuggling illicit of drugs within China are vigorously prosecuted (Zhao

2022). Abroad, the synthetic opioid crisis is a non-attributable threat below the threshold of war; the illicit drug trade is a non-governmental organized crime phenomenon, prosecuted at home but tolerated—and perhaps enabled—when it affects other states. In this, it is like cybercrime.

Like cybercrime and narcotics, disinformation can be profitable as well as dangerous. Algorithms that drive views and advertising models that reward viewer clicks combine to spread outrageous stories far and wide while rewarding those who concoct and host them (Braun and Eklund 2019). This may be as dangerous as the targeted political disinformation described above because it prepares the ground for further attacks.

Defending against disinformation that erodes cohesion and accentuates polarization requires a legal-technical fix for the platforms that enable it and a social-psychological understanding of how and why disinformation works. Jacques Elul draws on an understanding of propaganda used in World War II and the emerging Cold War:

> Loneliness inside the crowd is perhaps the most terrible ordeal of modern man… Propaganda is the true remedy for loneliness. It also corresponds to deep and constant needs… the need to believe and obey, create and hear fables, and communicate in the language of myths.
>
> (Ellul 1965:148)

Loneliness makes people vulnerable to propaganda and disinformation. The more people live in a society they cannot control, the more they need propaganda, or conspiracy and disinformation, to give them a sense of agency (Pomerantsev 2024:11–12).

Social cohesion is ultimately about willingness to live together. The ultimate threat to national social cohesion is organized secession. Many states today have boundaries that encompass alternative ethnic identities. Wherever growing states or receding empires have left mixed-language groups, there are potential secessionist movements—French and English in Canada and Cameroon, and Russian-speaking enclaves in every successor state of the Soviet Union. Identities that precede the modern state create another secessionist potential: Basques and Catalans in Spain, nationalist movements in Belgium, South Tyrol, Italy, Corsica, and France (Müller-Plotnikow 2017). A map of these movements illustrates their potential to disrupt the European project—a tempting target for hostile manipulation.

Transnational ideological movements have similar disruptive capacities. The leftist movements of the 1980s—Red Brigades and the Baader-Meinhof gang—appealed to a wide group of sympathizers. Iran's axis of resistance creates a regional network of influence. The Jihadist movement supported by the United States in Afghanistan to resist Soviet control became a global movement after the war. These movements carried out acts of violence intended to provoke a reaction. Reactions by the government and society can alienate some groups, create tensions within others, or generate avoidable costs. In democracies, good information and clear public debate can help to maintain cohesion and preserve focus on the public interest, human security, and national security. It is important to provide decision-makers and the public with an accurate understanding of events and motivations. Media outlets and scholars who contribute to war hysteria and fail to explore causal factors accurately can leave the state vulnerable to dangerous overreactions (Kellner 2002).

Technological Vulnerabilities

Technology creates social vulnerabilities through hardware, software, and consequent social interactions. Societies increasingly depend on digital devices for communication, information storage, and management of physical assets like public utilities—water, electricity, and transport.

Governments, businesses, and housing use digital security cameras and access controls. This hardware is vulnerable to technological interference. Fifth-generation (5G) networks allow more data to be transmitted and received faster by devices. Artificial intelligence (AI) allows for the sorting and interpreting of data more effectively than humanly possible. New software creates social unease among some, and emerging vulnerabilities are continually explored and exploited. The third level is the social interaction that accompanies engagement with technology. Already almost two decades old, social media are now combining with these hardware and software vulnerabilities. Social media like Facebook, Twitter (X), Telegram, and WeChat are characterized by two-way communication; users can transmit and receive information within a social network. The membership of that network and the messages within it are controlled by the users and the algorithms provided by the platform. They are vulnerable to manipulation by both.

Many countries with inadequate infrastructure have leapfrogged traditional telephone networks and moved fast toward 5G, aided by permissive regulatory environments (Ahmad et al. 2024). Because 5G networks are defined by software, their vendors have persistent access to the networks for updates and patches (Bogdanova 2022). Both the United States and China seek to expand markets and influence through technology. China's Digital Silk Road has technology agreements with 16 countries, including six NATO members (Triolo et al. 2020:3). China has also used trade agreements and legal strategies to pursue more favorable access to control 5G networks (Bogdanova 2022). Before crossing the threshold of war, comprehensive access to national information infrastructure can enhance cyber disruption, as Russia did in its attacks on Ukraine after thorough reconnaissance (Willett 2023:11). Hardware vulnerability merges with software risks and social vulnerabilities, affecting societies facing hostile manipulation.

Technological threats evolve with the interaction of these three levels of vulnerability. "Deep fakes" use AI to deceive audiences about the authenticity of audio-visual materials. Access to personalized big data can identify vulnerabilities and nudge groups and individuals toward beliefs and actions. The anonymous Dark Web can spread information and manage illicit networks; it is inherently transnational and difficult to police without new tools (Chertoff and Jardine 2021). Dependence on satellites and undersea cables for normal communications creates new vulnerabilities that cannot be contained within state boundaries (Bratton 2016).

Information technology that enabled the Arab Spring against autocratic governments in 2011 has been adapted for more centrally directed operations in the cognitive domain. The cognitive domain has become "the main battlefield for other countries conducting ideological penetration and is an important domain for both sides in a war to fight for or destroy troop morale and cohesion, as well as forming or deconstructing operational capabilities" (Beauchamp-Mustafaga 2023:33). The tools are being developed to manipulate emotions through direct access to the population (Wrzosek 2018). Algorithms help spread disinformation, encourage defections, and corrupt communications. Blurring the lines between truth and falsehood creates alternative realities that undermine the foundations of a cohesive society and erode trust in its institutions. How people search the internet and interpret results uncritically and emotionally is a vulnerability for open societies (Pomerantsev and Weiss 2014).

Images are powerful tools to manipulate emotions. Websites and web searches are a significant source of popular knowledge about events (Schwalbe 2006). Manipulating them with AI and targeting them precisely with advertising techniques and social media can generate groundswells of emotional response. The emotional mechanism of violence can start with bitterness and loss of self-respect arising when people feel inferior or dominated by another group. Narratives of historical hostility, perfidy, and violence between groups can be mobilized to generate inter-group distrust (Vitaljić 2013:20). Organized campaigns armed with technology for the mass production of images and text intended to deceive and manipulate can augment diplomatic, economic, and military subversion (Bradshaw and Howard 2017; Harold et al. 2021). Such campaigns can even attempt to reshape collective memory. Communication technologies can help reconstruct collective memories by changing the meaning of physical and intellectual symbols surrounding people.

The importance of symbols is not lost on those in power, and legal systems may be mobilized to protect flags, monuments, institutions, and other symbols of national identity (Tóth 2022; Chegeni et al. 2020). But legal tools can also be used to undermine national cohesion.

Legal Tools to Undermine Social Cohesion

A state's legal system includes its laws, lawyers, judges, enforcement mechanisms, and the rules and norms that make it function. An independent legal system not only helps to reconcile conflicts between citizens, market forces, and governments but also provides opportunities to pursue malign interests within the state. Lawfare is a general label for exploiting legal systems to achieve hostile objectives. It is easy to adapt to the adversarial legal process of an open society. Confidentiality, privacy rights, and lawyer-client privilege impede detection and defense against legal subversion. Chapter 14 addresses the intersection of lawfare and information war. Here, the focus is on using legal instruments to undermine social cohesion.

Open societies typically have within them tensions and conflicts of value and opinion that events can exacerbate. Companies seek commercial advantage; news stories lead to protests; diasporas mobilize to support homelands in conflict; political entrepreneurs mobilize conflicts over values and perceptions of reality. The law is a mechanism for resolving these tensions but can be manipulated by divisive or vexatious court cases. State and non-state actors can use a law to achieve results. International agreements on human rights, civil rights, or Indigenous rights can be manipulated or subverted (Zanin, Martins and Valim 2022:16). The US Legislation makes the Foreign Intelligence Surveillance Act and the Foreign Corrupt Practices Act complementary tools for legal attacks on, or manipulation of, foreign enterprises. China's defensive strategy involved a law to prohibit cooperation or disclosure of information by Chinese citizens and companies with foreign authorities (Zanin, Martins and Valim 2022:22). In the pursuit of commercial advantage, manipulation of jurisdictions, claims of intellectual property theft, and mobilizing public opinion can have a disruptive impact on economy and society.

The mass protests of the Arab Spring were occasions for the state's use of legal tactics to blunt the political impact of the protests. Preventive incarceration removed key leaders, extracted confessions, and delegitimized enemies by falsely incriminating them. Minor offenses were subject to more serious charges. People on the margins of the movement were subject to "carrot and stick" tactics: cooperate with authorities or face serious charges. Lawyers working for the opposition faced arbitrary state limitations, obstruction, and intimidation. Legal action silenced or intimidated those opposing the regimes (Zanin, Martins and Valim 2022). All these tactics undermine the rule of law but preserve authority and the *status quo*. Criminal investigations, lawsuits, the release of evidence or confidential information, and accompanying information campaigns can also generate public disillusionment that can undermine sitting governments. Public pressure following messy court cases involving corruption and mismanagement was involved in the fall of Dilma Rousseff's government in Brazil (2016), Park Geun-Hye's presidency in South Korea (2017), and Mariano Rajoy's government in Spain (2018).

Diaspora populations may be well-connected within host societies. Second-generation immigrants will be fluent in several languages and often highly educated. Sympathetic non-governmental organizations may assist them. Even when they lose in court, they may win allies or fellow travelers amongst the public or commercial enterprises, or they may drive away the allies of their enemies. Supporters in the Palestinian diaspora, for example, have waged three different types of legal warfare against Israel (Kittrie 2016). The first is in direct support of Palestinian statehood. The second is through non-governmental allies pursuing lawsuits. The third is a supporting role in what might be labeled strategies of atrocity. Hamas's terror attacks take tactical advantage of civilian populations, which are often harmed by Israeli responses. The orchestrated outrage about the violation of laws of war effectively leverages international support in response to terror attacks, which themselves are illegal.

Environmental Impacts on Social Cohesion

The world is getting hotter, and large parts are becoming less habitable. Climate change is increasingly linked to conflict and displacement (UNHCR 2024:4). In some countries, data show that violent crimes, including homicides, rise with the temperature (Xu et al. 2020). Conflict multipliers include migration, urban crowding, and the unequal impact of environmental shocks. Scapegoating, blaming, and anti-immigrant sentiments are common responses to the increased pressure that environmental stress puts on society.

Factors Increasing Vulnerability

For a socially progressive audience, it is convenient to label poverty, ignorance, and inequality as factors increasing vulnerability to social disruption. It is not always so. Poor and uneducated populations can be stable and cohesive under the influence of narratives purveyed by elite groups. Religion, nationalism, and ideology can provide powerfully stabilizing narratives, but conscious information operations may undermine those narratives. Social media and translation software make more of the world's linguistic communities accessible. Evolving culture, state over-reach, and state under-performance are factors increasing the vulnerability of social cohesion.

The late Roman Empire was marked by distracting the Roman mob with bread and circuses. Some see echoes in today's capitalism, where market failures create opportunities for authoritarian alternatives. Effective urban governance helps prevent the social exclusion that leads to resentment and violence (Andersen 2019). Rather than relying on national councils reflecting a single religion, municipal consultative bodies with various religious representations help to define acceptable and unacceptable public religious expressions (Martínez-Ariño 2019). However, even well-designed institutions will not integrate everyone. Violence and disorder will occur at the margins of democracies, and political entrepreneurs will be ready to mobilize this violence. Democracies degenerated through the disempowerment of groups from political and economic inequality, the failure to include groups in political and economic life, and hyper-partisanship at the expense of the public interest (Taylor 2022).

States can overreact to decaying social cohesion by cracking down on dissent, eliminating rivals to governing regimes, and driving opposition forces underground. Enough repression will buy stability in the short term. Underreactions can fail to identify hostile interference intended to weaken a state's cohesion.

Conclusion

Societies contain within them the potential for conflict and fragmentation. This is especially true of modern liberal democratic market states. Leaders should not assume that all threats to social cohesion originate with nefarious foreign actors, although it may be convenient to label them as such. The social cohesion that enables a collective response in the public interest can be eroded through political interference, economic attacks, technology and infrastructure, legal maneuvers, and environmental stresses.

References

Ahmad, I. A. I., Dawodu, S. O., Osasona, F., Akagha, O. V., Anyanwu, A. C. and Onwusinkwue, S. (2024), "5G Deployment Strategies: Challenges and Opportunities: A Comparative Review for Africa and the USA", *World Journal of Advanced Research and Reviews,* vol. 24, no. 1, pp. 2428–2439.

Andersen, H. (2019), *Governing European Cities: Social Fragmentation, Social Exclusion and Urban*, Routledge, London and New York.

Beauchamp-Mustafaga, N. (2023), *Chinese Next-Generation Psychological Warfare: The Military Applications of Emerging Technologies and Implications for the United States*, RAND Corporation, Santa Monica, CA.

Bogdanova, I. (2022), *Politicization of the 5G: Divergent Policy Responses and Huawei's Litigation Strategy*, Policy Brief, 4 January, Word Trade Institute. https://www.wti.org/research/publications/1385/policy-brief-politicization-of-the-5g-divergent-policy-responses-and-huaweis-litigation-strategy/

Bradshaw, S. and Howard, P. N. (2017), *Troops, Trolls and Troublemakers: A Global Inventory of Organized Social Media Manipulation*, Working Paper No. 2017.12. Computational Propaganda Research Project, Oxford Internet Institute, pp. 1–37.

Brattberg, E. and Maurer, T. (2018), *Russian Election Interference: Europe Counter to Fake News and Cyber Attacks*, 23 May. Carnegie Endowment for International Peace, Washington, DC. https://carnegieendowment.org/research/2018/05/russian-election-interference-europes-counter-to-fake-news-and-cyber-attacks?lang=en

Bratton, B. H. (2016), *The Stack: On Software and Sovereignty*, MIT Press, Cambridge, MA.

Braun, J. A. and Eklund, J. L. (2019), "Fake News, Real Money: Ad Tech Platforms, Profit-driven Hoaxes, and the Business of Journalism", *Digital Journalism*, vol. 7, no. 1, pp. 1–21.

Carlin, J. P. and Graff, G. M. (2018), *Dawn of the Code War: America's Battle Against Russia, China, and the Rising Global Cyber Threat*, PublicAffairs, New York.

Chegeni, M., Masuminia, G. A. and Sorkhvandi, H. (2020), "A Comparative Study of Protecting National Identity and its Symbols in Criminal Law", *Quarterly Journal of the Macro and Strategic Policies*, vol. 8, no. 2, pp. 316–339.

Chertoff, M. and Jardine, E. (2021), *Policing the Dark Web: Legal Challenges in the 2015 Playpen Case,* CIGI Papers 259 Centre for International Governance Innovation. https://www.cigionline.org/publications/policing-the-dark-web/

Chrustie, C. and Luna, D. M. (2023), *The Growing Harms of Cross-Border Illicit Trade Vectors and Threat Convergence to Canada's National Security: Urgent Attention and Action Are Needed*, International Coalition Against Illicit Economies, Washington, DC. https://icaie.com/wp-content/uploads/2023/11/ICAIE-Canada-FINAL-Policy-Brief-Illicit-Trade-Vectors-Threat-Convergence-Harms-to-Canadas-National-Security-27-November-2023.pdf

Ellul, J. (1965), *Propaganda: The Formation of Men's Attitudes*, Knopf Doubleday Publishing Group, New York.

Fonseca, X., Lukosch, S. and Brazier, F. (2019), "Social Cohesion Revisited: A New Definition and How to Characterize It", *Innovation: The European Journal of Social Science Research*, vol. 32, no. 2, pp. 231–253.

Gilmore, J. (2015), *The Cosmopolitan Military: Armed Forces and Human Security in the 21st Century*, Palgrave Macmillan, Basingstoke, UK.

Greenwood, L. and Fashola, K. (2021), "Illicit Fentanyl from China: An Evolving Global Operation", *US-China Economic and Security Review Commission*. Issue Brief (24 August 2021). https://www.uscc.gov/research/illicit-fentanyl-china-evolving-global-operation

Hanes, W. T. and F. Sanello (2002), *Opium Wars: The Addiction of One Empire and the Corruption of Another*, Sourcebooks, Naperville, IL.

Harold, S. W., Beauchamp-Mustafaga, N. and Hornung, J. W. (2021), *Chinese Disinformation Efforts on Social Media*. RAND Corporation, Santa Monica, CA.

Ibrahim, A. M. (2021), *Social Cohesion: Concept and Significance*, Zayed House for Islamic Culture. https://zhic.gov.ae/en/articles/social-cohesion-concept-and-significance-1

Ikenberry, G. J. (2018), "The End of Liberal International Order?", *International Affairs*, vol. 94, no. 1, pp. 7–23.

Kaldor, M. (2012), *New and Old Wars*, Polity Press, Cambridge.

Kellner, D. (2002), "September 11, Terrorism, and Blowback", *Cultural Studies-Critical Methodologies*, vol. 2, no. 1, pp. 27–39.

Kittrie, Orde F. (2016), *Lawfare: Law as a Weapon of War*, Oxford University Press, Oxford.

Kortukov, D. (2020), "Sovereign Democracy' and the Politics of Ideology in Putin's Russia", *Russian Politics*, vol. 5, no. 1, pp. 81–104.

Kubicek, P. (2022), "Illiberal Nationalism and the Backlash against Liberal Cosmopolitanism in Post-communist Europe", *Nationalism and Ethnic Politics*, vol. 28, no. 3, pp. 332–350.

Manca, A. R. (2014), "Social Cohesion", in A. C. Michalos (ed.) *Encyclopedia of Quality of Life and Well-Being Research*. Springer, Dordrecht, Netherlands, pp. 6026–6028.

Martínez-Ariño, J. (2019), "Governing Islam in French Cities: Defining 'acceptable' Public Religiosity Through Municipal Consultative Bodies", *Religion, State & Society,* vol. 47, nos. 4–5, pp. 423–439.

Morris, I. (2014), *War! What Is It Good For? Conflict and the Progress of Civilization from Primates to Robots*, Farrar, Straus and Giroux, New York.

Müller-Plotnikow, S. (2017), "European Independence Movements", *Deutsche-Welle: Society,* 30 September. https://www.dw.com/en/beyond-catalonia-separatist-movements-in-western-europe/a-40761144

Nemno, Y. (2015), "Concept and Nature of Technology of Corporate Raids of Enterprises," *Evropský politický a právní diskurz*, vol. 2, no. 1, pp. 331–335.

Ong, J. C. and Cabañes, J. V. A. (2018), *Architects of Networked Disinformation: Behind the Scenes of Troll Accounts and Fake News Production in the Philippines*, Report. Newton Tech4Dev Network, University of Massachusetts, University of Leeds, pp. 1–74. https://newtontechfordev.com/wp-content/uploads/2018/02/Architects-of-Networked-Disinformation-Executive-Summary-Final.pdf

Patnaik, P. (2017), "Caste, Community and Belonging: The Indian Case", *Social Scientist*, vol. 45, no. 1/2, pp. 73–79.

Pomerantsev, P. and Weiss, M. (2014), "The Menace of Unreality: How the Kremlin Weaponizes Information, Culture and Money", *The Interpreter: Institute of Modern Russia.* https://www.interpretermag.com/the-menace-of-unreality-how-the-kremlin-weaponizes-information-culture-and-money/?utm

Pomerantsev, P. (2019), *This Is Not Propaganda: Adventures in the War Against Reality*, PublicAffairs, New York.

Pomerantsev, P. (2024), *How to Win an Information War: The Propagandist Who Outwitted Hitler*, PublicAffairs, New York.

Realuyo, C. B. (2019), "The New Opium War", *Prism*, vol. 8, no. 1, pp. 132–142.

Rid, T. (2020), *Active Measures: The Secret History of Disinformation and Political Warfare*, Farrar, Straus and Giroux, New York.

Schneider, S. (2020), *Money Laundering in British Columbia: A Review of the Literature*. The Cullen Commission of Inquiry into Money Laundering in British Columbia. https://cullencommission.ca/files/reports/CullenCommission-FinalReport-LOT_TOC.pdf

Schwalbe, C. B. (2006), "Remembering Our Shared Past: Visually Framing the Iraq War on US News Websites", *Journal of Computer-Mediated Communication*, vol. 12, no. 1, pp. 264–289.

Taylor, C. (2022), "Degenerations of Democracy", in C. Calhoun, D. P. Gaonkar, and C. Taylor (eds), *Degenerations of Democracy,* Harvard University Press, Cambridge, MA, pp. 18–46.

Tilly, C. (1985), "War Making and State Making as Organized Crime" Chapter 5, in P. B. Evans, D. Rueschemeyer, and T. Skocpol (eds), *Bringing the State Back In,* Cambridge University Press, Cambridge, pp. 169–191.

Tóth, Z. (2022), "The Protection of State and National Symbols Across Europe: An Overview of Constitutional Law and Criminal Law Regulations", in Z. J. Tóth (ed.), *Constitutional and Legal Protection of State and National Symbols in Central Europe*, Central European Academic Publishing, Miskolc, Budapest, pp. 13–53.

Triolo, P., Allison, K., Brown, C. and Broderick, K. (2020), "The Digital Silk Road: Expanding China's Digital Footprint", *Eurasia Group*, 8 April. https://www.eurasiagroup.net/files/upload/Digital-Silk-Road-Expanding-China-Digital-Footprint.pdf

Turchin, P. (2016), *Ultra Society: How 10,000 Years of War Made Humans the Greatest Cooperators on Earth*, Beresta Books, Chaplin, CT.

UNHCR (2024), Focus Area Strategic Plan for Climate Action. 2024–2030. https://digitallibrary.un.org/record/4062910?v=pdf

Vitaljić, S. (2013), "*War of Images-Contemporary War Photography*" (*Rat slikama-Suvremena ratna fotogratvija)*, Algoritam, Zagreb.

Willett, M. (2023), "The Cyber Dimension of the Russia–Ukraine War", *Survival: October-November 2022,* vol. 64, no. 5, pp. 7–26.

Williams, P. (2001), "Crime, Illicit Markets, and Money Laundering", in P. J. Simmons & C. de Jonge Oudraat (eds.), *Managing Global Issues: Lessons Learned,* Carnegie Endowment for International Peace, Washington, D.C., pp. 106–150.

Wrzosek, M. (2018), *Wojny Przyszłości: Doktryna, Technika, Operacje Militarne (Wars of the Future: Doctrine, Technology, Military Operations)*, Wydawnictwo Fonda, Warszawa.

Xu, R., Xiong, X., Abramson, M. J., Li, S., and Guo, Y. (2020), "Ambient Temperature and Intentional Homicide: A Multi-city Case-crossover Study in the US", *Environment International*, vol. 143 (105992), pp. 1–10.

Zanin, C., Martins, V. and Valim, R. (2022), *Lawfare: Waging War through Law*, Routledge, London.

Zhao, M. (2022), "The Illicit Supply of New Psychoactive Substances within and from China: A Descriptive Analysis", *International Journal of Offender Therapy and Comparative Criminology*, vol. 66, no. 5, pp. 495–516.

Zygar, M. (2023), *War and Punishment: Putin, Zelensky, and the Path to Russia's Invasion of Ukraine*, Scribner, New York.

17 Social Organization and Strengthening the State

Jowita Brudnicka, Marzena Żakowska, and David Last

Introduction

The state remains central to the pursuit of national security, maintaining external boundaries and protecting national institutions for autonomy and decision-making. Beyond the apparatus of states, and regardless of the form of regime, states are responsible for the human security of their citizens, including social cohesion (Chapter 16). Stasis rarely exists in social organizations; the pursuit of equality or advantage by different groups is elusive, so the potential always exists for destructive changes to social organizations. Within the protected boundaries of a state, this creative destruction of social forces can be positive. Women's rights, universal suffrage, and the retreat of cruelty would not be possible without social conflict and eliminating some ways of organizing social behavior. The focus here is not the conflict of positive social change within states but the defense against attacks by hostile actors to weaken and disrupt society. This is the defense against Grey Zone threats and modern war.

Concepts for Defense of the Social Order

Social order implies an organized and structured society, including a hierarchy of power, roles, norms, and relationships that govern the interactions of individuals and groups, including political, economic, social, and legal relations. Social order does not afford everyone equal benefits or protection. This can lead to conflict within society or vulnerability to external interference. Early sociologists referred to studying a "larger mind" that governed social relationships (Cooley 1909:4–10,38). The idea of a coherent whole responding to threats has been elusive. Understanding Grey Zone threats and responses demands thought about states, nations, societies, and social organization.

States are neither fixed nor equal. They are nominally equal political and legal entities, but their power, boundaries, and identities change over time. Putin's Russia has attacked the very idea of Ukraine as a state. Ambivalence is captured in the label "nation-states," which combines the political and legal characteristics of the state with the social and cultural characteristics and sense of identity of the nation, many of which lack states. There are Kurdish, Basque, and Oromo nations without states. Efforts to "build" states have led to rethinking the conditions under which stable and peaceful states can emerge. Hostile efforts to attack social cohesion and defensive efforts to strengthen states and nations demand an understanding of social organization.

There is no origin point for the state. Rather, different forms of localized and enlarged polities have co-existed in which claims to sovereignty have varied (Graeber and Wengrow 2021:359). This is important because our modern concept of the state is not the only way to provide for human security. Many states today are constrained from above by weak governance

DOI: 10.4324/9781003428701-26

or foreign domination and from below by strong alternative sources of identity. Focusing on social organization is a path to strengthening resistance to Grey Zone threats, either by the state or social networks within or transcending states.

A state is, by definition, endowed with government, territory, population, legitimacy conferred by other states, and crucially, a monopoly on legitimate violence (Brown et al. 2018:508). States are also the nexus of territory, government, and society; society cannot be separated from the state (Buzan 1991:67). A strong state has a strong idea of the state; this idea is grounded in its society. The national idea is a source of identity and legitimacy for the state (Breede 2017:30,33). The idea of nationalism or national identity and the extent to which it can be mobilized to strengthen the state requires more explanation.

Nations predate the modern state, emerging from kin-culture communities transformed by modernity. Nations may reflect fundamental human preferences for their own kin (Gat 2013:386). The rationalists of the French Revolution argued people could choose their civic identity. In contrast, Germanic romanticism held that national identity rested on blood and soil (Renan 2002:55). Language, religion, and culture within territorial boundaries shaped early forms of ethnic and national consciousness (Armstrong 1982:8). A typology of nationalism includes ethnonationalism drawing on shared identity, civic nationalism drawing on shared citizenship, and cultural nationalism seeking to promote a common national identity through culture (Gellner 1983). Ethnic and cultural nationalism are more exclusive but potentially stronger in their expressions of core identity. This explains the resilience of the national identities of the Kurds, Armenians, and Ukrainians, deprived of statehood for generations. Czech Historian Miroslav Hroch describes a progression from awakening cultural-ethnic identity to political mobilization to achieve national goals and institutions and eventually to establishing and consolidating a nation-state within defined boundaries (Hroch 2004:96). National identity also waxed and waned within Europe's multinational empires, leading ruling elites to interpret minority claims as challenges to national security (Halász 2023:327).

Civic nationalism is more suited to capitalist economies, which rely on loose economic ties (Putnam's social lubricant). Common rights and acceptance of the rule of law help to support industry, commerce, employment, and the pursuit of positive peace within the modern state. Arguably, today's "imagined communities" exist because of mass literacy, state-sponsored education and socialization, and a common mass-produced culture (Anderson 1983:24,39–41).

What could be assembled in European states over hundreds of years from the Reformation to the 20th century can now be disassembled. Globalization induces fragmentation as individuals and groups are exposed to new ideas and pressures. Communications technology and social media allow direct appeals to fragmented and atomized identities. Mobilizing a population to defend the state is difficult when the people see no connection between the state and their security or economic interests in a global economy.

The state's power can also use coercion to dominate its citizens and maintain control internally, often with the pretext of external enemies (Rummell 1997:203). State repression can erode trust and weaken secondary social groups and civil society, forcing greater reliance on primary social groups like families and clans. These, in turn, can be disrupted by state-controlled socialization and the dislocation of families and communities. Kampuchea under the Khmer Rouge and North Korea today reflect these extreme forms of state control over society. Although they can be durable, state-dominated societies can be brittle and subject to rapid collapse, as the Khmer Rouge regime did, despite efforts to dominate society.

Social arrangements can strengthen or undermine the nation-state's identity and cohesion. People find safety and security in many social settings, including primary social arrangements like families, kinship, marriage, clan, and close living arrangements (Brudnicka-Żółtaniecka 2022).

Even in advanced industrial countries and major cosmopolitan cities, the state's ability to regulate or alter these primary social arrangements is limited. Sociologist Robert Putnam describes two forces: the superglue of strong in-group bonding and the lubricant of bridging social capital that transcends groups (Putnam 2000:23). Tight networks of social trust are important to stability and human security just as looser trust networks are essential to the commerce and political discourse of any advanced society. Research on business networks confirms this view. Any given individual has a network of close friends in regular contact and a wider network of business or professional acquaintances who are not in regular contact. Each of these acquaintances, however, has a similar close network of friends and regular contacts (Granovetter 1983:202). Both tight close ties and weak bridging ties can be targets for Grey Zone attacks on the coherence of society. Both are important for preserving social cohesion and a functioning society and economy.

Primary social groups are based on long-term, face-to-face interactions; the family or extended family is the typical example. Secondary social groups are based on shared goals or interests. Members of secondary groups like political parties or trade unions may never see each other face-to-face but collaborate across networks to pursue common goals (Colman 2015:273). Loose networks of secondary social groups may cut across international boundaries or operate mainly within a state. Organized religious groups and multinational corporations have identifiable offices in many countries and operate in a common language with common concepts, values, attitudes, and beliefs. National businesses and cultural clubs operate mainly within a state and in their vernacular language.

Strong states are characterized by primary and secondary social groups that reinforce the idea of the state. National narratives play a role in binding together primary and secondary social groups. A shared history, common purpose, and belief in common values combine personal and collective interests in the present with a common understanding of the past and a vision of the future. Religions and political ideologies, as collective belief structures, can be mobilized to support political action. Still, despite deep political and economic divisions, there may be common acceptance of myths, national symbols, historical narratives, and origin stories. Anthony Smith's historical ethno-symbolic approach describes nationalism over the long duration of groups, encompassing the many generations of past, present, and future but resting on ethnic foundations of language, culture, and phenotype. Both group members and outsiders typically have an identifying name or emblem, a myth of common ancestry, elements of a common culture, a link to a historic homeland or territory, and a measure of solidarity, at least among elites (Smith 1999:6). Ideas that transcend ethnic nationalism, like cosmopolitan service to universal values (Gilmore 2015; Nussbaum 2019), the ideals of European unity (Smith 1999:Chapter 9), or world socialism (Nielsen 2003:437) compete with more local loyalties to primary and secondary organizations and ethnic concepts of the nation. Large multinational states strive to create new origin myths and national symbols and may treat competing nationalisms as threats to the state.

Attacking Social Cohesion

Social cohesion consists of solidarity, connectedness, and trusted relationships within society. The World Bank's research on civil wars and the connections between violence and development in the 1990s demonstrated that social cohesion is an essential prerequisite for the many forms of social interaction that enable economic development (Colletta and Cullen 2000:35). It is also an essential prerequisite for national defense against concerted hostile action. The will to work together to defend national institutions can be undermined by violence, subversion, and propaganda.

Violent conflicts within a state—like those in Cambodia (1975–1979), Rwanda (1994), Guatemala (1960–1996), and Somalia (1981–present)—don't deplete the stock of social capital

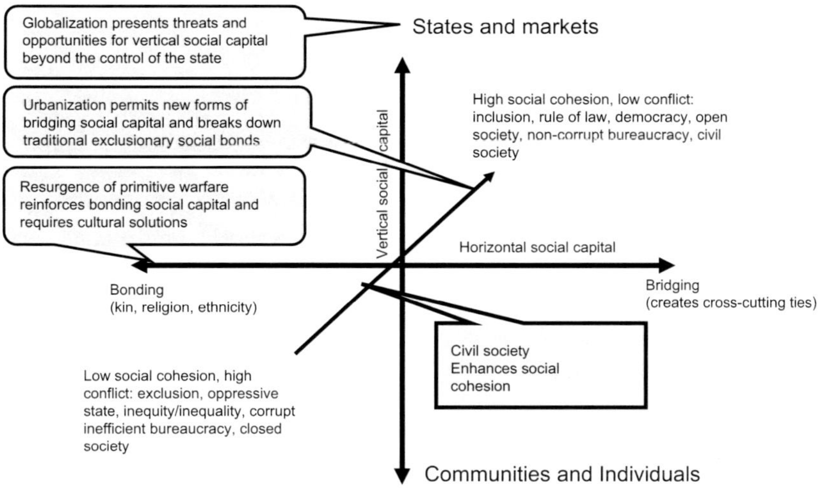

Figure 17.1 Violence and Social Cohesion.

Source: Adapted from Colletta and Cullen (2000:14).

(trusted relationships). Rather, they transform it, reducing trust in strangers and increasing trust in close relations. This is reflected in movement from the upper right toward the lower left quadrant in Figure 17.1. In an effective and well-functioning state, citizens have a justified trust in the probity of government and public services, the honesty of businesses and financial establishments, and the efficacy of law. Urbanization brings new opportunities to live with mutual tolerance in diverse communities, engendering trust in those from different linguistic, religious, and ethnic groups and breaking down exclusionary social bonds. Social media and new communications technologies blur the boundaries between urban and rural communities. Loose ties are broken, and trust in strangers weakens as primitive warfare spreads through a society. People fall back on the tight ties of kin, religion, and ethnicity. Social media can accelerate the breakdown of trust, but also the building of new communities of interest among individuals not physically in contact.

Subversion is a strategy that seeks to undermine the power and authority of established leadership and governance systems. The target for subversive strategies is how people think or the cognitive domain. Manipulation, deception, and direct action can be mobilized to change people's thinking. Finding a way into a community may entail mimicking or impersonating trusted messengers or co-opting community members. Outsiders with limited local knowledge cannot easily disrupt strong in-group bonds, but tensions between identity groups may be easier to instigate. Strong in-group bonds can generate out-group antagonism. Competing identities play a role in initiating and escalating violence between groups (Seul 1999:556). One tactic for Grey Zone attacks is to manipulate group identities to create tensions between tightly bonded groups in opposition. A few trusted contact points within an identity group can spread distrust and hostility to out-groups. There is a long history of appeals to co-religionists and ethnolinguistic populations living under foreign rule. The expulsion and starvation of minorities during the waning years of World War I were tactics for the consolidation of new territorial boundaries (Tooze 2014:165). Political entrepreneurs or elites have used foreign agents or trusted intermediaries to spread rumors, stir hostility, or engage in acts of sabotage as low-cost ways to stir division between groups.

External disruption of loose ties and social networks has accelerated with the age of social media, which facilitates direct communication with identified groups. The changing information landscape and influence operations have generated online glossaries to distinguish between disinformation, misinformation, malinformation, and more (Wardle 2018; Sessa 2023). External disruption is most effective when it capitalizes on existing social divisions and conflicts. During the first Cold War, the Soviet Union transmitted malinformation about America's race conflicts and civil rights struggles to African audiences as an effective distraction from their own problems of internal dissent (Rid 2021:136). During the 2016 US Presidential election, Russia used fake social media accounts, bots, and websites to stimulate hostility between pro-Black and white supremacist groups; "The purpose of this strategy was to manipulate public opinion on racial issues and disrupt the political process" (Aceves 2019:178). In one example, two separate Russian-sponsored Instagram accounts promoted competing pro-Beyoncé and anti-Beyoncé rallies at the headquarters of the National Football League in New York on the morning of 16 February 2016 in response to Beyoncé's support for Black Lives Matter. This kind of information manipulation has always been labor-intensive. Automated programs (bots) and artificial intelligence (AI) generation of content make such strategies more accessible and cost-effective. The broad range of active measures is presented in Table 17.1.

The techniques of active measures are less important than their targets. Hostile foreign powers must understand a society and its government well to weaken the bonds that link government, society, and security apparatus. Hostile operations are most successful when they can mobilize tensions within the society without the actors being aware that they are manipulated:

> It is very hard to distinguish—for an activist, for the target of an active measures campaign, even for a large organization running its own active measures—between a cunning influence agent on the one hand, and a genuine activist on the other. In theory, on an individual basis, one person is either a genuine activist or a controlled agitator, but this worldview applies only in the abstract. In practice, one individual can be both genuine and an exploited asset, a witting and unwitting collaborator at the same time.
>
> (Rid 2021:432)

The state plays a critical role in preserving human and national security through its power to regulate social relations. Much of this power depends on its agents' policing.

Policing and the Violent Transformation of Social Capital

Policing encompasses two distinct approaches. High policing is political; it focuses on maintaining the status quo and safeguarding established institutions and interests against threats of subversion or political unrest. Domestic intelligence services like MI5 (internal) and MI6 (external) in the United Kingdom, the Canadian Security Intelligence Service, and the French *Direction de la Surveillance du Territoire* are all engaged in high policing, as are the other intelligence services described in Chapter 9. Low policing, in contrast, is focused on criminal and anti-social elements, protecting not state institutions but its citizens. Low policing is concerned with maintaining order as a foundation of governance (Brodeur 1983:508). A characteristic of Grey Zone threats and modern wars is that the lines between policing and military responses are difficult to draw. High and low policing, civilian and military measures, and public and private control of force overlap.

Grey Zone threats weaken the state by transforming social capital; active measures are designed to undermine trust across the loose ties of secondary social organizations and throw

Table 17.1 Summary of Active Measures

Label	*Description or example*
Active measures	Methodical output of large bureaucracies; always include an element of disinformation; always purposive, usually to weaken an adversary
Bots	"Bots are social media accounts that are operated entirely by computer programs and are designed to generate posts and/or engage with content on a particular platform. In disinformation campaigns, bots can be used to draw attention to misleading narratives, to hijack platforms' trending lists, and to create the illusion of public discussion and support" (Wardle 2018).
Deepfakes	"Deepfakes is the term currently being used to describe fabricated media produced using artificial intelligence. By synthesizing different elements of existing video or audio files, AI enables relatively easy methods for creating 'new' content, in which individuals appear to speak words and perform actions, which are not based on reality" (Wardle 2018).
Dezinformatsiya	Russian for disinformation, Office established by Artuzov in USSR, 1923 (Rid 2020:25).
Disinformation	"Disinformation is false information that is deliberately created or disseminated with the express purpose to cause harm" (Wardle 2018).
Disinformation, first wave	1920s "political warfare," print, radio, forgeries, a weapon of the weak against the dominant capitalist world order.
Disinformation, second wave	Post World war II, professionalized, US intelligence leading, aggressive, unscrupulous, well-funded: "exacerbate existing tensions and contradictions within the adversary's body politic, by leveraging facts, fakes, and ideally a disorienting mix…" (Rid 2020:7).
Disinformation, third wave	1970s, USSR leading, "active measures" a global operational science, supported by a large, well-oiled bureaucracy, growth up to USSR disintegration.
Disinformation, fourth wave	2010s, technology and internet culture dominate, high tempo, low skill, remotely delivered, disjointed (Rid 2020:7) "attacks against a liberal epistemic order, or a political system that places its trust in essential custodians of factual authority." (Rid 2020:10) augmented by covert action to target machines (Rid 2020:13).
Doxing	"The act of publishing private or identifying information about a person or organisation online, with malicious intent" (Sessa 2023).
False flag operations	Operations designed to be attributed to a non-existent group established for the purpose, e.g. the *Kampfverband* in 1957 (Rid 2020:121) or to an innocent existing group.
Forgery	Generating materials attributed to another source; may entail elaborate cover stories, fabrication of background, e.g. the Czech Black Lake forgeries, 1964. (Rid 2020:156–165).
Leaked information	Broadcasting confidential information that is not intended for the public.
Malinformation	"Information that is based on reality but is used to harm or threaten a person, an organization, or a country (e.g., see "doxing").
Misinformation	"Information that is false but believed to be true by those disseminating it. It differs from disinformation in the absence of an intention to mislead or harm" (Sessa 2023).
Operational play	Active engagement with opposing factions to change their perception of reality (Rid 2020:21).
Propaganda	"Propaganda is true or false information spread to persuade an audience, but often has a political connotation and is often connected to information produced by governments. It is worth noting that the lines between advertising, publicity, and propaganda are often unclear" (Wardle 2018).

(*Continued*)

Table 17.1 (Continued)

Label	*Description or example*
Sock puppet	"A fictitious online identity created specifically to deceive, i.e., a fake persona. Sock puppet accounts differ from catfishing as the former are short-lasting, not very detailed, and not necessarily conceived for malign intent" (Sessa 2023).
Targeted leaks	Conveying confidential information to specific audiences to manipulate their perception and actions.

Source: Compiled by Authors from Wardle (2018), Rid (2020), and Sessa (2023).

people back on the tight bonds of primary social organizations, distrusting strangers and outsiders. Many people are separated from extended families in individualistic, developed capitalist economies. Clans and kin groups are left behind by modernity. Many states have weak or contested national narratives, which leave parts of society without a sense of belonging. People are lonely and isolated. French sociologist Jacques Ellul described loneliness inside the crowd as "the most terrible ordeal of modern man" (Ellul 1965:148). Loneliness makes people vulnerable to propaganda and the sense of belonging to an in-group that recognizes conspiracies.

Values, attitudes, and beliefs—in a word, ideas—are the target of active measures seeking to erode social cohesion by transforming social capital to weaken an enemy state. The policing of ideas that threaten the state is a classic function of high policing. Tolerance of divergent ideas is a strength of liberal democracies, while creating strong narratives that defy facts but satisfy emotions is a strength of totalitarian regimes and demagogues (Pomerantsev 2023:93; Snyder 2018:279). This creates a dilemma for police strategies to support social cohesion and build resilience in the face of foreign pressure. High policing of subversive or socially damaging ideas is the direct approach: shut down or discredit the extremists, cranks, sock-puppets, and anti-government narratives that exacerbate divisions. Doing so may paradoxically lower the acceptance of the responsible authorities. If active measures are well-crafted, they will have built on genuine inequalities and social divisions, which will not be resolved by attacking the messengers.

Real divisions are more dangerous than fake or fabricated ones. Consistent and credible low policing provides the foundation of order on which good governance is built. Policing and the rule of law create a predictable framework of daily life for all citizens. High policing can contribute by protecting against external and internal threats, but the nonviolent transformation of social capital is also needed to build social resilience.

Nonviolent Transformation of Social Capital

Social resilience might be visualized as the ability of a society under threat to remain in or bounce back to the upper right quadrant in Figure 17.1, where the state is better able to provide the advantages of positive peace: good governance, a sound business environment, equitable distribution of resources, acceptance of rights, neighborly relations, free flow of information, and low levels of corruption (IEP 2023:7). Civil society organizations—public voices not controlled by government—play an important role. These may be primary or secondary social groups. Three kinds of capabilities help a society to remain cohesive. *Coping* capabilities allow groups and individuals to cope with and overcome adversity. *Adaptive* capacities help them to learn from experiences and adjust to future challenges in their everyday lives. *Transformative* capacities allow them to build new institutions that foster individual welfare and sustainable social robustness to address future crises (Keck and Sakdapolrak 2013:6–8). Individuals who excel in society have these capabilities and may use them with others to gain control, set rules, ensure their children succeed, and sometimes oppress other groups. Building social resilience is, therefore, political, not technical. This means that the battleground in Grey Zone conflicts and modern war is as much between socially constructed elites, minorities, and majorities as between states.

Violence and active measures can transform social capital by eroding trust and pushing society into the lower left quadrant in Figure 17.1. Nonviolent measures—coping, adapting, and transforming—can help push society back into the upper right quadrant. Empires, colonialism, and neoliberal globalization have all been adaptive vehicles for building loose ties that empower some groups at the expense of others. If families and extended families are the essential primary social organizations, we can see how a minority of families benefit from empire, colonialism, or globalization while a majority are marginalized. Those marginalized tend to fall back on tighter bonds for survival and become potential sources of anti-systemic hostility, protest, and mobilization for change. The promise of nonviolent social change for the common good weakens the appeal of extremist or manipulative messages (IEP 2024:41).

Nationalism and national narratives can also expand the loose ties that foster cohesion and preserve identity. As conservative nationalist Ernest Renan wrote in 1882, "For hundreds of years to come, France, England, Germany, and Russia will be still…be historical individuals" (Renan 2002:54). The new digital age has made borders more permeable but has also provided new tools for nationalists. If nationalism is an invented tradition to enhance national solidarity (Hobsbawm 1983:307), then digital tools can accelerate its manufacture. From a critical theory perspective, digital capitalism and nationalist narratives are mutually supportive. Both are shaped today by social media, big data employed to manipulate consumer behavior, fake news, targeted advertising, bots, and algorithmic politics. The German and Austrian Federal elections of 2017 privileged right-wing parties with capitalist money and technology behind them. However, the nationalist narratives were deployed by political influencers and right-wing media from above to exacerbate social tensions (Fuchs 2019:248). In an earlier era, mass media permitted cultural homogenization, but this can no longer be assumed with social media. Because the internet is a key technology for keeping nations and other abstract communities together, it has the potential to strengthen national narratives (Eriksen 2007:16). When a state backs a national narrative, it can drown out competing claims online, particularly in national digital ecosystems dominated by a local language. Without state engagement, global communications ecology tends to drive diversification, fragmentation, and commodification of narratives (Mihelj and Jiminez-Martínez 2021), particularly in international languages like English, French, Spanish, and Arabic. Nations may survive, but their character and cohesion can be eroded online from within and by hostile external actors. How states benefit or suffer from this erosion will vary.

Liberal and autocratic regimes alike feel the pressure of new transnational social communications. Putin saw the "color revolutions" as a Western-sponsored threat to Russian sovereignty (Zygar 2023:271). Liberal democracies see international right-wing collaborations as menacing (Caiani 2018) as once they saw risk in communist conspiracies. Nationalist politicians and political entrepreneurs, sometimes mobilized by special interests, challenge the engagement of their governments in international institutions (De Vries et al. 2021:306). While autocratic regimes seem to have better tools to manage internet incursions on sovereignty, liberal democracies have better protections embedded in an educated and critically thinking citizenry. Distinct language and culture may help insulate a small country from international pressure, but AI and automatic translations may erode this advantage.

A strategy for social resilience must identify and counter messages from adversarial political and economic systems. Embassies, whose physical safety and confidential communications are guaranteed under international law, may be used to support the preferred world orders of their home countries. China has often relied on military diplomats who swear loyalty to the Communist Party rather than the government or state (Martin 2021:49). Major power embassies typically include large staffs with intelligence and security assets. Most embassies are unabashed in pursuing trade advantage, and the United States has long pursued an open door for trade and investment, predicated on global capitalism (Van Apeldoorn and de Graaff 2016:236). This can seriously affect small states' social cohesion if they are adversely affected by liberalism and the Washington Consensus. Migration and diaspora politics actively stirred by hostile embassies, social media based abroad, and market economies without safety nets can all threaten resilience.

Social organizations looking after people in need can grow stronger if they respond effectively to disruptions. They will weaken if they fail. When state organizations fail to respond to unemployment, insecurity, starvation, or suffering, people fall back on primary social organizations to meet their needs. State capacities erode, and primary social organizations benefit. A principle of state resilience is to be seen to look after people. Each marginalized or dissatisfied group is a potential source of opposition and disruption and a target for hostile action. Electoral politics provide an imperfect feedback mechanism to guide government action. Governments also need more regular and consistent action—social and economic policies responsive to well-being indicators.

Conclusions

Modern wars span hybrid, asymmetrical, and Grey Zone conflict. They undermine social stability and attack the foundations of trust and community solidarity within a state. To meet their national and human security responsibilities, states must be aware of domestic and international factors eroding social cohesion. Understanding primary and secondary social organization, bridging and bonding capital, and state responsibilities for social cohesion are essential elements of the new security environment.

References

Aceves, W.J. (2019), "Virtual Hatred: How Russia Tried to Start a Race War in the United States", *Michigan Journal of Race and Law*, vol. 24, no. 1, pp. 177–250.

Anderson, B. (1983), *Imagined Communities: Reflections on the Origin and Spread of Nationalism*, Verso, London.

Armstrong, J.A. (1982), *Nations Before Nationalism*, University of North Carolina Press, Chapel Hill.

Breede, H. (2017), *The Idea of Failed States: Community, Society, Nation, and Patterns of Cohesion*, Taylor and Francis, Routledge, London.

Brodeur, J.P. (1983), "High Policing and Low Policing: Remarks about the Policing of Political Activities", *Social Problems*, vol. 30, no. 5, pp. 507–520.

Brown, G., McLean, I. and McMillan, A. (2018), *The Concise Oxford Dictionary of Politics and International Relations* (Oxford Quick Reference), Oxford University Press, Oxford.

Brudnicka-Żółtaniecka, J. (2022), *Afganistan. Budowanie państwa* (*Afghanistan. State-building*), Polskie Towarzystwo Geopolityczne, Kraków.

Buzan, B. (1991), *People, States and Fear,* Harvester Wheatsheaf, Hemel Hempstead.

Caiani, M. (2018), "Radical Right Cross-National Links and International Cooperation", in J. Rydgren (ed.), *The Oxford Handbook of the Radical Right*, vol. 1, Oxford University Press, New York, pp. 394–411.

Colletta, N.J. and Cullen, M.L. (2000), *Violent Conflict and the Transformation of Social Capital: Lessons from Cambodia*, Rwanda, Guatemala, and Somalia: World Bank Group, Washington DC. https://documents.worldbank.org/en/publication/documents-reports/documentdetail/799651468760532921/violent-conflict-and-the-transformation-of-social-capital-lessons-from-cambodia-rwanda-guatemala-and-somalia

Colman, A.M. (2015), *A Dictionary of Psychology*, 4th edition, Oxford University Press, Oxford.

Cooley, C.H. (1909), *Social Organization: A Study of the Larger Mind*, Charles Scribner's Sons, New York.

De Vries, C.E., Hobolt, S.B. and Walter, S. (2021), "Politicizing International Cooperation: The Mass Public, Political Entrepreneurs, and Political Opportunity Structures", *International Organization*, vol. 75, no. 2, pp. 306–332.

Ellul, J. (1965), *Propaganda: The Formation of Men's Attitudes*, Vintage Books, New York.

Eriksen, T.H. (2007), "Nationalism and the Internet", *Nations and Nationalism*, vol. 13, no. 1, pp. 1–17.

Fuchs, C. (2019), *Nationalism on the Internet: Critical Theory and Ideology in the Age of Social Media and Fake News*, Routledge, New York.

Gat, A. (2013), *Nations: The Long History and Deep Roots of Political Ethnicity and Nationalism*, Cambridge University Press, Cambridge.

Gellner, E. (1983), *Nations and Nationalism*, Cornell University Press, Ithaca.

Gilmore, J. (2015), *The Cosmopolitan Military: Armed Forces and Human Security in the Twenty-First Century*, Palgrave Macmillan, Houndmills, Basingstoke, UK.

Graeber, D. and Wengrow, D. (2021), *The Dawn of Everything: A New History of Humanity*, Penguin Books, London.

Granovetter, M. (1983), "The Strength of Weak Ties: A Network Theory Revisited", *Sociological Theory*, vol. 1, pp. 201–233. https://doi.org/10.2307/202051

Halász, I. (2023), "'National and Ethnic Minorities' Legal Position in East Central Europe Between 1789 and 1989", in P. Sáry (ed.), *Lectures on East Central European Legal History*, 2nd edn, Legal Studies on Central Europe, Central European Academic Publishing, Miskolc, Budapest, pp. 311–342.

Hobsbawm, E.J. (1983), "Mass-producing Traditions: Europe 1870–1914", in E.J. Hobsbawm and J. Ranger (eds.), *The Invention of Tradition*, Cambridge University Press, Cambridge, pp. 263–307.

Hroch, M. (2004), "From Ethnic Group toward the Modern Nation: The Czech Case", *Nations and Nationalism*, vol. 10, nos. 1–2, pp. 95–107.

Institute for Economics and Peace (IEP) (2023), *Positive Peace Implementation Guide: A User's Guide to the Positive Peace Framework*, Institute for Economics and Peace, Sydney. https://www.economicsandpeace.org/wp-content/uploads/2021/03/Positive-Peace-Implementation-Guide-web.pdf

Institute for Economics and Peace (IEP) (2024), *Positive Peace Report 2024: Analysing the Factors that Build, Predict and Sustain Peace*, Institute for Economics and Peace, Sydney. https://www.economicsandpeace.org/research/positive-peace-report/

Keck, M. and Sakdapolrak, P. (2013), "What Is Social Resilience? Lessons Learned and Ways Forward", *Erdkunde*, vol. 67, no. 1, pp. 5–19.

Martin, P. (2021), *China's Civilian Army: The Making of Wolf Diplomacy*, Oxford University Press, Oxford.

Mihelj, S. and Jiménez-Martínez, C. (2021), "Digital Nationalism: Understanding the Role of Digital Media in the Rise of 'New' Nationalism", *Nations and Nationalism*, vol. 27, no. 2, pp. 331–346.

Nielsen, K. (2003), "Toward a Liberal Socialist Cosmopolitan Nationalism", *International Journal of Philosophical Studies*, vol. 11, no. 4, pp. 437–463.

Nussbaum, M.C. (2019), *The Cosmopolitan Tradition: A Noble but Flawed Ideal*, Harvard University Press, Cambridge.

Pomerantsev, P. (2023), *How to Win an Information War: The Propagandist Who Outwitted Hitler*, PublicAffairs, New York.

Putnam, R.D. (2000), *Bowling Alone: The Collapse and Revival of American Community*, Simon & Schuster, New York.

Renan, E. (2002), "What Is a Nation?", in S. Woolf (ed.), *Nationalism in Europe: From 1815 to the Present,* Routledge, London and New York, pp. 54–66.

Rid, T. (2020), *Active Measures: The Secret History of Disinformation and Political Warfare*, Farrar, Straus and Giroux, New York.

Rummel, R.J. (1997), *Power Kills: Democracy as a Method of Nonviolence*, Transaction Publishers, New Brunswick, NJ.

Sessa, M.G. (2023), *Disinformation Glossary: 150+ Terms to Understand the Information Disorder*, EU DisinfoLab. https://www.disinfo.eu/publications/disinformation-glossary-150-terms-to-understand-the-information-disorder/

Seul, J.R. (1999), "'Ours Is the Way of God': Religion, Identity, and Intergroup Conflict", *Journal of Peace Research*, vol. 36, no, 5, pp. 553–569.

Smith, A.D. (1999), *Myths and Memories of the Nation*, Oxford University Press, Oxford.

Snyder, T. (2018), *The Road to Unfreedom: Russia, Europe, America,* Tim Duggan Books, New York.

Tooze, A. (2014), *The Deluge: The Great War, America and the Remaking of the Global Order, 1916–1931*, Penguin, London.

Van Apeldoorn, B. and de Graaff, N. (2016), *American Grand Strategy and Corporate Elite Networks: The Open Door Since the End of the Cold War*, Routledge, London and New York.

Wardle, C. (2018), *Information Disorder: The Essential Glossary*, Harvard Kennedy School Shorenstein Centre on Media, Politics, and Public Policy, Cambridge, MA.

Zygar, M. (2023), *War and Punishment: Putin, Zelensky and the Path to Russia's Invasion of Ukraine*, Scribner, New York.

Design Problem 4

Solution Evaluation with Convergent Thinking and Empathy Mapping

Archipelago of Design (AoD)

Oliver Jones, Kassie Miedema, and Philippe Beaulieu-Brossard

Introduction

Section IV of *Modern War and Grey Zones: Design for Small States* explores the social and human dimensions of national security in the face of modern wars and Grey Zone conflict. The state plays a vital role in social resilience, cohesion, and the legal frameworks required to defend against Grey Zone threats and modern warfare. However, the social element may be underemphasized in traditional security studies.

Design Problem 1 used divergent thinking to scan the horizon and find weak signals indicating emerging problems. Design Problem 2 used convergent thinking and rich pictures to define complex problems more coherently. Design Problem 3 explored solutions with divergent thinking by applying causal loops to encourage systems thinking. Design Problem 4 leverages convergent thinking through empathy mapping. Empathy mapping is well-suited to elicit innovative solutions that address complex human and social dimensions of security.

How can small states enhance their social resilience and state integrity to manage Grey Zone conflicts effectively? Preceding design exercises highlighted key actors and stakeholders in the problem space and helped identify leverage points for interventions. This exercise explores the impact of a potential solution on stakeholders by *empathy mapping* or trying to imagine what stakeholders might think and feel and how they might react to specific interventions or messaging. This exercise aims to test out interventions and understand the consequences these changes will have on other people. Empathy mapping may also highlight aspects of a proposed intervention or policy that need to be rethought to avoid negative or capitalize on positive impacts. Empathy mapping might be employed by a red team, a group directed to think differently to test decisions or tactics. Mimetic red teaming involves role-playing and imitating the actions of an opponent. Decision support red teaming produces a careful analysis of the real world from the perspective of other players. Empathy mapping is an important tool for designing appropriate responses and can be a helpful tool for understanding the deep implications of decisions (UFMCS 2018).

Method

For the next exercise, pick a stakeholder (person, group, or organization) significantly affected by one of the interventions identified in the previous exercise. As a way of modeling how to apply empathy mapping when developing security interventions or policies, this example shows how to apply the tool when assessing the impact of approaches to stabilizing the Sahel region and improving human security. For this example, suppressing rebel groups in Darfur is one of the leverage points.

DOI: 10.4324/9781003428701-27

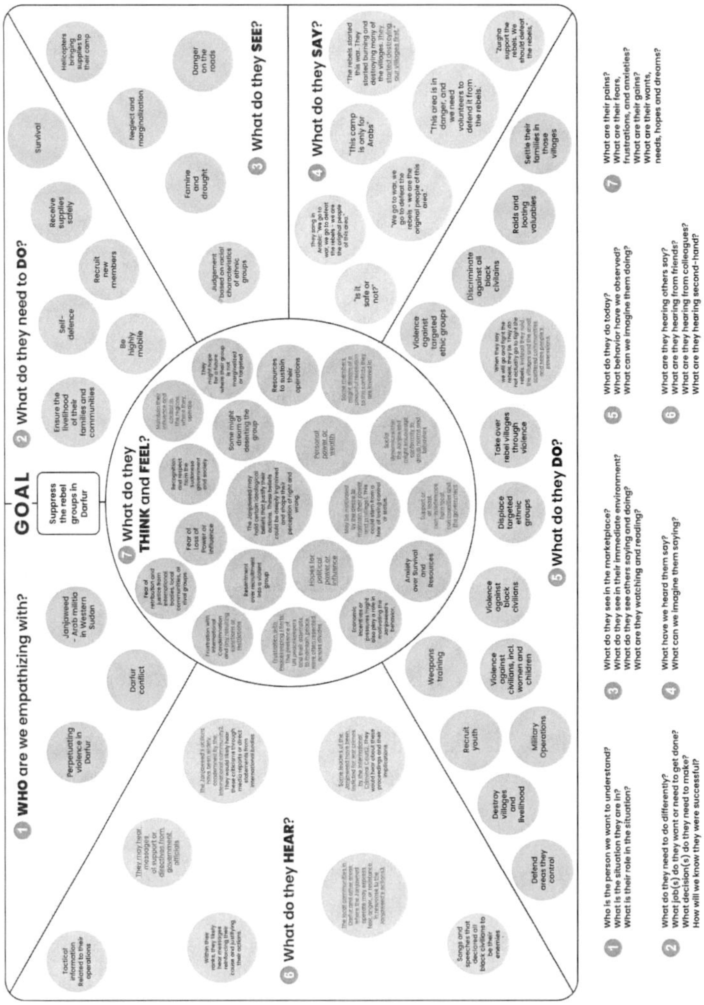

Figure DP4 Empathy Map Depicting Aspects of the Mindsets of Armed Groups in the Darfur Conflict.

The Empathy Map Canvas (Gray 2017) illustrates a diagram around which a goal and six questions lead to an understanding of what a key stakeholder—in this example, the Janjaweed militia—might think and feel. This suggests ideas about how they might be influenced. Group work will probably produce a random creative walk through the various steps with implications examined in the order that ideas occur to participants, and this aligns with the intent of using the tool. Use sticky notes on a whiteboard or wall to cluster ideas around a circle representing the target group or individual.

The empathy map in this example targets the Janjaweed as a significant party to the Darfur conflict (see Figure DP4). They are an ethnic militia of Arabized indigenous people in Western Sudan. Their goal is the suppression of rebel groups opposed to the Sudanese government. They have multiple needs and perspectives related to martial purpose and collective and individual survival. They must be mobile, recruit new members, and provide for their families.

The Janjaweed sees the ongoing normalization of violence, resource scarcity, famine, and drought. They see dangerous roads and routine use of resupply by government helicopters. They broadcast a mix of messages. Some are political exhortations and calls to arms. They issue harsh judgments based on racial characteristics. They make claims to territory based on ethnic identity and spread messages (including through songs) affirming the identity of preferred groups and attacking others. But they voice questions and concerns about their safety. They engage in martial tasks and rent-seeking activities to sustain themselves and accomplish their purposes. From the Sudanese government, they hear messages of support, orders, and intelligence reports. They hear condemnation from the international community (which is likely filtered or indirect). They likely hear internal messaging that reinforces their cause, but they also probably hear some anger and resistance from affected parties, alongside information about the prosecution of their leaders in the International Criminal Court.

Having mapped the surroundings of the target group or individuals, we can represent some of their thoughts and feelings to understand their motivations better. These categories can be simplified as "pains" (sources of frustration and grievance) and "gains" (satisfaction of their needs and desires). Their "pains" arguably stem from a fear of losing their power and influence. They genuinely fear retribution not only from international bodies, including sanctions, but also from rival groups and local communities. In other ways, they are simply anxious about survival amid scarce resources. On the "gains" side, the Sudanese government grants them token legitimacy and protection. They gain a modicum of security and some influence in the regions they control. Their leaders can use that security and influence to amass wealth and power. At a deeper level, they likely want a future in which they are not marginalized, and they may dream of a peaceable resolution of the conflict. Some may even dream of deserting. Outside of these calculations, some may resent the conditions under which they were recruited. Economic factors may motivate them. There may be deep ideological commitments to justify their beliefs to themselves, and those ideologies might have unobserved influences on group dynamics.

By understanding how potential recruits or dispossessed organization members receive messages, an intervening agency could develop a communication strategy to dissuade individuals from joining the Janjaweed or promote alternative messages of social cohesion or legitimacy. This approach might require some refinement or prototyping to implement and target accurately. If a proposed solution seems unfeasible or counterproductive, revisit the empathy map and target another potential intervention. The exercise aims to identify human elements of a conflict that are within an actor's power to influence, develop, and test potential solutions. For example, empathy mapping might be applied to Russia's Africa Corps (successor to the Wagner Group) and its clients. Understanding mindsets and motivations is essential in formulating strategies for the complex world of modern war and Grey Zone conflicts.

The chapters in Section IV focus on the state's fragmentation and cohesion and challenges to state authority from both within and outside the state. How might an empathy map help to identify potential solutions that capitalize on state power or enhance the strength and resiliency of state institutions?

Group Exercise—Empathy Mapping and Its Implications

Task. This exercise has four parts. First, identify a target for the exercise. Consider a disaffected group, disruptive leader, or political movement. It might be a minority language group with ties to a foreign state, a fringe ideological group, or an insurgent leader. Look for a group, movement, or leader actively supported by social media messages originating outside the country or a domestic leader playing a disruptive political role on the margins of political acceptability. What's acceptable will vary depending on the regime and political climate in the state. Domestic dissent is usually less tolerable when external threats increase, but some regimes have lower tolerance for dissent.

Second, articulate the aims of the state or agency conducting the exercise. Aims will reflect the range of tools available. Security agencies will have more latitude if the risk of violence is high or if it combines with external threats. The goal may be maintaining public order and confidence in the government and security services. Authoritarian governments have the latitude to crush opposition, while democratic regimes may focus on the reintegration and appeasement of disaffected groups.

Third, conduct the empathy mapping exercise, following the example above. Finally, what deductions can be drawn about tactics or approaches to achieve the aims of the state?

Consider. The questions posed in the example about the target.

- Who are we empathizing with?
- What do they need to do (from their perspective)?
- What do they see?
- What do they say?
- What do they do?
- What do they hear?
- What do they think and feel?
- What are their fears, frustrations, and anxieties (pains)?
- What are their wants, needs, and aspirations (gains)?

After completing an empathy mapping exercise, how might different actions or strategies impact the target's perceptions? Will appeasement lead to escalating demands? Will suppression create martyrs or delegitimize the government in the group's eyes?

Consider

- Targeted communication campaigns
- Blocking or countering disruptive social media
- Psychological or social community support for alternative groups
- Community engagement initiatives, capitalizing on group loyalty
- Work on the next generation with education, socialization, and youth programs
- Satire, humor, and ridicule as tactics to address bizarre beliefs (QANON)

Implications. Empathy mapping close to home may lead to the realization that there is insufficient information to do the exercise effectively. What is known about marginal groups, movements, and leaders, and how is it known? Who is the enemy, and how is the enemy like or unlike us in their motivations and thought processes? Can security organizations access specialized media channels and language communities to understand them? Have a lack of understanding and communication failures contributed to the problems this exercise is intended to address? What deductions may be drawn about intervention strategies, information gathering before intervention, or community engagement as a strategy for democratic resilience and national cohesion?

References

Baker, J. G. (2017), *Imagining Empathy: Counterfactual Methods and the US-Iran Security Dilemma* (PhD Thesis) University of Birmingham, Edgbaston, UK.

Gray, D. (2017), "Updated Empathy Map Canvas", *Medium*, 15 July. https://medium.com/@davegray/updated-empathy-map-canvas-46df22df3c8a

Further Readings

Galli, C., Minervini, A., and Sitze, A. (2009), "On War and on the Enemy", *CR: The New Centennial Review*, vol. 9, no. 2, pp. 195–219.

Packer, G. (2006), "Knowing the Enemy", *The New Yorker*, 10 December. https://www.newyorker.com/magazine/2006/12/18/knowing-the-enemy

Conclusion

Professional Military Education, Modern War, and Grey Zones

David Last and Marzena Żakowska

There is no front line or declaration of war in the war nexus. Kinetic violence by flag-waving regular troops of a recognized hostile state is a flawed trigger for defensive efforts. They may come after the battle is lost or never come. Modern war is waged in the Grey Zone and cannot be won by uniformed troops alone. Those in military, police, and gendarme uniforms may be the first to notice that they are at war without the knowledge, skills, and allies they need to defend their vital ground, but lawyers and judges, business and political leaders, and ordinary citizens will recognize something amiss. Against the background of a deteriorating planet and accelerating technology, major powers are driving conflict that penetrates small states' political, social, and economic fabric. Not all states, societies, and economies feel the pressure equally. Some global neighborhoods are better than others; proximity to disruptive major powers can be an indicator.

It is important to remember what is being defended. The goals are good governance, a sound business environment, equitable distribution of resources, acceptance of the rights of others, good relations with neighbors, free flow of information, high levels of human capital, and low levels of corruption (IEP 2024). Achieving these goals makes states more resilient. Positive peace improves national security as well as human security. Bad governance, bad business, inequity, oppression, exclusion, secrecy, and corruption weaken a state and penalize its people, whether under external threat or not. This concluding chapter reviews the main themes and explores their implications for professional security education, policy, and strategy.

Revisiting Key Themes

The war nexus, dependency on the global commons, the need to apply design thinking, society as a target, and the breadth of tools for responding to Grey Zone threats are themes across the four sections of this book.

The War Nexus

In Chapter 1, Simonyi describes a new ontology of war in which mobilization for victory is not the objective. Continued conflict below the threshold of full effort confers advantages on a major power disrupting the established order. The war nexus for disruption inverts the policing necessary to preserve international order and therein lies the incentive for small states to support an international rule of law, tolerance of social differences, and open economies. We can disagree about the details but agree that modern war in the Grey Zone is to be resisted in our common interests. There are potential failures in small-state responses to modern war and Grey Zone threats.

DOI: 10.4324/9781003428701-28

The first failure is to overlook external threats: to be oblivious to electoral interference, manipulation, foreign purchase of assets, and disinformation. Governments and citizens may naively think all is well or have a vague unease without being able to attribute hostile acts to any actor. A state can be defeated without battles, and a nation can lose its sense of itself.

The second failure is seeing (or claiming) external threats when they are part of domestic culture. It is difficult to know the difference in practice between a domestic actor of conscience and a paid foreign agent (Chapter 17). Opposition is built into open societies' adversarial political, legal, and economic systems. Party fights, acrimonious lawsuits, hostile takeovers, direct foreign investment, and conflicting media viewpoints strengthen open societies.

The third failure is to focus on symptoms but be distracted from underlying causes. Survival migration is spurred by war and famine, not only hostile intent. Receptiveness to disinformation rises with dissatisfaction, loneliness, and poor education, not just foreign trolls. Chapter 2 reminds us that our deteriorating planet is an urgent priority. Small states must pool their efforts to restrain great powers from destructive conflicts that distract from common priorities without succumbing to the first failure.

Mutual Dependency and Global Commons

Climate, oceans, space, and technology are all global commons on which every state depends. Technology's advantages result from being a global system (Bratton 2016:198). Its vulnerabilities also stem from dependence on connections, including satellites, undersea cables, remote server farms (the "cloud"), and energy consumption. One of the least vulnerable states—North Korea—is also one of the least connected. States can maximize their benefits from connection and technology while remaining wary of technological dependence and developing low-technology alternatives.

All states need space and rely on markets. Both should be guarded as global commons, like the climate and oceans. For small states, protecting global commons implies conventions, laws, and international norms. It implies active engagement in the institutions discussed in Chapter 11, and caution about the disruptions is described in Chapter 14. Self-help is also in order. Small states can enhance their position through cooperative arrangements like the European Space Agency and protecting national start-ups in niche areas, the way Luxembourg and Monaco have.

Protecting the global commons of market systems might be more difficult. Bretton Woods is in tatters, and the US dollar's status as a global reserve currency cannot last forever. Repeated use of the banking and exchange mechanisms for punishing sanctions and the weaponization of the SWIFT mechanism will eventually produce alternatives. Supporting the rules of an economic system that benefits some over others will generate opposition. The collective actions of small states will influence whether the global financial system evolves or collapses and the speed of that process.

Need for Design Thinking

A consistent theme has been the need for design thinking to understand complex systems and rethink strategies to address them. Considering the array of Grey Zone threats in Appendix, how can security leaders know when they are interconnected, strategic, hostile, or coincidental and locally generated? Treating a serendipitous group of unrelated threats as a hostile campaign could reduce society's tolerance of security responses. Missing the connections could increase vulnerability in the prelude to a more aggressive campaign. Horizon scanning and looking for weak signals (Design Problem 1) can help.

Design thinking can help select alternative strategic objectives and evaluate the relative importance of political, social, and economic objectives. Our authors have considered actors, factors, threats, battlefields, and responses. These are worth emphasizing. In the 20th century, staff college exercises were typically limited to considering regular military actors from states. Today, planners must consider state and non-state actors, as well as regular, irregular, and civilian actors. Who has standing? Who are the threats, and who are the stakeholders? Can organized crime groups be flipped from enemies to assets?

We have used PESTLE—political, economic, social, technological, legal, and environmental—as a convenient mnemonic to organize factors or categories for Grey Zone threats. Just as there is a list of threats (see Appendix), a parallel list of defenses might be developed in each category. Central banks defend against currency attacks by sales or purchases according to need. Hostile sponsorship of organized crime can be countered with policing, but it might also be met with counteroffers of alliance and information sharing, depending upon the motivation and pressures affecting the group. Empathy mapping might reveal new strategies. Romans hired barbarians into their legions; could a country mobilize gangs for border security without setting a fox to guard the hens? We don't know. Alternative decision analysis (red teaming) can be combined with design thinking to set up policy experiments to determine what works. Rapid assessment process brings a variety of policy expertise together to triangulate what works under different circumstances and according to different professional expertise (Beebe 2001). Something often omitted is post-experiment data collection and evaluation to determine whether a policy or strategy was effective. Grey Zone threats may have a long fuse because stable authoritarian governments can be patient and relentless.

Society as Strength and Target

The word "battlefield" conjures images of muddy fields and contending men-at-arms. The fields of contention in the war nexus are less obvious. They include space, cyber, markets, institutions, and above all society—perhaps the central battlefield. It is an asset for achieving the goals of positive peace, and it is the target of many of the activities of modern war in the Grey Zone. The "Vancouver Model" was described by a public inquiry in Canada's British Columbia. Drug trafficking linked to international cartels generated illicit funds for organized crime. Profits were laundered through casinos, real estate, and legitimate businesses. Government revenues benefited, reducing the incentive to police the source of funds. Real estate prices skyrocketed, pushing locals out of the housing market. Municipal and provincial officials were corrupted. Lawsuits were used to intimidate witnesses and whistleblowers (Cullen 2022; Cooper 2021). Society seemed to be under attack from below via growing damage from drug trafficking and gang activities and from above by the corruption of law, policing, and governance.

Canada's strengths include its ability to conduct independent inquiries, engage lawmakers, discuss problems in public, experiment with alternative policies, and trust citizens and institutions to produce solutions. However, each of these strengths is a potential target. Community-level foreign-language media outlets can be intimidated or taken over, lawmakers can be sued, and witnesses can be disparaged and harassed. Gangs cooperate with foreign governments. Mutual interests in the short term, like gambling taxes and rising property values, convey long-term strategic advantages to hostile actors, like economic erosion and unlivable cities, which are more vulnerable to influence. Governments and security planners must think strategically and holistically about how society is affected by hostile forces and domestic fellow travelers.

Tools for Response

The acronyms DIME (diplomatic, information, military, and economic), DIMEFIL (add financial, intelligence, and law), and PESTLE (political, economic, social, technological, legal, and environmental) have been used in this book to describe instruments of state power and factors for consideration in Grey Zone conflicts and the war nexus. The culture of security emphasizes service in uniform. Military, gendarmes, police, and intelligence services refer to "civilians" outside the community. Considering the range of DIMEFIL or PESTLE tools, militarizing or securitizing social functions may be less useful than drawing the circle larger to include whole-of-government and whole-of-society defenses—bringing society into the confidence military, police, and intelligence.

Awareness of Grey Zone threats must be concrete but nuanced. An amorphous sense of unease about threats can be socially corrosive; cynical political entrepreneurs use it to mobilize fear, target scapegoats, and justify violence. Concrete information about hostile actors, strategic industries, security intelligence functions, and public responsibilities must be balanced by action against racist stereotypes and exclusions. If Iran or Russia are hostile actors, Farsi- or Russian-speaking citizens are assets. This implies strong counterintelligence and monitoring.

Strong independent media, investigative journalism, public information, and active government communications strategies are essential for an open society. This demands constraints on transnational platforms like TikTok, X, and Facebook, which erode independent journalism (Public Policy Forum 2017). New media monitoring services like Critical Mention (www.criticalmention.com) help companies monitor and respond to their media coverage. Small states can also benefit from greater awareness of the information environment about their policies and interests. Civil society organizations, government departments, and private sector contracts can help.

Much better awareness of the cyber domain is needed. The University of Toronto's Citizen Lab (www.citizenlab.ca) is a good model for independent monitoring and public awareness, but more observation posts are needed. States must support cyber observation and response, preservation of information and electoral integrity, and policing of bad actors. This implies stronger public prosecution. If court victories always go to the side that spends the most on lawyers, small states can be overwhelmed by well-funded hostile states or corporations.

Authoritarian states have an advantage in controlling large bureaucracies and developing long-term strategies to yield disinformation and active measures. Liberal democracies can mobilize civil society to counter hate and disinformation. Advances in natural language processing and artificial intelligence may give non-governmental organizations new tools to identify online hate and develop and deliver counter-messages (Chung et al. 2021).

Defense against Grey Zone threats cannot stop at borders. Diplomatic initiatives to call out offending governments, sanction individuals, and share information need resources and consistent commitment. Strategies are needed to police illegal markets, counterfeiting, smuggling, the drug trade, money laundering, and state-sponsored erosion of national economies. Financial tracking works with intelligence-sharing, network infiltration, and imaginative policing. Europol, the European Union Agency for Law Enforcement Cooperation, ran Operation DisrupTor, successfully targeting vendors and buyers on the dark web with deception to penetrate encryption. It is hard to determine how extensively organized crime contributes to hostile state strategies. Gangs in major ports and transport hubs can be self-serving until paid to disrupt logistics support in a crisis. Entrepreneurial criminals can steal information, assassinate leaders, disrupt communities, intimidate enemies, and derail economies. Money is a universal motivator, but anti-regime animus, ethnic or religious motivations can be mobilized by manipulation and deception. Targeted intelligence operations to understand motivations may be as important as understanding activities.

Few tools exist to respond to Grey Zone threats and modern war that are purely within a single sphere. Like all-hazards intelligence, strategies and defensive tools must combine traditionally separate expertise.

Professional Security Education for Modern Wars in the Grey Zone

There is a broad international constituency for professional military education (PME) and a vested interest in defense universities, staff colleges, and military academies to preserve the institutions that specialize in military education. Modern wars in the Grey Zone suggest revisiting this ecosystem. Small states should think more broadly about professional security education while preserving hard military and policing skills. This means more cross-appointments and mixed careers, rewarding rather than penalizing unorthodox career paths. A holistic approach highlights gaps in PME and the risks and rewards of fighting dirty.

Holistic Education, Cross-Domain Employment, and Common Training

The smaller the state, the greater the advantage of combining and sharing institutional expertise. Small size can be advantageous as services learn to leverage their assets rather than copy the practices of larger and better-resourced allies, who invest more resources in inter-agency coordination and internal communications. A whole-of-government effort implies that all government departments and agencies contribute to security objectives. A whole-of-society effort extends this concept to the profit and non-profit private sectors. In a clear life-or-death struggle, cooperation should be possible to elicit. In the Grey Zone of ambiguous conflict and political and economic contention, many actors will not be easy to harness to a national security effort. National government leadership may be painted as antidemocratic or demagogic as it strives to mobilize a whole-of-society response.

A broad understanding of security stakes, spread throughout different sectors of society, will help stay on track. For military, paramilitary, police, and intelligence officers, early career exposure to other sectors, including the private sector and civil society organizations, will help. Exchanges and problem-solving exercises with selected members of civil society, the private sector, and officers of multinational corporations will build capacity to mobilize a whole-of-society response when it is most needed. Preserving institutional memory of who's who and networks of informal communications can improve resilience and identify vulnerabilities.

A comprehensive and adaptive educational framework that goes beyond specialized training helps security leaders think comprehensively and creatively about new threats and challenges. A monoculture can emerge in leadership cadres who all receive the same education from the same institution. On the other hand, socialization and commitment in a community of trust are essential for the security professions. The traditions of service and leadership developed in military academies can be expanded by inviting diverse student bodies committed to national security through different career paths across government. Local and international exchanges, sabbaticals, cross-appointments, and secondments should be career-enhancing. Beyond multi-year bachelor-master programs (usually four plus one or three plus two years), mid-career master programs, short courses, certificate programs open to civilians, and mixed military-police-civilian boards for accreditation and examination can broaden the focus of professional security education.

Military, paramilitary, police, intelligence, and public service are all examples of inverted professions. While professions like law, engineering, and medicine begin with a general body of knowledge from which professionals become increasingly specialized, inverted professions begin with specialized fields, which must be integrated as professionals advance. Mid-career staff colleges teach army, navy, and air force officers to plan operations together. To deal with the challenges of modern war in the Grey Zone, at least some security leaders need more generalization earlier in their careers.

Gaps in Security Education and Recruitment

Joint education is difficult, even for the uniformed services. Army, navy, and air force officers are comfortable in their tribes and build esprit with narratives about their superiority to other services and civilians. Budgets can compel, and for small countries, the path may be easier than for large powers. Mid-career staff colleges are typically joint in smaller countries, while they are single service in larger ones. The picture painted above of civil-military, public-private, and multinational collaborative education is a long way from the current reality in any country. The gap between that picture and the present creates vulnerabilities to Grey Zone threats and the war nexus. Small states will be ill-situated if the best cyber talent emigrates, if multinational corporations hire away the educated youth, and if talented immigrants are excluded from government service.

The necessary talent doesn't have to be in uniform. Knowing where the talent is, in country or in a diaspora, and being able to call on it in a crisis is one thing. Exercising those calls regularly and deliberatively can help to build networks to respond to national needs as Grey Zone threats blend with accelerating violence. Military, paramilitary, police, and intelligence services cannot be slow to adapt. They need rapid situational understanding and learning in very short cycles—within minutes and hours in special operations that may last days or weeks (Björklund and Tillberg 2024). But they also need a historical sense and retention of learning both because some threats evolve over years and because history provides clues to meaning and context for combining the talents of a whole-of-society response to Grey Zone threats. Greater civilian involvement and historical context for security planning will help security establishments to avoid the second trap, to make choices about the kinds of society and values people want before making choices about defense and styles of defense.

Playing Dirty

Grey Zone threats might be labeled "playing dirty"—breaking the norms for state conduct, breaking domestic and international law, justifying sanctions and retaliation. The Grey Zones faced by small states usually originate with major powers and aggressive neighbors. Are small states justified in retaliating with dirty tricks or Grey Zone threats? Are they better keeping the moral high ground? Deterrence works best under conditions of certainty (Chapters 12 and 13). Offensive cyber operations and unattributable dirty tricks do not deter. They are used to achieve desirable results at a minimum cost. Among the costs that should be considered is the damage done to international norms. Canada alleges that India paid gangsters to murder Sikh nationalists in Canada; India denies the charge and accuses Canada of welcoming separatist criminals in exile. When successful, murder creates martyrs. When detected, as Indian actions were in the United States, dirty tricks damage relations, but how badly? The answer seems to be, not sufficiently to prevent their use.

The United States, China, Russia, and India are large enough and have enough diplomatic capital with other states not to be labeled rogue regimes and shunned, even for repeated and egregious breaches of law and norms. They continue to enjoy normal diplomatic relations;

transgressions are routinely ignored. Iran and North Korea have more assertive Grey Zone strategies and suffer proportionately more international hostility, but also retain normal status as states in the international community.

Any state attacked or threatened by state or non-state actors may opt for self-defense using the most expedient means. Grey Zone tactics, including the full range of threats listed in Appendix and more, can be applied usefully in the service of small countries, seemingly with little risk of serious consequences. Consulting allies beforehand might be prudent, but many of the targets for small states playing dirty are likely to be individuals, corporations, and non-state actors, lowering the risks of extra-legal international actions undertaken by proxies.

Implications for Policy and Strategy

The war nexus and modern war in the Grey Zone are both more than war and less. It is more, because there are threats beyond the traditional forms of military threat to states. Adversaries are trying to win without fighting battles. This requires vigilance and a flexible, resilient strategy. Small states must be ready to turn to a total conventional defense when necessary, as Ukraine did in February 2022. Modern war in the Grey Zone is also less than war; most small states will not see the intensity of conventional conflict in Ukraine or Gaza but may still experience an erosion of their national and human security and a diminution of positive peace.

States cannot assume peace based on their image of the threshold of war. Regardless of bandwagoning, balancing, hedging, or self-sufficiency strategies, small states make continual adjustments to national strategies (Edström, Gyllensporre and Westberg 2019:183). Beyond the halls of power, these adjustments must occur in the streets, public institutions, and among the people. Strategy and tactics in the Grey Zone matter to society. Clever actors may use Grey Zone threats to drive a wedge between government, society, and security forces, rendering the relationships unable to respond to the demands of war as threats escalate.

The chapters in this book indicate a need for foresight, all-hazards intelligence, and continuous reassessment of the level of threat faced across political, economic, social, technological, legal, and environmental factors. Facing changing circumstances, governments and security leaders should use creative design thinking to develop policies and strategies, alternative decision analysis to assess their robustness, and policy experiments to test them. The safety net for policy experiments is the democratic decision-making process coupled with the professionalism of security organizations, sometimes pilloried as the "deep state." Professionalism, education, and adaptation of the police, paramilitary, military, and intelligence thinking are the safety nets that permit experimentation with confidence to confront modern war and Grey Zone threats.

References

Beebe, J. (2001), *Rapid Assessment Process: An Introduction*, Altamira Press, Walnut Creek, CA.

Björklund, L. and Tillberg, V. L. (2024), "Police and Military Tactics for the Unpredictable – Epistemological Perspectives on Knowledge in Motion". *Presented at the 17th Biennial Conference of the European Research Group on Military and Society, Working Group on Military and Police Relations, 1-5 July 2024*, Stockholm, Sweden.

Bratton, B. H. (2016), *The Stack: On Software and Sovereignty*, MIT Press, Cambridge.

Chung, Y. L., Tekiroğlu, S. S., "Tonelli, S. and Guerini, M. (2021), Empowering NGOs in Countering Online Hate Messages", *Online Social Networks and Media*, vol. 24. https://doi.org/10.1016/j.osnem.2021.100150

Cooper, S. (2021), *Willful Blindness: How a Network of Narcos, Tycoons, and CCP Agents Infiltrated the West*, Optimum Books, Toronto, Canada.

Cullen, A. F. (2022), *Commission of Inquiry into Money Laundering in British Columbia, Final Report*, Province of British Columbia, Victoria BC, Canada. ISBN 978-1-7774125-2-4. https://canadacommons.ca/artifacts/2470962/cullencommission-finalreport-full/3492968/#:~:text=This%20is%20a%20report%20published%20by%20Commission%20of,Property.%20It%20was%20written%20by%20Austin%20F.%20Cullen.

Edström, H., Gyllensporre, D. and Westberg, J. (2019), *Military Strategy of Small States: Responding to External Shocks of the 21st Century*, 1st ed. Routledge, New York.

Europol (2020), "*International sting against dark web vendors leads to 179* arrests", European Union Agency for Law Enforcement Cooperation, 22 September. https://www.europol.europa.eu/media-press/newsroom/news/international-sting-against-dark-web-vendors-leads-to-179-arrests

IEP (2024), The Eight Pillars of Positive Peace, Institute for Economics and Peace, https://www.visionofhumanity.org/the-eight-pillars-of-positive-peace/

Public Policy Forum (2017), *The Shattered Mirror: News, Democracy and Trust in the Digital Age*, Public Policy Forum, 26 January. https://ppforum.ca/publications/shattered-mirror-news-democracy-trust-digital-age/

Appendix

Tabular Summary of Grey Zone Threats

David Last

The table is organized according to the primary instrument or tactic relating to types of Grey Zone threats, recognizing that in most cases, these instruments or tactics are interconnected, and the target of the threat may be related to second-order effects in other fields. For example, economic actions may have political motivations or targets, or political actions may support actions on economic intentions or targets. Grey Zone threats are inherently complex and concealed by design, so any tabular summary is a necessary simplification, and these labels and examples are offered only for illustrative purposes. Examples can be found from any state, but mostly from major powers. While our attention is mainly on authoritarian threats to liberal democracies—like Russia, China, Iran, and even North Korea—some of the most successful Grey Zone attacks go back to the days of colonialism and post-Cold War neoliberal globalization, in which major Western countries achieved significant foreign policy goals using social and economic hegemony, establishing rules in their favor, and then enabling commercial sidestepping of many of those rules. Many of these forms of market attack are now being mobilized by Russia, China, and India. Authoritarian states have the advantage of long-term planning and executive stability, while liberal states (particularly the United States) probably have an advantage in manipulating market forces, although this might be a declining asset.

Categories and labels	*Elaboration*
Political threats	Description/example
Elite penetration	Make key elites beholden to foreign interests, e.g., cabinet ministers on foreign boards, or relying on information or services provided by foreign power, election, and foundation donations are a good start
Elite family strategies	Employing family members, or offering benefits or opportunities, power-marriages with reliable insiders, and *kompromat* strategies
Family influence strategies	Influence key appointments through threats or promises to family members, often easily identifiable through social media presence
Insider threat	Use well-placed insiders to alter processes or decisions if favor of foreign interests
Corruption strategies	Corruption is the use of public position for personal gain; it applies to any position of trust. Corrupting people in positions of trust opens potential for influence while invisible and also erodes trust in institutions when it becomes visible, so corruption can be used strategically in phased operations
Funding parties and activists	Can be tailored to the political system, e.g., supporting vote-splitting parties or
Suborning diplomats	Use of any *kompromat,* intimidation, or other strategy

(*Continued*)

(Continued)

Categories and labels	*Elaboration*
Hostage-taking (low-profile)	Can involve relative unknowns for minor or stepping-stone objectives, part of *kompromat*, intimidation or manipulation of other political players or processes
Hostage-taking (high-profile)	e.g., the Two Michaels for Meng Wanzhou
Intimidation	Threatening to achieve objectives; may be direct (we'll hurt you) or indirect (we'll hurt your family or friends), can be subtle threats (we know where you live), or demonstrative (we did this to your neighbor). Demonstrating ruthlessness is helpful for enforcement
Targeting leaders via constituents	Mobilize constituents to support or oppose action by elected leaders; depends on "tipping-point" or "squeaky wheel" dynamic, may be combined with
Kompromat	Smear campaigns (reputational damage), compromising information collected or fabricated for use in blackmailing, discrediting, or manipulating someone for political purposes
Fellow travelers and "useful idiots"	Fellow travelers are those with congruent aims; useful idiots are those who can be persuaded or manipulated to undertake actions whose consequences they do not fully understand
Indirect "clean hands" strategies	Ensuring that no direct connection to "dirty tricks" can be traced to national policies; entails the use of proxies, NGOs, and often fellow travelers or useful idiots
Astroturfing	A fake grass-roots organization established to push a particular interest; often related to corporate profits (e.g., overturning environmental regulations in favor of jobs)
Targeted assassination	Removing a key individual, either for that individual's activities, or to provide an example to others, e.g., murder of a family member as part of an intimidation strategy
Electoral manipulation	Dependent on the electoral system and degree of penetration, may involve gerrymandering, focus on executive, legislative, or (sometimes) judicial elections or appointments, voter-suppression
Direct electoral interference	Usually through disinformation or malinformation (true info but used maliciously), but could also be through intimidation, bribery, targeting politicians
Support for political extremists	Training, funding, and supporting extreme political factions or terrorists, exacerbating polarization
Support for armed action	Training, *arming*, funding, and supporting terrorists or insurgent groups, forcing state security reaction or over-reaction, or justifying direct intervention
Economic threats	Description/example
Currency attack	Use of market mechanisms and financial institutions to erode national currency value
Credit rating attack	Use of market mechanisms, cyber-attacks, information, financial manipulation to undermine credit ratings
Erosion of foreign reserves	Force a draw-down of foreign reserves, e.g., by demanding payment in a particular currency
Supply chain domination	Monopolizing or dominating means of transport, routes, sources, distributors, or components to achieve influence over-dependent actors
Supply chain disruption	Target distribution or supply networks to disrupt the economy or critical imports/exports; may be done strategically at a distance, including point of origin

(*Continued*)

(Continued)

Categories and labels	*Elaboration*
Boycott-Divest-Sanction (BDS) movements	Networking of NGOs and social movements to undermine economic well-being, limit access to markets and finance, and attack public identification with political actors
Infiltration/dominance of an economic sector	In Europe, energy was Russia (RU) instrument of choice; in Africa, China (CN) is dominating small business retail sector and supply chains, strangling African enterprise, as former colonial powers did
Critical resource dominance	Rare earth metals, strategic minerals, energy, water, food—any resource can be critical at times and places and under the right market conditions
Alliance with organized crime	Typical in smuggling, counterfeiting, money laundering, and cybercrime. Criminal networks open doors for political objectives
Sponsorship of organized crime	Typical of cyber-attacks, robberies, assaults, intimidation, and any action where the state wishes to use illegal activities to pursue intelligence or security interests
Disrupting government revenues (grey markets)	Smuggling and grey markets (unregulated, untaxed) can disrupt sources of government revenue like alcohol, tobacco, and gasoline
Infiltrating key industries	Telecommunications, infrastructure, energy, and private security are all possible targets for infiltration through commercial or personal networks and may be stepping stones to other objectives
Manipulation of financing/ banking	Can create cascading banking failures, crisis of confidence (e.g. Bank of Credit and Commerce International (BCCI), 1991, PricewaterhouseCoopers PwC Sandstorm report and can serve as a tool for concentrating illicit funds for other objectives
Use of shell companies	Concealing the actual beneficiaries and lowering reputational risk or opposition to mergers and acquisitions, or support for illicit activity, or acquire intellectual property
Shareholder manipulation	Shareholders may be manipulated with respect to their stock, e.g., to support mergers and acquisitions, or in relation to other activities, like political donations, voting, or others
Mergers and acquisitions	May target specific technology, Intellectual Property (IP), or contractual relationships, e.g., Canada's Norsat International Inc. sale
C-suite intimidation	Impacts commercial decision-making, helps with mergers and acquisitions
Money laundering	Moves money illicitly
Counterfeiting and credit card fraud	Provides funds for local operations, debases currency, erodes confidence
Asset stripping	Purchasing and dismantling a firm to remove a capability, hobble a competitor, and provide short-term profit
Social threats	Description/example
Penetration of institutions	Getting friendly or compliant individuals in key positions in important institutions
Cultural affinity institutions	British Council, Alliance Française, Göethe Institute, Confucius Institute
Supporting or opposing publications	Ownership or manipulation of media, changing rules of ownership or usage, e.g., concentration of ownership may be advantageous or may be a barrier to manipulating information flow
Suppressing or manipulating education	Entry points may be local government (6% rule) national regulation, key influencers, parties, or mass and social media memes; TikTok has had a measurable impact on primary and secondary education

(Continued)

(Continued)

Categories and labels	*Elaboration*
Mass media ownership or manipulation	Purchase, infiltration, or influence over media for conveying messages
Cultural hegemony	Dominating cultural memes and narratives through a successful entertainment industry, e.g., Hollywood for much of the world during the Cold War and up to the present.
Counterculture	Find marginal cultural movements and support insurgency against mainstream culture, targeting disaffected youth, e.g., fascist punk
Social media manipulation	Creation, manipulation, and exacerbation of divisive narratives or memes online, use of bots and algorithms to manipulate perceptions of reality
Propaganda and information war	Information intended to persuade, may be false or misleading, combination of many different means to achieve specific objectives
Agitprop	intentional vigorous promulgation of ideas, including "propaganda of the deed" which are actions intended to foster revolution or dramatic change
Media influence	Use of editorial boards, ownership
Use of religion, religious infrastructure	Use of religious leaders or channels of communication
Use of ideology, political parties, and groups	Use or manipulation of existing ideological platforms or organizations
Discredit organizations	Smear or discredit NGOs that are doing useful or supportive work, to undermine social capacity
Generate schisms	Strategic exaggeration of differences, distrust of others, scapegoating, minority attacks
Support extreme views	Find and support extreme and hateful activists likely to erode trust in government, e.g, Alternative für Deutschland (ADF) in Germany
Weaponized drug trade	From the Opium Wars (1839, 1853) to the Fentanyl epidemic, tactics have changed, but social disruption, imposing costs, and extracting profit are still the basic motivations
Weaponized refugee flows	Enabling large numbers to seek asylum in a neighboring country, sometimes transporting them from a third country, may include infiltration of refugee flows
Technological threats	Description/example
Theft of intellectual property	As a national strategy, systematic theft of Intellectual Property (IP) can weaken the global market position of the target country.
"Deep fakes"	Use of Artificial Intelligence (AI) to deceive about the source of information by creating convincing audio, video, and written simulations.
Access to personalized big data	Sophisticated mining of big data for manipulation potential—Cambridge Analytica case of use for electoral manipulation
Social nudging using individualized big data	Nudging individual behavior based on personal information profiles
Dark web use	Use of dark web to spread information, manage illicit networks
Offshoring	Moving jobs or other functions outside national boundaries to evade laws, regulations, or reporting requirements that restrict nefarious actions
Offensive Cyberspace Operations	cyber-enabled espionage and exploitation, non-destructive operations to collect intelligence
Cyberspace attacks (within cyberspace)	Operations to disrupt, deny, degrade, or destroy information resident in computers, computer networks, or computers and networks themselves

(*Continued*)

(Continued)

Categories and labels	*Elaboration*
Cyber-enabled attacks (outside cyberspace)	Use of cyber connections to enable physical damage to infrastructure and other targets
Disruption of electromagnetic spectrum	Disruption of radio waves necessary for normal operations may be localized or broad spectrum
Software manipulation	Leaving "backdoors" or security loopholes in commercial software; developing and sharing cyber-attack tools
Data pollution	Disrupting the advantage of those with access to large datasets by corrupting or contaminating them, reducing confidence that they can be used effectively
Space disruption	Disrupting any part of the system which permits routine exploitation of space-based assets

Legal threats	Description/example
"Law fare"	A general label for exploitation of the legal system to achieve hostile objectives; recognizing that the legal system is inherently adversarial, and international law is an alternative to costly kinetic options for dispute management
Masking	Using lawyer-client privilege to conceal actors, intentions, and nefarious activities sub-contracted by law firms to private investigators and others
Exploitation of justice system	Delays, appeals, disclosures, preliminary hearings to explore strategies
Manipulation of justice system	Surveillance, intimidation, false witness, fabricated documents
Imposing legal obligations on citizens	Citizens obliged to cooperate with intelligence services when living abroad; descendants of citizens unable to renounce citizenship, and pressured through relatives in the home country; citizens required to report on relatives and acquaintances
Extraterritorial application of law	Citizens are subject to national law when living abroad; special branches of police responsible for enforcement
Use of international laws and agreements	Suing under international agreements, seeking compensation, using these suits as bargaining chips for other objectives (see political and economic threats)
Extraterritorial enforcement of law	laws extending beyond boundaries, courts enforcing jurisdiction beyond territorial boundary, seizing or freezing assets
Manipulation of tribunals	Pressure to get favorable outcomes in tribunal processes, e.g. trade decisions, may entail anything from signaling to bribery, extortion, family pressures (similar to manipulation of justice system, but with different targets)
Suits for reputational damage	International Criminal Court (ICC) cites Putin for war crimes in Ukraine under international law; Putin government cites ICC Chair for war crimes under Russian statutes
Internationalizing disputes	Palestinian supporters sue Israel in international courts over restriction of movement with separation barrier
Erosion of boundaries	United Nations Declaration on the Rights of Indigenous Peoples (UNDRIP), and lawsuit for free movement of indigenous across Canada-United States international boundary

(*Continued*)

(Continued)

Categories and labels	*Elaboration*
Environmental threats	Description/example
Evading local regulations	An offshoring strategy that transfers the costs of pollution to another country
Hazardous waste disposal	A strategy to reduce waste disposal costs and evade national/ international regulations
"Poisoning wells"	An ancient meme deployed to vilify an out-group, often Jews, creating panic and social tension
Land purchase or lease	Long-term agreements for the use of large plantations—increasingly important in the context of food security, may alienate states from capacity to feed themselves
Cash-crop market dominance	Similar to land purchase or lease, market forces may be used to reduce state capacity for self-sufficiency
Down-stream deterrence	Threats to cut off supply to down-stream water users; by analogy, other exclusion strategies
Lifeboat strategies	Any strategy that excludes others from a safe environment on the grounds of capacity
Survival migration	Mass movements of people driven by environmental change, which can reach inundation levels
Heat deaths	Wet bulb temperatures over 35 degrees causing large numbers of deaths
Water loss	Depletion of potable groundwater by excess use, salination, fracking
Air pollution	Deterioration of air quality by industrial activity
CBRN spill-over	Spreading effects of chemical, biological, radiological, or nuclear weapons tests or experiments, e.g. strontium 90, 1959; lab-leak sources of pandemics (hypothetical?), intergenerational impact of chemical warfare (Agent Orange in Vietnam)

Index

Note: **Bold** page numbers refer to tables and *italic* page numbers refer to figures.